# Towards Impact and Resilience

# Towards Impact and Resilience: Transformative Change In and Through Agricultural Education and Training in Sub-Saharan Africa

Edited by

Frans Swanepoel, Zenda Ofir
and Aldo Stroebel

CAMBRIDGE
SCHOLARS
PUBLISHING

Towards Impact and Resilience:
Transformative Change In and Through Agricultural Education and Training
in Sub-Saharan Africa
Edited by Frans Swanepoel, Zenda Ofir and Aldo Stroebel

This book first published 2014

Cambridge Scholars Publishing

12 Back Chapman Street, Newcastle upon Tyne, NE6 2XX, UK

British Library Cataloguing in Publication Data
A catalogue record for this book is available from the British Library



ISBN (10): 1-4438-5680-0, ISBN (13): 978-1-4438-5680-5

Cover design: Sun Media Stellenbosch

Cover images:
*A farmer's field in Western Kenya with maize intercropped with beans.*
Courtesy of Wandera Ojanji © CIMMYT/2011

*Mozambiquan farmer Amelia Ubisse, in Marracuene, near Maputo, with goats and bread.*
Courtesy of ILRI /Mann © Creative Commons/2008/
http://creativecommons.org/licenses/by/2.0/deed.en

*Woman in maize field. © CIMMYT/*undated
*Research visit. © CIMMYT/*undated
*Test tubes.* © Eric Cuthbert/Creative Commons/2007
http://creativecommons.org/licenses/by/2.0/deed.en

# TABLE OF CONTENTS

**Part Two: International Insights**



**Part Three: Building African Capacities**

**Part Four: Critical Issues**



**Part Five: Conclusion**

# List of Tables

# LIST OF FIGURES

# CONTRIBUTORS

**Teshome Alemneh**, Programme Officer for Africa, Higher Education for Development (HED), Washington D.C., USA. *talemneh@hedprogram.org*

**Ponniah Anandajayasekeram**, Independent consultant and Adjunct Professor, School of Environmental and Rural Sciences, University of New England, Australia. *anandaponniah@yahoo.com*

**Irene Annor-Frempong**, Director, Capacity Strengthening, Forum for Agricultural Research in Africa (FARA), Accra, Ghana. *ifrempong@fara-africa.org*

**Ralph Christy**, Professor and Director, Charles H. Dyson School of Applied Economics and Management, Cornell University Ithaca, New York, USA. *rdc6@cornell.edu*

**Canagasaby Devendra**, Consulting Tropical Animal Production Systems Specialist, JalanAwanJawa, Kuala Lumpur, Malaysia. *cdev@pc.jaring.my*

**Luc D'Haese**, Emeritus Professor, Department of Bioscience Engineering, Faculty of Science, University of Antwerp, Antwerp; Department of Agricultural Economics, Faculty of Bioscience Engineering, Ghent University, Ghent, Belgium. *luc.dhaese@ua.ac.za*

**Maureen Bandama**, Agribusiness Programme Coordinator, Market Matters Inc, Ithaca, New York, USA. *mbandama@marketmattersinc.org*

**Carl Eicher**, University Distinguished Professor Emeritus, Michigan State University, East Lansing, Michigan, USA. *ceicher@msu.edu*

**Marcus Hakutangwi**, Extension and Communication Consultant, Swedish Regional Cooperative Centre, Harare, Zimbabwe. *marcus.hakutangwi@gmail.com*

**Monty Jones,** Ambassador-at-Large and Advisor to the President of Sierra Leone, Sierra Leone. *Monty.p.jones@gmail.com*

**Johan Jordaan**, Senior Lecturer, School for Natural Resource Management, Saasveld Campus, Nelson Mandela Metropolitan University, George, South Africa. *Johan.Jordaan@nmmu.ac.za*

**Margaret Kroma**, Programme Officer, Gender Alliance for a Green Revolution in Africa (AGRA), Nairobi, Kenya. *mkroma@agra.org*

**Edward Mabaya**, Research Associate, Charles H. Dyson School of Applied Economics and Management, Cornell University, Ithaca, New York, USA. *em37@cornell.edu*

**John Makina**, Former Director, MHO Centre for Staff Development and International Linkages, Tshwane University of Technology, Pretoria, South Africa. *john@motheowathuto.co.za*

**Zenda Ofir**, International Evaluator, Switzerland; Fellow, Stellenbosch Institute for Advanced Study (STIAS), Stellenbosch, South Africa. *zenda@evalnet.co.za*

**Daniel Sherrard**, Provost, Earth University, San José, Costa Rica. *dsherrar@earth.ac.cr*

**Aldo Stroebel**, Senior Research Fellow, Centre for Sustainable Agriculture, Rural Development and Extension, University of the Free State, Bloemfontein, South Africa; Executive Director, International Relations and Cooperation, National Research Foundation (NRF), Pretoria, South Africa. *aldo.stroebel@nrf.ac.za*

**Frans Swanepoel**, Director, African Doctoral Academy (ADA), Stellenbosch University, Stellenbosch, South Africa; Visiting Fellow, Institute for African Development (IAD), Cornell University, Ithaca, New York, USA. *fjcswanepoel@sun.ac.za*

**Ranga Taruvinga**, Director, Mananga Centre, Ezulwini Valley, Swaziland. *director@mananga.sz*

**Glen Taylor**, Senior Director, Innovation and Business Development, University of the Free State, Bloemfontein, South Africa.
*taylorgj@ufs.ac.za*

**Johan van Rooyen**, Director, Standard Bank Centre for Agribusiness Mentorship and Leadership, Stellenbosch University, Stellenbosch, South Africa. *cjvr@sun.ac.za*

**Akke van der Zijpp**, Emeritus Professor, Animal Production Systems Group, Department Animal Sciences, Wageningen University, Wageningen, The Netherlands, *akke.vanderzijpp@wur.nl*

**Larry Zuidema**, Former Associate Director, International Agriculture Programme, Cornell University, Ithaca, New York, USA.
*lwz1@cornell.edu*

# BIOGRAPHIES OF EDITORS

**Frans Swanepoel** is Director of the African Doctoral Academy (ADA) at Stellenbosch University, South Africa. He is a Visiting Fellow at the Institute for African Development (IAD) at Cornell University, USA. He is the former Senior Director of Research Development and Professor of Sustainable Agriculture at the University of the Free State, South Africa. He holds a PhD in Animal Breeding. Previously he was Dean of Agriculture at the University of Fort Hare, South Africa, where he is now an Adjunct Professor, and Professor and Director of the Postgraduate School of Agriculture and Rural Development at the University of Pretoria, South Africa. He serves on the Board of the South African Agricultural Research Council. During 2008/2009, he was a Fulbright Visiting Fellow at the Cornell International Institute for Food, Agriculture and Development at Cornell University (CIIFAD), USA. He is a member of the Academy of Science of South Africa (ASSAf), and has recently been appointed as Panel Chair for the national consensus study to revitalise agricultural education and training. He has successfully supervised more than thirty Masters and Doctoral students, and is the author/co-author of more than 150 journal articles, monographs, reports and book chapters. His fields of specialisation include African smallholder agriculture, research and innovation management, agricultural education and training, and capacity development.

**Zenda Ofir** is a South African international evaluation specialist. With a PhD in Chemistry, she initially worked as manager in a South African science council, responsible for *i.a.* national higher education/industry research programmes on biotechnology and on food production and security, and later as Director Research at the University of Pretoria. Since then she has worked on assignments in nearly 40 countries on four continents, conducting evaluations, facilitating the design of monitoring and evaluation systems and helping programmes and organizations clarify their change logic ('theories of change') towards impact. She has done extensive work in the agriculture sector for *i.a.* Rockefeller Foundation, the International Fund for Agricultural Development (IFAD), the Alliance for a Green Revolution in Africa (AGRA), the Consultative Group on International Agricultural Research (CGIAR) and AWARD, a Bill and Melinda Gates Foundation funded programme cultivating African women leaders in agricultural research and development. Zenda is a former President of the African Evaluation Association (AfrEA) and former Vice-President of the International Organization for Cooperation in Evaluation (IOCE). She also served on the first steering committee of the Network of Networks on Impact Evaluation (NONIE) - a global network representing hundreds of donor organizations and professional evaluation networks - and the first Board member living outside North America of the (now) nearly 8,000 member American Evaluation Association (AEA).

**Aldo Stroebel** is Executive Director of International Relations and Cooperation at the National Research Foundation (NRF), South Africa, and Affiliated Associate Professor at the Centre for Sustainable Agriculture, Rural Development and Extension, at the University of the Free State (UFS), South Africa. He is the former Director of International Academic Programmes in the Office of the Vice-Chancellor, UFS. He is a well-known agriculture- and rural development specialist, focusing on smallholder farming systems in Africa. He qualified with a Masters degree from the University of Ghent, Belgium before researching for his PhD at UFS and Cornell University, USA. He was a Postdoctoral Research Fellow at Wageningen University, The Netherlands. He has successfully supervised more than fifteen Masters and Doctoral students, and is the author/co-author of more than eighty journal articles and publications. At the University of Pretoria, he was the Programme Coordinator of the Integrated Rural Development Programme (IRDP) in partnership with the W.K. Kellogg Foundation, directing the establishment of a model for human capital and economic development in nine geographical areas in six southern African countries, including Botswana, Lesotho, Mozambique, South Africa, Swaziland and Zimbabwe. He was recently elected as one of twenty Founding Members of the South African Young Academy of Science (SAYAS). His fields of specialisation include smallholder farming systems, human capital development and capacity building, and internationalisation of higher education.

# FOREWORD

Africa is growing - economic growth has continued to rise steadily at an average of five percent per annum for the past decade. At the same time the continent's population continues to expand and it is expected that Africa's population will quadruple within the next century. In the decades to come, this continued expansion will require a concomitant increase in employment opportunities, the acceleration of economic transformation and a substantial upsurge in agricultural production in order to ensure food and nutrition security for this ever-growing population.

As a sector, agriculture has the potential to contribute significantly to food and nutrition security whilst simultaneously driving inclusive economic growth. More than two-thirds of the African population rely on agriculture for their livelihood (including income) and the sector represents a third of the continent's GDP. From this vantage point, efforts to improve farmer productivity and incomes can drive demand and growth in other economic sectors. In fact, the World Bank has calculated that growth in the agriculture sector is 2.5 times more effective at reducing poverty than growth in other sectors, and more recently research has shown that in sub-Saharan Africa agriculture is eleven times more effective than any other sector at reducing poverty. Although almost all African countries could realize greater potential from agriculture, wide disparities in the extent to which countries prioritise investments and policy reform remain pervasive. Despite its abundance of natural resources – if Africa is to realize the potential of its agricultural sector – intervention is needed to overcome inadequate infrastructure, the insufficient high-level human and research capacity, the paucity of enabling policy and business environments, and the need for increased public, private and donor funding.

Ten years ago, at the African Union summit held in Maputo, African leaders made a bold commitment to reverse the underinvestment that continued to hold back the sector by pledging to allocate at least ten percent of national budget to agriculture, adopt sound agricultural development policies and achieve at least six percent agricultural growth. Regardless of the country-specific plans developed through the Comprehensive Africa Agriculture Development Programme (CAADP), progress on the Maputo declaration targets has been inconsistent and erratic with many countries falling short of their intended targets. The absence of proper governance structures – including the lack of accountability structures and tracking mechanisms – continue to stunt progress and development, particularly in priority areas such as bolstering human capital development and addressing the gender gap.

In the above context, the need to transform agricultural education and training to improve its effectiveness and efficiency is now clearly recognized by African leaders and development partners. The Ministerial Conference on Higher Education in Agriculture (CHEA) held in Uganda (2010) and the creation of the Tertiary Education for Agriculture Mechanism for Africa (TEAM-Africa) in September 2012 represent both the urgency of and the unified response to the need for substantial shifts in AET. The support of prominent African institutions, for example the African Union (AU), the New Partnership for Africa's Development (NEPAD), the Forum for Agricultural Research in Africa (FARA) and the Regional Universities Forum for Capacity Building in Agriculture (RUFORUM) attest to this.

It is an opportune time during 2014, the AU Year of Agriculture and Food Security, to reconsider the continent's approach to AET. At the upcoming AU summit in July 2014 African leaders will review the Maputo Declaration to make new policy commitments for the next decade. At the same time, CAADP is undergoing a reform process to ensure that it can provide more effective guidance and deliver an improved results framework. In addition, as we move forward to finalise the continent's vision for the next 50 years - Agenda 2063 - it is encouraging that this book *Towards Impact and Resilience: Transformative Change in and through Agricultural Education and Training in sub-Saharan Africa* is published at this time.

The authors recognize that AET should be a top priority in the continent's development through the education of human capital. The book is broadly divided into four parts: the African context, particularly sub-Saharan Africa (highlighting FARA's contribution in capacity development and strengthening); relevant global experiences (including Europe, Asia and South America); lessons in AET from sub-Saharan Africa; and critical success factors including systems approaches, mainstreaming gender, quality assurance, financing, agricultural innovation and entrepreneurship. The concluding chapter synthesizes these ideas, experiences and evidence in order to highlight critical areas and proposes potential solutions.

As African leaders are planning for Africa's response to the post-2015 African agenda that will replace the Millennium Development Goals a vibrant, resilient and effectively functioning AET system is needed more than ever. I commend the editors of this book for their relentless commitment to an African AET system which can deliver a strengthened platform from which African agriculture can continue to grow and contribute to the continent's development and prosperity.

—*Professor Monty Jones*
Special Advisor to the President of Sierra Leone
Immediate past Executive Director, FARA and Chairperson GFAR
World Food Prize Laureate 2004

# PREFACE

The University of the Free State (UFS), South Africa, received a grant from the W.K. Kellogg Foundation (WKKF) to establish partnerships, interrogate agricultural education and training (AET) models, share information, and develop together with partners and key actors on the continent a proposal for a project that could transform the agricultural education and training sector in sub-Saharan Africa. Good progress was made in some aspects, in particular in cultivating partnerships and networks, and gaining and sharing insights into what may work in a long-term effort to develop the sector. However, due to changes in WKKF's representation in southern Africa, the project did not continue as intended. Instead, the project energies and resources were redirected to produce a high quality, highly relevant and tangible output in the form of this book. The book has been developed to ensure that the knowledge obtained through the work done is consolidated and enriched by the expertise of leading figures in the field of AET in Africa and worldwide.

The global and continent-wide trends affecting agriculture, and the extraordinary opportunity offered by Africa's increasingly critical role in the world, highlight the need for a dramatic increase in agricultural expertise, as well as a shift in focus, over the coming decades. Skills for farm productivity and employment in the public sector need to expand to include those in post-farm value-addition, related policy and regulatory issues, and private sector employment. Entrepreneurs, exporters, importers, and suppliers have also become more interested in African agriculture since the agricultural commodity "price bubble" of 2007–2008, drawing attention to the need for a similar upgrade of skills in areas such as business and financial management, marketing, and communications.

Appropriate and adequate education and training at all levels and in many spheres are therefore essential for a revitalised agricultural sector. Agricultural development underpins all other development efforts. For example, the World Bank, in a 2006 synthesis of an institutional, regional, and international review, argues that the contributions of agricultural education and training lead to a variety of significant productivity increases. The importance of agriculture for development, and the role of AET within agriculture, have also been reinforced by USAID and other prominent actors. A mechanism—TEAM-Africa (Tertiary Education for

Agriculture Mechanism)—has been established and endorsed by the African Union, the Forum for Agricultural Research in Africa, and the NEPAD Planning and Coordination Agency to define and promote transformation of the agricultural education sector, as well as to facilitate, convene, and coordinate stakeholders and reform initiatives on the continent.

More specifically, in a highly competitive, innovation-driven, and knowledge-intensive world, development strategies need to include as a top priority the education of human capital through quality higher education. Development takes place when effective institutions and productive enterprises are supported and advanced by the collective generation and use of knowledge and the expansion of social and technological capabilities that lead to—and build on—transformative change.

Twenty-two esteemed international and African scholars contributed to this manuscript, which aims to transform the AET sector for the benefit of agriculture and development in sub-Saharan Africa. Consisting of seventeen chapters, the book focuses on clarifying the challenges, issues, and priorities of agricultural education and training in SSA and provides suggestions for practical solutions that can help guide organisations interested in furthering AET for agricultural development on the continent.

The book is divided into four sections that focus on: 1) predicting and illuminating the context in Africa, and particularly in SSA, within which AET will have to flourish over the coming decades; 2) highlighting relevant experiences from around the world—from Europe, North America, Latin-America, and Asia (especially South and Southeast Asia)—in developing human potential through agricultural education and training; 3) drawing lessons in agricultural education and training from relevant experiences in sub-Saharan Africa; and 4) addressing critical issues for success—systems, quality, gender, financing, innovation, and entrepreneurship. The concluding chapter synthesises the ideas, experiences, and evidence from the chapters in order to highlight critical issues as well as possible solutions.

The book is uniquely positioned to support a call to action on AET, to pull together state-of-the-art knowledge from within and outside SSA, and to advance out-of-the-box thinking about the principles, values, and character of AET for development, with an emphasis on the models that can help to cultivate leaders and change-makers at all levels of the agricultural sector.

# ACKNOWLEDGMENTS AND PEER-REVIEW PROCESS

The Editors would like to acknowledge the valuable support and participation of all contributing authors, the Editorial Advisory Committee, individuals who provided editorial assistance, and organisations for financial and other support.

## Editorial Advisory Committee

| | |
|---|---|
| Prof. Izak Groenewald | Director, Centre for Sustainable Agriculture, Rural Development, and Extension, University of the Free State, South Africa |
| Ms. Nancy Hart | International Editor, Italy and USA |
| Prof. Monty Jones | Executive Director, Forum for Agricultural Research in Africa (FARA); Chair of the Global Forum on Agricultural Research (GFAR); and World Food Prize Laureate, Ghana |
| Dr. Vuyo Mahlati | Private Consultant and Commissioner, National Planning Commission in the Presidency, South Africa |
| Prof. Lindela Ndlovu | Vice Chancellor, National University of Science and Technology, Zimbabwe |
| Prof. Willis Oluoch-Kosura | Programme Director, Collective Masters Programme in Agricultural and Applied Economics (CMAAE), Kenya |
| Ms. Myra Wopereis-Pura | Director for Access to Knowledge Technologies, Forum for Agricultural Research in Africa (FARA), Ghana |

## Editorial Assistants

| | |
|---|---|
| Dr. Melody Mentz (Leader) | Director, Melody M Consulting, South Africa |
| Dr. Hendrik Swanepoel | Veterinarian, Swans Veterinary Services, Australia |
| Ms. Esley van der Berg | Research Assistant, Melody M Consulting; and Independent Consultant, South Africa. |
| Ms. Liana Stroebel | Provincial Coordinator, Farmer Development, GrainSA, South Africa |


## Financial and Other Contributions

| | |
|---|---|
| WK Kellogg Foundation, USA | University of the Free State, South Africa |
| Forum for Agricultural Research in Africa, Ghana | Stellenbosch University, South Africa |
| Institute for Africa Development, Cornell University, USA | SunMedia, South Africa |
| National Research Foundation, South Africa | |

The Editors are greatly indebted to the Institute for African Development (IAD), Cornell University, USA, for accepting this publication as part of the IAD/Cambridge Scholars Book Series. Special mention must be made of Prof. Muna Ndulo, Director of IAD, Ms. Evangeline Ray, Publications Manager, and Ms. Mihret Tamrat, Student Assistant.

Special mention is due to Prof. Mandi Rukuni, former Director of the WKKF Africa Office, who conceptualised this project, for his unwavering support and commitment throughout.

We gratefully acknowledge the Stellenbosch Institute for Advanced Study (STIAS) for their support through appointment of one of the Editors as a Fellow to focus on this work.


## Peer Review Process

The chapters in this book have been reviewed by persons chosen for their diverse perspectives and technical expertise in accordance with procedures required by the Department of Higher Education and Training, South Africa. In order to maintain and ensure the highest quality, a three-stage review process was followed:

1. A group of expert international and African authors were invited to prepare and present draft outlines of their chapters at a workshop held in Cape Town in 2010. The presentations were assessed over a three-day period by these peers as well as the Editorial Advisory Committee. The focus and content of the chapters and the book were critically assessed, refined, and sharpened.
2. Informed by the workshop consensus, all authors reviewed their chapters and submitted to the Editors. Each chapter was independently reviewed by the Editors and in addition, where necessary, by an acknowledged expert selected to provide candid and critical comments to further improve the content and to ensure scientific soundness. In addition, all chapters were reviewed by an internationally-recognised, independent highly authoritative professional editor.
3. Finally, the manuscript has been peer reviewed by the IAD/ Cambridge Scholars Publishing Editorial Board and anonymous reviewers to meet standards of the IAD/Cambridge Scholars Book Series and to warrant international expectations of objectivity, evidence, and responsiveness.

# ACRONYMS

| | |
|---|---|
| 2IE | International Institute for Water and Environmental Engineering (Burkina Faso) |
| AABSA | Association of African Business Schools Agribusiness Consortium |
| AAC | Agribusiness Consortium |
| AAPP | African Agriculture Productivity Programme |
| AAU | Association of African Universities |
| ABC | Agricultural Business Chamber |
| ACBF | African Capacity Building Foundation |
| ACCI | African Centre for Crop Improvement |
| ACE | African Centres of Excellence |
| ACE | American Council on Education |
| ACT | Agricultural Consultancy Training |
| ACU | Association of Commonwealth Universities |
| ADA | African Doctoral Academy |
| ADB | Asian Development Bank |
| ADC | Agricultural Development Council |
| ADEA | Association for the Development of Education in Africa |
| ADII | Agriculture, Development and Innovation Index |
| AEA | American Evaluation Association |
| AET | Agricultural Education and Training |
| AET4D | Agricultural Education and Training for Development |
| AEZ | Agro-Ecological Zone |
| AFAPP | African Food and Agricultural Policy Platform |
| AfrEA | African Evaluation Association |
| AfriQAN | African Quality Assurance Networks |
| AFSIP | Agricultural Food Security Plan |
| AFSRN | Asian Farming Systems Research Network |
| AgGDP | Agricultural Gross Domestic Product |
| AGRA | Alliance for a Green Revolution in Africa |
| AGROCURI | Agricultural Open Curriculum and Learning Initiative |
| AIAT | Assessment Institutes for Agricultural Technology |

| | |
|---|---|
| AIBP | Agro-Industrial By-Products |
| AIDS | Acquired Immunodeficiency Syndrome |
| AIS | Agricultural Innovation System |
| AKIS | Agricultural Knowledge and Information System |
| ALO | Association Liaison Office for University Cooperation and Development |
| AMUL | Anand Milk Union Limited |
| ANAFE | Agroforestry and Natural Resources Education |
| ANK | Ansanm Nou Kapab (Together We Can) |
| APLU | Association of Public and Land-grant Universities |
| ARD | Agricultural Research and Development |
| AR4D | Agricultural Research for Development |
| ARFSN | Asian Rice Farming Systems Network |
| ASARECA | Association for Strengthening Agricultural Research in Eastern and Central Africa |
| ASSAF | Academy of Science of South Africa |
| ASTI | Agricultural Science and Technology Indicators |
| ATAI | Agricultural Technology Adoption Initiative |
| ATVET | Agricultural Technical and Vocational Education and Training |
| AU | African Union |
| AUCC | Association of Universities and Colleges of Canada |
| AUF | *Agence Universitaire de la Francophonie* |
| AWARD | African Women in Agricultural Research and Development |
| BA | Bachelor of Arts |
| BBBEE | Broad Based Black Economic Empowerment |
| BE | Business and Enterprise |
| BEAT | Barefoot Education for Africa Trust |
| BECA | Biosciences Eastern and Central Africa |
| BI | Bridging Institutions |
| BMGF | Bill and Melinda Gates Foundation |
| BRAC | Bangladesh Rural Advancement Committee |
| BRICS | Brazil, Russia, India, China and South Africa |
| BSc | Bachelor of Science |
| CAADP | Comprehensive Africa Agriculture Development Programme |
| CAMES | *Conseil Africain et Malgache pour l'Enseignement Supereur* |
| CASREN | Crop-Animal Systems Research Network |

| | |
|---|---|
| CBO | Community-Based Organisation |
| CD | Compact Disc |
| CGIAR | Consultative Group on International Agricultural Research |
| CHE | South African Council on Higher Education |
| CHEA | Conference on Higher Education in Agriculture in Africa / Council of Higher Education Accreditation |
| CIAT | *Centro Internacional de Agricultura Tropical* |
| CIFOR | Center for International Forestry Research |
| CIIFAD | Cornell International Institute for Food, Agriculture and Development |
| CIMMYT | *Centro Internacional de Mejoramiento de Maiz y Trigo* |
| CIP | *Centro Internacional de la Papa* |
| CLA | Community Livestock Auxiliary |
| CMAAE | Collaborative Masters of Agricultural and Applied Economics |
| CORAF | *Conseil Ouest et Centre Africain Pour la Recherche et le Développement Agricoles* / West and Central African Council for Agricultural Research and Development |
| COS1 | Convergence of Sciences Programme |
| CREST | Centre for Research on Evaluation, Science and Technology |
| CRP | Consultative Group on International Agricultural Research Research Programme |
| CRSP | Collaborative Research Support Programme |
| DANIDA | Danish International Development Agency |
| DBSA | Development Bank of Southern Africa |
| DFID | Department for International Development |
| DGIS | Ministry of International Cooperation |
| DoE | Department of Education |
| DP | Development Plan |
| DP | Development Partner |
| DS | Diploma Supplement |
| E&SG | Equitable and Sustained Growth |
| EAAP | European Association of Animal Production |
| EAC | European Consortium for Accreditation |
| EACEA | Education, Audiovisual and Culture Executive Agency |

| | |
|---|---|
| *EACI* | European Association for Creativity & Innovation |
| EAPP | East Africa Productivity Programme |
| EAQA | European Association for the Quality Assurance |
| EAQAHE | European Association for the Quality Assurance in Higher Education |
| EARTH University | *Escuela de Agricultura de la Region Tropical Humeda* |
| ECA | European Consortium for Accreditation |
| ECHAE | European Conference on Higher Agricultural |
| ECTS | European Credit Transfer Accumulation System |
| EE | Enabling Environment |
| EHEA | European Higher Education Area |
| EI | Education International |
| EIWR | Ethiopian Institute of Water resources |
| ELLS | Euroleague for Life Sciences |
| EMBRAPA | *Empresa Brasileira du Pesquisa Agropecuária* (Brazilian Agricultural Research Corporation) |
| ENLASA | *Enlace Agropecuario S.A.* |
| ENQA | European Association for Quality Assurance in Higher Education |
| ENSA | *Ecole Nationale Superieure Agronomique* |
| EQAR | European Quality Assurance Register |
| ERASMUS | European Community Action Scheme for the Mobility of University Students |
| ESAMI | Eastern and Southern African Management Institute |
| ESAP | Economic Structural Adjustment Programme |
| ESCAP | Economic and Social Commission for Asia-Pacific |
| ESG | European Standards and Guideline |
| ESIB | National Unions of Students in Europe |
| ESU | European Students' Union |
| EU | European Union |
| EUA | European University Association |
| EU-RA | European Research Associates |
| EURASHE | European Association of Institutions in Higher Education |
| EUROP GAP | Euro retailer produce / Good Agricultural Practice |
| EURYDICE | Information on Education System and Policies in Europe |
| FAAP | Framework for African Agricultural Productivity |
| FAC | Future Agricultures Consortium |

| | |
|---|---|
| FANRPAN | Food, Agriculture and Natural Resources Policy Analysis Network |
| FAO | Food and Agriculture Organization |
| FAOSTAT | Food and Agriculture Organization Statistical Database |
| FARA | Forum for Agricultural Research in Africa |
| FDI | Foreign Direct Investment |
| FEF | Farmer Extension Facilitator |
| FETAD | Farmer Empowerment through Agribusiness Development |
| FFS | Farmer Field School |
| FLS | Farmer Life School |
| FSR | Farming Systems Research |
| GCHERA | Global Consortium for Higher Education and Research in Agriculture |
| GDP | Gross Domestic Product |
| GEM | Global Entrepreneurship Monitor |
| GER | Gross Enrolment Ratios |
| GFAR | Global Forum for Agricultural Research |
| GIMPA | Ghana Institute of Management and Public Administration |
| GIZ | *Gesellschaft für Internationale Zusammenarbeit* |
| GM | Genetically modified |
| GNP | Gross National Product |
| GPI | Gender Parity Index |
| GTZ | *Gesellschaft Technische Zusammenarbeit* / German Technical Corporation |
| H&R | Hit and Run Development |
| HACCP | Hazardous Analyses Critical Control Points |
| HED | Higher Education for Development |
| HEI | Higher Education Institution |
| HEQC | Higher Education Quality Committee |
| HIV | Human Immunodeficiency Virus |
| HLPE | High Level Panel of Experts |
| HOORC | Harry Oppenheimer Okavango Research Centre |
| IAASTD | International Assessment of Agricultural Knowledge, Science and Technology for Development |
| IAC | Inter Academy Council |
| IAD | Institute for African Development |
| IADB | Inter-American Development Bank |

| | |
|---|---|
| IAR4D | Integrated Agricultural Research for Development |
| IARC | International Agricultural Research Centre |
| ICA | Interfaculty Committee Agraria |
| ICARDA | International Center for Agricultural Research in the Dry Areas |
| ICP | International Course Programme |
| ICRAF | World Agroforestry Centre |
| ICRISAT | International Crops Research Institute for the Semi-Arid Tropics |
| ICT | Information and Communication Technologies |
| IDA | International Development Association |
| IDAT | Institutional Development for Agricultural Training |
| IDRC | International Development Research Centre |
| IFAD | International Fund for Agricultural Development |
| IFAMA | International Food and Agribusiness Management Association |
| IFC | International Finance Corporation |
| IFPRI | International Food Policy Research Institute |
| IFSA | International Farming Systems Association |
| IITA | International Institute of Tropical Agriculture |
| ILRI | International Livestock Research Institute |
| IMF | International Monetary Fund |
| IMRD | International Master Programme in Rural Development |
| INREF | Interdisciplinary Research and Education Fund |
| IOCE | International Organization for Cooperation in Evaluation |
| IP | Investment Plan |
| IPM | Integrated Pest Management |
| IPMS | Improving Productivity & Market Success of Ethiopian Farmers |
| IRDP | Integrated Rural Development Programme |
| IRRI | International Rice Research Institute |
| ISNAR | International Service for National Agricultural Research |
| ISO | International Standardization Organisation |
| JFFLS | Junior Farmer Field and Life School |
| JiT | Just in Time |
| JKUAT | Jomo Kenyatta University of Agriculture and Technology |

| | |
|---|---|
| JU | Jimma University |
| K4D | Knowledge for Development |
| KAM | Knowledge Assessment Methodology |
| KE | Knowledge and Education |
| KEI | Knowledge Economy Index |
| KVKs | *Krishivigyankendras* |
| KVL | Agricultural and Veterinary University of Copenhagen |
| LAN | Local Area Network |
| LEA | Local Enterprise Authority |
| LFS | Livestock Farming Systems |
| LGU | Local Government Unit |
| LPI | Logistics Performance Index |
| MA | Master |
| MARDI | Malaysian Agricultural Research and Development Institute |
| MBA | Master in Business Administration |
| MCS | Maluti Crop Share |
| MDFT | Multi-Donor Trust Fund |
| MDG | Millennium Development Goal |
| MF | Master Farmer |
| MSc | Master of Science |
| MSME | Micro, Small and Medium Enterprise |
| MSU | Michigan State University |
| MVA | Manufacturing Value Added |
| NAADS | National Agricultural Advisory Services |
| NAES | National Agricultural Extension System |
| NAETS | National Agricultural Education and Training System |
| NAIS | National Agricultural Innovation System |
| NARI | National Agricultural Research and Innovation |
| NARS | National Agricultural Research Systems |
| NASULGC | National Association of State Universities and Land-Grant Colleges |
| NCFR | Non-Conventional Feed Resources |
| NEPAD | New Partnership for Africa's Development |
| NERICA | New Rice for Africa |
| NGO | Non-Governmental Organisation |
| NONIE | Network of Networks on Impact Evaluation |
| NOQA | Nordic Quality Assurance Network in Higher Education |
| NORAD | Norwegian Agency for Development Cooperation |

| | |
|---|---|
| NPCA | NEPAD Planning and Coordinating Agency |
| NRF | National Research Foundation |
| NRM | Natural Resource Management |
| NSF | National Science Foundation |
| NUFFIC | Netherlands Universities Foundation for International Cooperation |
| NWO | Dutch National Science Foundation |
| OECD | Organisation for Economic Cooperation and Development |
| OFR | On-Farm Research |
| OSU | Ohio State University |
| PAID | Pan African Institute for Development |
| PAU | Punjab Agricultural University |
| PEARL | Partnership for Enhancing Agriculture in Rwanda through Linkages |
| PEST | Policy, Economic, Social, Technology |
| PhD | Doctor of Philosophy |
| PHEA | Partnership for Higher Education in Africa |
| PLAAS | Institute for Poverty, Land and Agrarian Studies |
| PNAS | Proceedings of the National Academy of Sciences |
| PPPs | Public-Private Partnerships |
| PQM | Programme and Qualification Mix |
| PRA | Participatory Rural Appraisal |
| PRDA | Provincial Rural Development Administration |
| PTD | Participatory Technology Development |
| R&D | Research and Development |
| R&E | Research and Evaluation |
| RED | Real Enterprise Development |
| REESAO | *Réseau pour l'Excellence de l'Enseignement Supérieur en Afrique de l'ouest* |
| RRA | Rapid Rural Appraisal |
| RUFORUM | Regional Universities Forum for Capacity Building in Agriculture |
| S&G | Stop and Go |
| SACCAR | Southern African Centre for Cooperation in Agricultural and Natural Resources Research and Training |
| SADC | Southern African Development Community |
| SADC-FANR | SADC Food, Agriculture and Natural Resource Directorate |

| | |
|---|---|
| SAFEX | South African Futures Exchange |
| SAQA | South African Qualifications Authority |
| SAREC | Sustainable Agriculture Research and Extension Center |
| SASACID | Strengthening Africa's Strategic Agricultural Capacity for Impact on Development |
| SAU | State Agricultural University |
| SAYAS | South African Young Academy of Science |
| SCARDA | Strengthening Capacity of Agricultural Research for Development |
| SCSU | South Carolina State University |
| SD | Sustainable Development Department |
| SDRE | Agricultural Education Group of the Extension, Education and Communication Service |
| SDRE | Extension, Education and Communication Service |
| SEAMEO | Southeast Asian Ministers of Education Organisation |
| SEARCA | Southeast Asia Regional Centre for Graduate Study and Research in Agriculture |
| SEDA | Small Enterprise Development Agency |
| SEMCIT | Sustainability, Education and Management of Change in the Tropics Project |
| SETA | Sector Education and Training Authority |
| SHF | Smallholder Farmer |
| SIDA | Swedish International Development Cooperation Agency |
| SIDA-SAREC | Swedish International Development Cooperation Agency and Sustainable Agriculture Research and Extension Center |
| SINAES | National Accrediting Agency of Costa Rica |
| SLP | Short Learning Program |
| SOFA | State of Food and Agriculture |
| SRO | Subregional Organisation |
| SRUPNA | Small Ruminant Production Systems Network |
| SSA | sub-Saharan Africa |
| SSI | Self-Sufficiency Index |
| STIAS | Stellenbosch Institute for Advanced Study |
| SUA | Sokoine University of Agriculture |
| T&V | Training and Visit |
| TAE | Tertiary Agricultural Education |

| | |
|---|---|
| TEAM | Training in Environmental and Agricultural Management |
| TEAM-Africa | Tertiary Education for Agriculture Mechanism in Africa |
| TEEAL | The Essential Electronic Agricultural Library |
| TIPCEE | Trade and Investment Programme for Competitive Export Economy |
| TLU | Tropical Livestock Unit |
| TSFS | Three-Strata Forage System |
| TTFP | Third Trimester Field Programme |
| TVET | Technical and Vocational Education and Training |
| UA | University of Antwerp |
| UB | University of Botswana |
| UDS | University of Development Studies |
| UFS | University of the Free State |
| UK | United Kingdom |
| UN | United Nations |
| UNAM | University of Namibia |
| UNCTAD | United Nations Conference on Trade and Development |
| UNDP | United Nations Development Programme |
| UNECA | United Nations Economic Commission for Africa |
| UNESCO | United Nations Educational, Scientific and Cultural Organization |
| UniBRAIN | Universities Business Research in Agricultural Innovation |
| UNICE | Union of Industrial and Employers' Confederations of Europe |
| UNIDO | United Nations Industrial Development Organisation |
| UOS | University of Stavanger |
| UPLB | University of the Philippines College of Agriculture at Los Baños |
| US | United States |
| USA | United States of America |
| USAID | United States Agency for International Development |
| USIU | United States International University |
| VLIR | *Vlaamse Interuniversitaire Raad* |
| VLIR/VLHO RA | *Vlaamse Interuniversitaire Raad / Vlaamse Hogescholen Raad* |

| | |
|---|---|
| VLIR-UOS | *Vlaamse Interuniversitaire Raad* and University of Stavanger |
| WAAPP | West Africa Agricultural Productivity Programme |
| WASH | Water Sanitation and Health |
| WECARD | West- and Central-African Council for Agricultural Research and Development |
| WKKF | W.K. Kellogg Foundation |
| WREM | Water Resources and Engineering Management |
| WUR | Wageningen University and Research Centre |
| YCFP | Young Commercial Farmers Education Programme |

# PART I

# CONCEPTUAL FRAMEWORK AND THE AFRICAN CONTEXT

# CHAPTER ONE

# AGRICULTURAL EDUCATION AND TRAINING FOR DEVELOPMENT IN SUB-SAHARAN AFRICA

FRANS SWANEPOEL, ALDO STROEBEL AND ZENDA OFIR

## Africa in the World

Africa is changing. It has become a continent of hope. While much of the developing world is mired in financial difficulties and economic stagnation, Africa's per capita income growth has recently been among the highest in the world. In sub-Saharan Africa (SSA), the Gross Domestic Product (GDP) growth accelerated from an average annual rate of 2 percent in the 1990s to 5.5 percent over the past decade (IMF, 2011); the number may even be higher. Key countries such as Ghana and Nigeria are revising their GDP estimates, which were based on outdated data and may have been too conservative (Jerven, 2013). Growth on the continent, which declined in the wake of the global financial crisis, recovered in 2010, supporting IMF predictions that Africa rather than Asia could be the fastest growing continent over the next few years. Contrary to conventional wisdom, commodities form only a third of the GDP increase; the rest comes from services, manufacturing, and construction (Renaissance Capital, 2012). Between 2000 and 2011, Africa's trade with the rest of the world increased by over 200 percent.

More than a fifth of the fifty-four countries in Africa grew at an annual rate of 7 percent or more, sufficient to double the economy in ten years. Nine of these are in SSA and three have been growing at more than 10 percent per year (South Africa, Niger, Sierra Leone) (IMF, 2011; Renaissance Capital, 2012; World Bank, 2013). Although some observers question Africa's development trajectory, a few countries appear to be performing better in certain key indicators than the Asian "tigers" did during their first decade of rapid economic development (Samman, 2012).

Forecasts for the future are generally positive, in particular as Africa's human capital potential continues to grow. The gross secondary school enrolment rate increased ten-fold, from 3 percent in 1960 to 29 percent in 2005. SSA also has a young labour force. It is the only region in the world positioned to experience 15–20 percent growth in the crucial fifteen to twenty-four year age range in coming decades (IAC, 2004; UNDP, 2012). Urbanisation, which is seen as increasing the efficiency and productivity of a country, is also set to advance. UN-Habitat (2008) forecasts that eleven African metropolitan cities may see their populations increase by over 50 percent between 2010 and 2025.

Importantly, over the past thirty years, the African middle class has tripled in actual numbers to nearly 350 million people, or 34 percent of the total population, in 2010. For the whole period between 1980 and 2000, the percentage remained stable at 27 percent before rising rapidly. For the first time in history, the growth of the middle class over the past decade has been faster than the population growth (AfDB, 2013).

The disposable income of the growing middle class is making a visible difference. For example, in countries such as Nigeria, Mauritius, South Africa and Botswana, the demand for passenger cars more than doubled in the five years between 2002 and 2007. The phenomenal growth in information and communication technologies has led to an estimated 0.5 billion mobile phones on the continent. There were 80.6 million Internet users in 2008, compared to just 4.5 million in 2000. Instead of one in 2007, seven fibre optic cables were connecting SSA to the world in 2012, and that number is soon to be increased to twelve (Renaissance Capital, 2012). A strong middle class not only raises economic and living standards, acting as a catalyst to start up entrepreneurial services, it also brings greater awareness of human and social rights (AfDB, 2013). With the growing middle class, prospects for better governance are also rising.

But there is another side to Africa's development. Africa is home to fourteen of the least competitive countries in the world, and development rates are very uneven across the fifty-four countries. In terms of industrialisation, a joint study by the United Nations Conference on Trade and Development (UNCTAD) and the United Nations Industrial Development Organisation (UNIDO) found that most African countries are either stagnating or moving backwards (UNCTAD, 2009). Only five countries achieved a manufacturing value added (MVA) per capita growth above 4 percent. Approximately one-third of people living in extreme poverty are found in Africa, and it is the only region where the expected absolute number of poor people in 2015—

estimated to grow to more than 400 million—is likely to be higher than in 1990 (World Bank, 2013). Although some countries have made good progress towards achieving the Millennium Development Goals (MDGs), disparities are increasing, and many countries will not achieve most of their targets.

Most disconcertingly, according to the African Development Bank (ADB), a quarter of the African population continues to "float" between the poverty and middle class income bands, and the percentage is rising (AfDB, 2013). Special strategies and support are needed to enable the poor to become permanent members of the middle class.

This situation is in sharp contrast to the remarkable rate of poverty reduction achieved in East Asia and the Pacific, where the proportion of people living in extreme poverty has decreased from 77 percent in 1988 to 14 percent in 2009 (Fosu, 2009). This means that the inspiring economic progress of Africa over the past decade is not yet sufficiently robust. This has important implications for development strategies on the continent.

## The Importance of Agriculture

Over the centuries, economies on a positive development trajectory have made transitions from primary agriculture or extractive industries to manufacturing and value-added services. Yet in Africa, the share of manufacturing value added in its GDP fell from 12.8 percent in 2000 to 10.5 percent in 2008, while the share of manufactured goods in its total exports fell from 43 percent in 2000 to 39 percent in 2008. The share of global manufacturing exports from Africa has risen only slightly, from 1 percent in 2000 to 1.3 percent in 2008 (Bolaky, 2011).

Some therefore argue that Africa's development path should leapfrog manufacturing entirely, moving directly from agriculture to services as its engine of growth. Yet this would mean that Africa would fail to focus on capitalising the one sector that can not only provide it with a major competitive advantage globally, but that is crucial to grow for the sake of the well-being of its people—agriculture. Rapid urbanisation, growth of human capital on the continent, increasing trade and other connections to many different parts of the world, and the entrepreneurship, disposable income, and high expectations of the growing number of middle class Africans all provide an opportunity for agriculture to be framed, planned, and executed in a manner that surpasses earlier attempts.

Africa imports US$20 billion in food annually, yet has two-thirds of the world's uncultivated arable land (Asenso-Okyere and Jemaneh, 2012). In an upbeat study published by Renaissance Capital (2012), a leading emerging markets investment bank, a group of respected African economists and analysts note that the rapidly growing population and accelerating urbanisation across Africa also mean that agriculture and food producers will benefit from the higher growth on the continent. These analysts also point out that Africa is charged not only with feeding itself, but increasingly with feeding the rest of the world as well.

These needs are becoming increasingly urgent. Hunger, poverty and disease are inextricably linked, each contributing to the presence and persistence of the other two; malnutrition and hunger are the number one risk factors for illness worldwide. In SSA, hunger still kills more children than the infectious diseases HIV and AIDS, malaria, and tuberculosis combined (UNDP, 2012). More than a quarter of the population on the continent remains chronically undernourished (FAO, 2012). The situation is particularly dire in the Sahel[1] region, where an estimated 19 million people are at risk of food insecurity (FAO, 2013).

The World Development Report (World Bank, 2007a) concluded that in the twenty-first century, agriculture will remain a fundamental instrument for poverty reduction, economic growth, and environmental sustainability. In addition, a number of very influential initiatives have pointed out the critical role of agricultural science and technology in this process (IAC, 2004; Conway and Waage, 2010; European Commission, 2012; UNDP, 2012; WEF, 2013).

The World Bank and the International Food Policy Research Institute (IFPRI) furthermore note that approximately 70 percent of the MDGs' target group, particularly in Asia and Africa, live in rural areas, and that agriculture is a critical component for the successful attainment of the MDGs for the majority of the rural poor (World Bank, u.d.). Agriculture is linked to the MDGs in several ways:

- Agriculture relates directly to two of the goals, namely halving the proportion of people living in extreme poverty and hunger (MDG 1), and promoting gender equality and empowering women (MDG 3).
- Two other goals, namely environmental sustainability (MDG 7), and a global partnership for development (MDG 8), have reciprocal links with agriculture. In other words, progress in agriculture will affect progress towards these goals, just as progress towards these goals will influence progress in agriculture.
- The remaining goals—to achieve universal primary education (MDG 2), reduce child mortality (MDG 4), improve maternal health (MDG 5), and combat HIV and AIDS, malaria and other diseases (MDG

[1] Burkina Faso, Cameroon, Chad, the Gambia, Mali, Mauritania, the Niger, and Senegal.

6)—have indirect but vital links with improved agricultural productivity.

Agriculture also has a comprehensive and decisive impact on all spheres of African society: economic, social, and political. Economically, Africa remains a predominantly agricultural continent, and the sector is still regarded as the engine of growth in spite of neglect over the past decades. In most African countries 60–80 percent of the population are employed in the agriculture sector (World Bank, 2007). Socially, agriculture continues to play a critical role in countries where the vast majority of the population is rural. Agriculture is vital to livelihoods and must also provide food for an expanding urban population. Politically, agriculture has long been largely neglected by most African governments and partly also by international donors. Politicians maintain their power base in urban areas, and for a long time most donors held an unrealistic belief in the rapid modernisation of African society, thereby somewhat neglecting the overarching importance of agriculture.

This situation has now changed. African heads of state have pledged to support an agriculturally-led continental strategy and to increase their political support and budgetary allocations to the sector. High profile reports outlining new visions, priorities, and directions for African agriculture have emerged. The most significant of these initiatives are the Comprehensive African Agriculture Development Programme (CAADP) of the New Partnership for Africa's Development (NEPAD), a programme of the African Union (AU), now known as the NEPAD Planning and Coordination Agency (NPCA) (CAADP, 2013).

In the first decade of the twenty-first century, African leaders have finally—at this crucial stage in the continent's development—embraced agriculture as an essential component of their development strategies.

## The Performance of the Agricultural Sector in SSA

Agriculture remains the most dominant economic sector in SSA, accounting for 40 percent of GDP, 25 percent of exports, and 60–80 percent of employment (World Bank, 2007). The World Bank (2008) notes that growth in the sector—especially productivity growth among the continent's millions of smallholder farmers—would make a major contribution to broad-based poverty reduction and economic growth on the continent.

Yet over the past five decades, agriculture in SSA has performed worse than in any other region of the world. Per capita food production has declined, and Africa is the only continent where the incidence of child malnutrition is increasing. Per capita production of cereals over the period has fallen by about 15 percent, while cereal yields have risen less than 50 percent compared with rises of 180–300 percent in Asia (FAO, 2012a). Meanwhile, Africa's share of world agricultural trade fell from 5–6 percent during the 1960s and 1970s to 2 percent in 2000 before recovering somewhat to 3.5 percent in 2008 (FAOSTAT, 2011).

African agriculture is less modernised and smaller in scale than agriculture in other parts of the world (Stroebel et al., 2011). Approximately 90 percent of African farmers—the majority of whom are women—are engaged in smallholder farming systems where diversification is carried out for various reasons, including spreading climate and market risks, exploiting crop synergies, and increasing land productivity (Swanepoel et al., 2010). Although some African countries are known for large-scale farming, largely based on a colonial plantation heritage, the continent still maintains a traditional structure of small-scale farming. This traditional structure has prevailed, partly on a communal basis with land tenure systems.

CAADP confirmed that larger investments in agricultural research, extension, and education systems are required to achieve the targeted increase in agricultural output of 6 percent per year for the next twenty years (NEPAD, 2002). As noted earlier, Heads of State have pledged their support to increasing budgetary allocations to the sector. However, during the *Political Economy of Agricultural Policy in Africa Conference* held in Pretoria in March, 2013, co-organised by the Future Agricultures Consortium (FAC) with the Institute for Poverty, Land, and Agrarian Studies (PLAAS) in South Africa, key figures in the sector questioned this commitment. Few African governments have met the 2003 Maputo Declaration's mandate for investing 10 percent budgetary allocation in agriculture (IFPRI, 2008). The political will, means, and capacity to make agriculture flourish are not yet in abundance on the continent.

## Key Global Trends Affecting Agricultural Development in SSA

Turning around the poor performance of agriculture is an urgent priority for the continent. This urgency is accentuated by several global trends.

*Firstly,* climate change is likely to make Africa's food security and agricultural productivity challenge greater, as much of the continent is expected to experience increased drought and extreme weather events.

*Secondly,* evidence suggests that, for the foreseeable future, agricultural commodity prices in world markets will remain well above levels of the past four decades (OECD-FAO, 2011). Since 2008, prices in nearly all African countries have persistently exceeded previous levels (HLPE, 2011; Tschirley and Démbele, 2011). These high prices present great opportunities for farmers who have the knowledge and resources to respond with increased production. Yet this new price environment will not automatically spur broad, substantial production increases. Using the new price environment for broad-based productivity growth that includes smallholder farmers will require greater investment in improved technology such as improved seed varieties, animal breeds better matched to agro-ecological zones and extreme weather conditions, better water and plant nutrient management, and rotational systems. It will also require greater accessibility and quality of support services such as agricultural advisory and extension services, as well as linking to input and output markets for smallholder farmers. In order to manage all of the above points effectively, the role of human capital with the relevant knowledge and skills will be crucial.

*Thirdly*, the unexplored potential of African agriculture, including its vast tracts of under-utilised land, has been attracting significant interest from investors. Africa may increasingly have to feed the world. This will require new paradigms and innovations to spur growth in the sector, as well as an understanding among policy and decision-makers of how best to deal with the potential exploitation of its agricultural resources.

*Fourthly,* as whole agricultural value chains worldwide increasingly come under the control of massive international industrial and financial concerns, African farmers and other parts of the value chain on the continent run the risk of being exposed to increases in input prices, barriers to accessing inputs, and manipulation of food prices for political and commercial gain. Signs of the latter were already visible during the 2007–2008 crisis, when agricultural commodity prices reached historic highs. African actors in the value chains will be increasingly vulnerable unless policies, regulations, and protocols are put into place to prevent such situations from taking hold in the sector.

## Key Continental Trends Affecting Agricultural Development

Within Africa, three main trends are influencing the nature and magnitude of the challenges in the agriculture sector.

*Firstly,* the continent is urbanising more rapidly than any other area of the world. This remarkable economic turnaround is the result of, among other things, increasing macro-economic stability, reforms that have shaped market imperfections, and most significantly, the increasingly competitive and rapidly growing global demand for the natural resource-based commodities exported by SSA countries (MGI, 2010).

Because much of the resulting income growth is occurring in urban areas,[2] these developments are driving towards an explosion in demand for marketed food and a rapid change in the types of foods that consumers want. These patterns make agricultural productivity growth even more urgent, not only at the farm level but also increasingly in the post-farm system of processing and marketing. For the education sector, it will be crucial to respond to the demand for newly required skills all along the value chains in the agriculture sector and to ensure that the human capital needed to address these new challenges will be available. This provides SSA with an extraordinary opportunity for transformation and sustained growth.

*A second trend*, linked to the first, is a substantial increase in foreign direct investment in African agriculture, forestry, and the rural tourism sectors.[3] Observers have characterised some of these investments as "land grabbing," criticising African governments for allowing such transactions to happen in an uncontrolled manner (Aliyu, 2012). Indeed, they threaten dispossession for some rural households, raise legitimate concerns about local authority, and impact on equitable growth. Yet they also present real opportunities for a sector and continent that have had, as documented in the Framework for African Agricultural Productivity (FAAP), enormous difficulties generating productivity-enhancing investment over at least the past four decades. Understanding the threats and opportunities posed by this phenomenon is of great importance if it is to be turned into a developmental advantage.

*Thirdly,* African leaders increasingly regard regional integration as the most critical ingredient needed to drive growth. This is dependent on accelerating the development of infrastructure, including for the agriculture sector, on the continent. For centuries, Africa has lacked rail, road, and water freight assets. Bottlenecks in infrastructure have greatly impeded growth in global exports while also hindering the development of intraregional trade. In many countries, the rapid rates of urbanisation are stretching existing infrastructure, resulting in congestion, poor delivery, and power outages, all with serious implications for productivity and public and private sector expansion.

---

[2] Exact data are lacking on rates of per capita income growth in rural and urban areas. However, the continuing higher rates of poverty in rural areas, stagnant agricultural productivity, and much higher rates of capital investment in urban areas suggest that most recent income growth is concentrated in urban areas.

[3] Most foreign direct investment (FDI) data is not sufficiently disaggregated to develop a detailed quantitative assessment of these trends. FAO (2009) suggests that domestic investment in rural areas has increased greatly over the past five years. UNCTAD (2009) provides more focus on agriculture and highlights growth (but only up to 2008).

The renewed and growing emphasis on infrastructure development over the past decade thus has the potential to greatly enhance economic growth in the region. In a breakthrough event at the 2013 African Summit of the World Economic Forum, a priority list of fifteen out of fifty-one proposed infrastructure projects was drawn up by African leaders (WEF, 2013). These are multi-government, cross-regional initiatives designed to catalyse regional integration, propel growth through private sector activity, and link African markets. The development of agricultural infrastructure will be critical in attracting further investments in the sector.

## AET for Development

The global and continent-wide trends affecting agriculture and the extraordinary opportunity offered by Africa's increasingly critical role in the world also highlight the fact that over the next decades agricultural expertise will not only need to increase in scope, but will also need to shift in focus. Among other things, skills for farm productivity and employment in the public sector need to expand to include those in post-farm value-addition, related policy and regulatory issues, and private sector employment. Entrepreneurs, exporters, importers, and suppliers became more interested in African agriculture after the agricultural commodity "price bubble" in 2007–2008, drawing attention to the fact that areas such as business and financial management as well as marketing and communications should also be included in the expansion of skills development.

Appropriate and adequate education and training, at all levels and in many spheres, are therefore essential for a revitalised agricultural sector. Education and training underpin all other development efforts. For example, a 2006 World Bank review synthesising institutional, regional, and international reports argues that the contributions of agricultural education and training (AET) lead to a variety of significant productivity increases. The importance of agriculture for development, and the role of AET within that, has also been reinforced by USAID (2010).

More specifically, in a highly competitive, innovation-driven, and knowledge-intensive world, development strategies need to include the education of human capital through quality higher education as a top priority. Development takes place when effective institutions and productive enterprises are supported and advanced by the collective generation and use of knowledge, and the expansion of those social and technological capabilities that lead to, and build on, transformative change (Chang, 2010).

NEPAD reinforces this notion for the agriculture sector in Africa: "the quality of tertiary agricultural education is critical because it determines the expertise and competencies of scientists, professionals, technicians, teachers, civil servants, and business leaders in all aspects of agriculture and related industries" (NEPAD, 2002). The higher education sector is at the core of a strong AET system that produces scholars and researchers—the "knowledge creators" responsible for innovations and their application—and the social understanding that can advance the African economy and ensure its competitiveness (Kapur and Crowley, 2008).

The higher education sector is also the primary source of human capital for agricultural research and development (R&D) agencies and for the academic staff of learning institutions. University graduates are influential in the staffing, management, funding, and operation of colleges, training centres, research centres, extension, and other rural services. In addition to building highly critical human resources, the higher education sector has a direct impact on agricultural productivity through its research, innovation, and community outreach programmes (Davis et al., 2008; Kingslow, 2007).

In spite of the significant constraints discussed later in this chapter, the potential is growing for the AET system to become central in transformative change and, in turn, to contribute to development impact on the continent.

## AET as International Priority

International evidence confirms the critical role of AET in agricultural development. There is now a wealth of information, based on research and studies, supporting and advocating the importance of higher education for accelerated development and transformation in Africa (African Union, 2006; Kellogg et al., 2008; Parker, 2012; UNESCO/World Bank, 2000; World Bank, 2008). Yet, according to Johanson and Saint (2007), though vital to development, AET has been much neglected in agricultural strategies and action plans in many SSA countries.

This situation is, however, improving. This was reinforced by, among others, the African Ministerial Conference on Higher Education in Agriculture in Africa (CHEA), hosted in November 2010 by the Government of Uganda and the Regional Universities Forum for Capacity Building in Agriculture (RUFORUM). This important meeting confirmed the African governments' commitment to a "renewed and vigorous emphasis" on restoring the quality of higher education in agriculture and to increased investment in agricultural education as part of CAADP Country Compacts and Medium Term Agricultural Productivity Programmes (CAADP, 2013).

AET has been an integral part of national strategies in countries such as India, Brazil, Malaysia, Chile, and the Philippines that have developed their agricultural sectors successfully (Staatz and Démbele, 2008). These countries have achieved notable successes in establishing productive and financially sustainable AET systems through consistently high levels of investment in AET, both by the countries themselves and by their development partners.

A World Bank review (2007) identified a set of critical factors that have contributed to building productive AET systems. These include:

- mobilising and sustaining political support for AET investments;
- supporting public investment in capacity building to create the scientific leadership needed to implement the agricultural and rural development strategies;
- building a system of core AET institutes that make sustained commitments over multiple generations to produce returns;
- undertaking massive campaigns to develop human capital;
- establishing closer links between research and higher education institutes, and recognising that the administrative separation of research and higher education in many African countries has crippled the development of national agricultural innovation systems; and
- creating incentives to retain the well-qualified and experienced staff in research, extension and educational institutions.

AET's contributions to agricultural development in selected countries are summarised in Table 1.1.

The analysis of these international examples of AET contributions to agricultural development identifies the following strategic pointers:

1. *Direct investment at country level.* Patterns across Asia and Latin America confirm that AET (and broader agricultural development) require strong country-level institutions.
2. *Pro-Smallholder Farmer (SHF) agricultural development as a national priority.* Without sustained commitment of national governments to continuously improve and invest in the agriculture sector, donor projects are likely to fail. In addition, development literature clearly shows that agricultural growth will not help SHFs in the absence of appropriate policy environments and supportive institutions.
3. *Expansion of existing and creation of new programmes.* It is important to prioritise postgraduate training to invigorate research and support national agricultural development. This is evidenced by the recent support of Rockefeller/USAID, where PhD graduates in agriculture from China have played a central role in transforming AET development in their country. Recent innovations show the potential of new approaches to rapidly upgrade AET, as illustrated by China's Open University system, which serves in excess of one million students per year.

4. *Improve quality and ensure alignment with SHFs' needs*. Curricular reform efforts have successfully improved linkages with agribusiness, SHFs, natural resource management, and practical skills, exemplified by, inter alia, EARTH University in Costa Rica.
5. *Ensure pro-SHF decision systems*. Agricultural development literature shows that the AET systems most beneficial to SHFs have actively incorporated their needs and aspirations throughout. This is particularly well illustrated by the land grant model in Japan and the success in building the SAU system in India. The Vietnamese agricultural extension system also directly engaged SHFs in the entire process and emphasised the potential role of farmer organisations and local NGOs in developing pro-SHF systems.
6. *Improve agricultural innovation systems (AIS)*. Development literature indicates that the most effective pro-SHF systems closely link education and training, research and extension at national and provincial levels. This is also supported by the 2007 World Bank review, as noted earlier, which emphasises the importance of integrated agricultural innovation systems to ensure productivity gains for SHFs. There is some evidence in SSA that land grant-style institutional structures, where AET is closely linked to research and extension, are more successful.
7. *Strengthening the funding base*. In Asia, the success achieved through significant investment in agricultural R&D and AET confirms the necessity of sustained public funding for AET to achieve impact. Development Partners (DPs) have been able to leverage strategic follow-up investments across agricultural innovation systems through partnerships and in collaboration with committed national governments.
8. *AIS and decentralisation of decision-making*. Land grant-style institutional structures have been critical and highly successful for AET transformation and agricultural development impact in the U.S., Japan, and India. The main reason for these successes is that the models in these countries have been demand-driven—and thus have developed "from scratch."

| Country | Key AET investments | Results |
|---|---|---|
| **Vietnam** | • In 2001, the government committed to spending 6.9 percent of GDP on education. Aims to grow participation in the higher education system four times by 2020.<br>• Vocational education is a major area for future investment; the aim is to train 40 percent of workforce (includes extension agents). | • Rapid, agriculture-led economic growth and substantial poverty reduction<br>• Tertiary growth: from 120 to 224 institutions in 10 years<br>• Significant tertiary enrolment growth, from 160,000 to 1.5 million in 15 years, with 30 percent currently in vocational programmes |
| **Malaysia** | • Massive human capacity building programme; sent thousands of researchers to U.S. universities for postgraduate training. | • Developed from a low- to a middle-income country.<br>• Established booming export industry: $6.4 billion palm oil/year. |
| **India** | • Established State Agricultural University (SAU) system (to be built over 40 years).<br>• Human capital investment to educate thousands of researchers, with focus on postgraduate education | • Substantial and sustained reductions in rural poverty<br>• Agricultural increases: high adoption rates, with crop yield increase of 1.6 percent/year for 30 years to 60 million tons grain surplus<br>• Built strong national AET network: 41 SAUs. |
| **Japan** | • Investment strategy began with explicit focus on building education system first<br>• Invested heavily in developing indigenous education systems focused on smallholder farmers | • Leveraged an agriculture-led development strategy which contributed to become one of the world's wealthiest countries<br>• System of technical colleges with strong links to smallholder farmers |
| **USA** | • Supplied permanent funding to build a decentralised, applied, and sophisticated AET system: 60 land grant universities (built over 60 years). | • Created one of the most complex but efficient agriculture sectors in the world, including the effective and globally recognised land grant model |

**Table 1.1. International examples of AET contributions to agricultural development. Source: Eicher, 2006; Staatz and Démbele, 2008; Stroebel et al., 2011; Lele et al., 2012.**

## Challenges and Constraints in the AET System in SSA

The sub-Saharan AET system has been largely neglected for at least three decades and, with some exceptions, is now inadequately prepared to address the enormous task resulting from the recognition of AET's critical role in development on the continent. The most significant challenges facing the system are summarised in Table 1.2.

Constraints to effectively building human capacity in support of agricultural development are also highlighted in the World Bank (2007) review, *Cultivating Knowledge and Skills to Grow African Agriculture*, summarised in Box 1.1.

| | |
|---|---|
| Misalignment between demographics of graduates and SHF population | • Women represent 70–80 percent of farmers, with limited numbers of women AET students and agricultural workers (e.g. only 10 percent of extension workforce in Mozambique is female).<br>• Students tend to come from urban backgrounds; the few from rural areas are not interested in returning to the rural areas; government and NGOs find it difficult to employ staff (researchers and extension agents) prepared to move or return to rural areas.<br>• Employers find it difficult to employ staff who speak the local languages required to communicate effectively with smallholder farmers. |
| Misalignment between type of training provided and skills needed in the workforce | • Students lack opportunity to develop technical competencies prior to seeking employment and require additional training before taking on their assignments.<br>• Employers identify gaps in communication skills, business management and planning, post-harvest processing and marketing.<br>• Curriculum and teaching methods include little practical training due to lecturers without appropriate skills, outdated course materials, poor facilities (particularly for research) and a shortage of high quality research projects to engage postgraduate students. |
| Misalignment between level of training provided and sector needs | • Lack of Africa-based postgraduate programmes coupled with a reduction in number of scholarships to pursue training abroad has led to fewer MScs and PhDs.<br>• Too few high-level technicians trained for productive work in the agriculture sector. |

| | |
|---|---|
| AET isolated from research and extension | • Administrative separation of agricultural research and higher education into two different ministries has hindered development of an effective innovation system and feedback loop from research to education.<br>• AET institutions fail to respond to the need for problem-solvers, facilitators, and practically-oriented graduates for extension services.<br>• Land grant-style institutional structures have been most beneficial inactively incorporating the needs of SHFs in the agricultural innovation system, i.e. Japan and India. |
| Poor feedback mechanisms | • Students in AET institutions are too rarely directly exposed to SHFs and are more engaged in theoretical rather than practical training.<br>• Information about labor market needs in agricultural employment is scarce; few AET institutions have tracer studies to understand how their graduates have done in the workplace. |
| Limited leadership | • Immigration has led to an exodus of senior academics (estimated 30 percent of all professionals trained in African universities live outside the continent).<br>• Over-stretched, under-paid lecturers often need to spend more time consulting to augment income than supporting research and postgraduate education.<br>• Many academics are nearing retirement age.<br>• HIV/AIDS has had broad, detrimental effects on human capital. |
| Inefficient allocation of scarce resources | • AET institutions have proliferated across the continent with duplication of efforts (e.g. six faculties in Cameroon, five in Ghana). There appear to be opportunities for consolidating some of these investments, and for developing a focus on more cost-effective, integrated approaches. |

**Table 1.2. Challenges to the AET system. Source: Rivera, 2006; Vandenbosch, 2006; World Bank, 2007; Stroebel and Swanepoel, 2008; Swanepoel and Stroebel, 2012.**

Despite these challenges and constraints, the situation has now changed. African leaders have finally, at this crucial juncture in the continent's development, embraced agriculture as an essential component of their development strategies. In addition, the 2010 Ministerial CHEA emphasised the need to strengthen AET (CHEA, 2010).

## Current AET Initiatives in SSA

Since the 2010 CHEA meeting, a number of efforts have been underway to develop models and a new approach, supported by renewed invest- ment, that will transform Africa's AET system for maximum impact in the context of the challenges and constraints noted earlier. This transformation envisions the emergence of an agricultural human resource/knowledge system that drives SHF-led sector development, innovation, adoption of new approaches, and adaptive problem solving.

The Ministerial Conference (CHEA, 2010) called for:

- development of an African strategic plan for tertiary agricultural education in Africa;
- increased investments in higher education in agriculture to be included as an integral part of the CAADP process; and
- renewed and vigorous emphasis on restoring the quality of higher education in agriculture by African governments.

Firstly, confirming the development partners' increased level of interest in response to the CHEA meeting, the Forum for Agricultural Research in Africa (FARA), with World Bank support, convened a CAADP Donor Retreat in Zurich in January, 2011, with a follow-up meeting in London in March, 2011, to consider how to implement these three imperatives. The main purpose of these working group meetings was to further discuss requirements for implementing and coordinating reforms to tertiary agricultural education in Africa within the framework of CAADP, and to support the establishment of an appropriate mechanism (now known as TEAM-Africa).

To initiate the process of integrating tertiary agricultural education into CAADP Investment Plans, FARA undertook a review of these plans in early 2011 to establish the degree to which tertiary agricultural education had been considered in devising the plans. In response to the findings from the review, FARA and the World Bank launched a process to address the apparent shortcomings with regard to agricultural education and undertook to further mainstream AET into CAADP processes and plans.

Secondly, responding to the call for scaling-up investments in tertiary agricultural education, the working group recommended that African tertiary agricultural education networks and their development partners design a mechanism to determine and articulate the preferred modalities, processes, and approaches through which support for reforming and strengthening tertiary agricultural education might be structured and implemented.

The resulting mechanism, referred to as the Tertiary Education for Agriculture Mechanism in Africa (TEAM-Africa), was conceived with the following objectives: to bring greater coherence to investments in tertiary agricultural education in Africa;

- to increase financial and technical support as well as attention to tertiary agricultural education; and
- to help facilitate and guide the reform and transformation of African AET institutions, all within the framework of CAADP.

The working group identified the need for a small secretariat to develop and coordinate implementation of TEAM-Africa. TEAM-Africa is housed at the RUFORUM Secretariat in Uganda, and jointly implemented by RUFORUM and the Agroforestry and Natural Resources Education (ANAFE) network. There is strong support from the AU's Commissioner of Rural Economy and Agriculture and the NPCA for the initiative.

TEAM-Africa was officially launched during RUFORUM´s bi-annual meeting in Uganda in September, 2012 and was confirmed in its defined role on a continental level by a wide range of stakeholders.

Thirdly, aligned with the vision of TEAM-Africa, the secretariat will identify the types of reforms in governance and incentive structures, new approaches to teaching, research, and outreach, and new investments needed to build African capacities and to make it a key contributor to rural transformation, economic growth, and poverty reduction.

This new vision and approach will be broader, going beyond what TEAM-Africa can undertake on its own. However, TEAM-Africa´s role will be to make a major contribution to defining and promoting this vision as well as to facilitating, convening and coordinating stakeholders and reform initiatives on the continent (TEAM-Africa, 2012).

The two working group meetings in Zurich and London referred to earlier generated strong momentum for African ownership behind the TEAM-Africa concept. In an effort to support this on-going Africa-led process of developing and eventually implementing an agenda for tertiary agricultural education, the World Bank worked with this working group to organise the first partnership workshop held in Ås, Norway, in October, 2011. The workshop was a first step in launching a broader, actively engaged partnership among development partners, academic experts, and African leaders to provide coordinated support to the African tertiary agricultural education agenda. This meeting was followed by a second partnership workshop held in Wageningen, Netherlands, in March, 2012, and by a third partnership meeting in Stellenbosch, South Africa in November, 2012—for the first time organised by the African partners, in Africa, with support from the World Bank.

**Box 1.1. Constraints in Building AET Capacity in SSA**

- **Distorted enrolment pattern**
  - Unbalanced technical education enrolment
  - Distorted enrolment pattern by gender
  - Waning interest of students in agriculture
  - Declining share of post-secondary enrolment in agriculture
  - Enrolments are often greater than the capacity of universities to handle
  - Failure to attract the best quality students from secondary schools
- **Isolated and fragmented AET institutions**
  - Lack of strategic alignment to national development priorities, declining relevance to national needs
  - Weak linkage to the AET systems with stakeholders, employers and the productive sector
  - Poor connections with international sources of knowledge
  - Over-centralised and divided administration
  - Weak connection with other parts of the agricultural education system—colleges, vocational schools, farmers' training networks, etc.
- **Curricula that are obsolete and disassociated from the economy**
  - Often outdated curricula
  - Narrow focus on production
  - Premature specialization
- **Crisis in staffing**
  - Eroded staff qualifications
  - Shortage of staff
  - High level of staff turnover
  - Brain drain
  - Staff attrition due to HIV and AIDS
  - Teaching and research staff are frequently graduates of the same university, thereby excluding the entry of new ideas from the wider world of academic research (inbreeding).
- **Inadequate teaching facilities and outdated teaching methods**
  - Overwhelmingly traditional teaching methods
  - Insufficient emphasis on practical instruction; poor practical skills
  - Deficient learning infrastructure
  - Insufficient budgets and over-dependence on public financing
- **Other**
  - Unsustainable patterns of expenditure for higher education
  - Decline in the quality of education
  - Too many universities in certain countries, unsustainable with existing budgets
  - Higher unemployment of graduates from the university, often due to lack of relevance of curriculum

Source: Adapted from World Bank (2007)

The second partnership meeting in Wageningen focused on principles and guidelines for coordination and alignment of African partner institutions to develop work plans, including actions, responsibilities, and time lines. The third partnership meeting in Stellenbosch brought together agricultural education experts and institutions from Africa, Europe, and the U.S. to work together to build a shared understanding of an African transformation process, clarify institutional arrangements for mainstreaming of AET and agricultural development across the continent, and review actions to be taken to align development partners in support of the TEAM-Africa transformation initiative.

Concurrently, there are a number of other efforts underway to revitalise AET in sub-Saharan Africa. This time, the process is different—it is a concerted, well-coordinated effort with increasing commitments aimed at aligning and pooling resources towards a common goal to strengthen agricultural development on the continent. The prospect of transforming AET to support this goal is framed, planned, and executed in a manner that surpasses all earlier attempts (TEAM-Africa, 2012).

## Purpose and Structure of the Book

The increasing attention paid to AET at global and regional forums, and the tone and focus of these discussions, all point to the need for urgent reconsideration of the entire AET system in SSA. Influential actors[4] have increasingly made commitments in an effort to align or pool resources towards common goals in the development of agriculture on the continent. AET must be an integral part of these efforts. In order to maximise the impact of such commitments, strategies for AET have to be informed by highly credible analyses and syntheses of issues and solutions, based on an understanding of practical AET experiences across the world and in Africa.

Yet with the exception of the World Bank report (2007), there has been a dearth of materials that can serve as a reference point for this purpose. This was confirmed in the lead context paper prepared for CHEA in Uganda (Blackie et al., 2012), and during the third TEAM-Africa partnership meeting in Stellenbosch (Swanepoel and Stroebel, 2012). Both emphasised that CAADP review teams from several countries had commented on the absence of sufficient quantitative and qualitative information on capacity needs to inform the AET system.

---

[4] Amongst others: GCHERA, World Bank, RUFORUM, ANAFE, FARA, USAID, AgriNATURA

TEAM-Africa will greatly benefit from highly credible analysis and synthesis of critical issues based on a global understanding and practical examples from SSA to inform its future strategies.

This book is designed to focus on clarifying the challenges, issues and priorities for AET in Africa, and to provide suggestions for practical solutions to those organisations interested in furthering AET and agricultural development on the continent. The book is uniquely positioned to support a call to action on AET, and to pull together state-of-the-art knowledge from within and outside of Africa. It also promotes and informs *out-of-the-box* thinking about the principles, values, and character of AET for development, with an emphasis on the concepts of change for systemic transformation of AET, drawn from the discussions in the chapters.

The book connects the context for AET in Africa to international and African experiences and lessons that can help decision- and policy-makers develop priorities and practices for modernising the AET system across SSA. It is therefore divided into four sections that focus on:

1. predicting and illuminating the context in Africa, and particularly in SSA, within which AET will have to flourish over the next decades;
2. highlighting relevant international experience in AET-related human capital development from which Africa can learn—from Europe, North America, Latin-America, and Asia, especially South and Southeast Asia;
3. drawing from experiences in SSA related to different parts of the AET system; and
4. addressing critical issues that will be crucial for success—systems, quality, gender, financing, and innovation and entrepreneurship.

Finally, the book provides a synthesis of principles for systemic change of AET to advance out-of-the-box thinking with an emphasis on the models that can help to cultivate leaders and change-makers at all levels of the agriculture sector.

## References

AfDB. 2013. *African Competitiveness Report 2013*. African Development Bank, Tunisia. http://www.afdb.org/fileadmin/uploads/afdb/Documents/Publications/The%20Africa%20Competitiveness%20Report%202013.pdf

African Union. 2006. *The Second Decade of Education for Africa (2006–2010)*. Addis Ababa, Ethiopia.

Aliyu, B. 2012. *Agricultural Development and "Land Grabs": The Chinese Presence in the African Agricultural Sector.* Consultancy Africa Intelligence.

Asenso-Okyere, K. and S. Jemaneh. 2012. *Increasing Agricultural Productivity and Enhancing Food Security in Africa: New Challenges and Opportunities.* International Food Policy Research Institute (IFPRI), Washington, DC.

Blackie, M., R. Blackie, U. Lele and N. Beintema. 2012. *Capacity Development and Investment in Agricultural R&D in Africa.* Lead Background Paper at the Ministerial Conference on Higher Education in Agriculture in Africa, November 15–19, Speke Resort Hotel, Munyonyo, Kampala, Uganda.

Bolaky, B. 2011. Fostering Industrial Development in Africa in the New Global Environment: Key Policy Recommendations. *Trade Negotiations Insights* 10: 9.

CAADP. 2013. *Implementing the CAADP Agenda*. NEPAD. Addis Ababa, Ethiopia. http://www.nepad-caadp.net/implementing-caadp-agenda.php

Chang, H. 2010. "Hamlet without the Prince of Denmark: How Development has Disappeared from Today's 'Development' Discourse." In Khan, S. and J. Christiansen, eds. *Towards New Developmentalism: Market as Means rather than Master.* Routledge, USA.

CHEA. 2010. *Ministerial Communiqué issued at the Ministerial Conference on Higher Education in Agriculture in Africa (CHEA).* November 15–19, Kampala, Uganda.

Conway, G. and J. Waage. 2010. *Science and Innovation for Development.* UK Collaborative on Development Sciences. London, UK.

Davis, K., J. Ekboir, and D. Spielman. 2008. Strengthening Agricultural Education and Training in sub-Saharan Africa from an Innovation Systems Perspective: A Case Study of Mozambique. *Journal of Agricultural Education and Extension* 14: 1, 35–51.

Eicher, C. 2006. *The Evolution of Agricultural Education and Training: Global Insights of Relevance for Africa.* Staff Paper 2006–26. East Lansing, MI, Department of Agricultural Economics, Michigan State University.

European Commission. 2012. *The Common Agricultural Policy Explained.* Directorate-General for Agriculture and Rural Development. Brussels, Belgium.

FAO. 2009. *Land Grab or Development Opportunity? Agricultural Investment and International Land Deals in Africa.* Food and Agriculture Organisation of the United Nations, Rome, Italy.

—. 2012. *The State of Food Insecurity in the World.* Food and Agriculture Organisation of the United Nations, Rome, Italy.

—, 2012a. *Statistical Yearbook.* Food and Agriculture Organisation of the United Nations, Rome, Italy.

—. 2013. *The FAO's Response to the 2012 SAHEL Food Crisis*. Food and Agriculture Organisation of the United Nations, Rome, Italy.

FAOSTAT. 2011. *Food and Agricultural Production*. Food and Agriculture Organization of the United Nations, Rome, Italy.

Fosu, A. 2009. "Inequality and the Impact of Growth on Poverty: Comparative Evidence for sub-Saharan Africa." *Journal of Development Studies* 24: 5, 726–745.

HLPE. 2011. *Land Tenure and International Investments in Agriculture and International Investments in Agriculture*. A Report by a High Level Panel of Experts on Food Security and Nutrition of the Committee on World Food Security. Rome, Italy.

Inter-Academy Council (IAC). 2004. *Realizing the Promise and Potential of African Agriculture.* Amsterdam: Royal Netherlands Academy of Arts and Sciences.

IFPRI. 2008. *The 10 Percent that Could Change Africa.* International Food Policy Research Institute, Washington, DC.

International Monetary Fund (IMF). 2011. *Regional Economic Outlook: sub-Saharan Africa, Sustaining the Expansion*. Washington, DC.

Jerven, M. 2013. *Poor Numbers: How we are Misled by African Development Statistics and What to Do about it.* Cornell University Press: Ithaca, NY.

Johanson, R. and W. Saint. 2007. *Cultivating Knowledge and Skills to Grow Africa's Agriculture: A Synthesis of an Institutional, Regional and International Review.* World Bank, Agricultural and Rural Development Department, Washington, DC.

Kapur, D. and M. Crowley. 2008. *Beyond the ABCs: Higher Education and Developing Countries.* Working Paper Number 139. Centre for Global Development.

Kellogg, E., A. Hervy, and T. Yizengaw. 2008. *Africa-U.S. Higher Education Initiative: Empowering African Higher Education for Africa's Transformation.* The Institute of International Education Magazine. IIE Networker: Fall 2008, 36–38.

Kingslow, E. 2007. *AET in Africa: A Survey of Progress.* A Report Prepared for the Kellogg Foundation. Kingslow Associates LLC.

Lele, U., M. Agarwal, P. Timmer, and S. Goswami. 2012. *Patterns of Structural Transformation and Agricultural Productivity Growth with Special Focus on Brazil, China, Indonesia and India.* Keynote Address at the

Global Conference on Agricultural Research for Development, October, Punta del Esta, Uruguay.
MGI. 2010. *Lion on the Move: The Progress and Potential of Africa's Economies*. http://www.mckinsey.com/insights/africa/lions_on_the_move
New Partnership for African Development (NEPAD). 2002. *Comprehensive Africa Agriculture Development Programme*. Addis Ababa, Ethiopia.
OECD-FAO. 2011. *Agricultural Outlook 2011*. OECD-FAO, Rome, Italy.
Parker, S. 2012. *Lessons from a Ten-year Collaborative: A Case Study of the Partnership for Higher Education in Africa (PHEA).* Clear Thinking Communications.
Rivera, W. 2006. *Transforming Post-Secondary Agricultural Education and Training by Design: Solutions for sub-Saharan Africa*. Unpublished paper. World Bank, Washington, DC.
Renaissance Capital. 2012. *The Fastest Billion: The Story Behind Africa's Economic Revolution.* http://www.fastestbillion.com/
Samman, E. 2012. *Measuring Wellbeing: Different Approaches: Their Implications and an Illustration.* Development Progress: UK. http://www.developmentprogress.org/sites/developmentprogress.org/files/measuring_wellbeing_different_approaches_their_implications_and_an_illustration.pdf
Staatz, J. and N. Démbele. 2008. *Agriculture for Development in sub-Saharan Africa.* Background Paper for the World Development Report. World Bank, Washington, DC.
Stroebel, A. and F. Swanepoel. 2008. *Cultivating Knowledge and Skills to Grow African Agriculture—An Analyses and Synthesis.* Southern African Agricultural Education Networking and Clarification Workshop, October 3–4, Manzini, Swaziland.
Stroebel, A., F. Swanepoel, and C. Eicher. 2011. *Agricultural Education and Training for Development (AET4D) in Africa*. Seventh Global Confederation of Higher Education Associations for the Agricultural and Life Sciences (GCHERA) Conference, June 27–29, Beauvais, France.
Stroebel, A., F. Swanepoel, and A. Pell. 2011. Sustainable Smallholder Livestock Systems: A Case Study of Limpopo Province, South Africa. *Livestock Science* 139, 186–190.
Swanepoel, F, A. Stroebel, and S. Moyo, Eds. 2010. The Role of Livestock in Developing Communities: Enhancing Multifunctionality. CTA Wageningen, The Netherlands.
Swanepoel, F., and A. Stroebel. 2012. *Capacity Building for African Agricultural Development: the Need for Transformation in Tertiary Agricultural Education.* Keynote address at the Third AET Partnership Workshop on

Tertiary Agricultural Education in Africa, November 13–15, Stellenbosch, South Africa.

TEAM-Africa. 2012. *Proceedings of the Third Partnership Workshop on Tertiary Agricultural Education in Africa*. Background and Synthesis on Processes, Workshop Objectives, and Results/Next Steps. November 13–15. Stellenbosch, South Africa.

Tschirley, D., and N. Dėmbele. 2011. *On Managing the New Food Price Environment in Countries with Food Insecure Populations*. Michigan State University.

UNCTAD. 2009. *World Investment Report 2009.* United Nations: New York and Geneva.

UNDP. 2012. *Africa Human Development Report: Towards a Food Secure Future.* United Nations Development Programme, New York, NY.

UNESCO/World Bank. 2000. *Higher Education in Developing Countries: Peril and Promise.* The Task Force on Higher Education and Society. World Bank, Washington, DC.

UN-HABITAT. 2008. *The State of African Cities: A Framework for Addressing Urban Challenges in Africa.* United Nations Human Settlements Programme, Nairobi, Kenya.

USAID. 2010. *Agriculture Long-Term Training: Assessment and Design Recommendations.* United States Agency for International Development, Washington, DC.

Vandenbosch, T. 2006. *Post-Primary Agricultural Education and Training in sub-Saharan Africa: Adapting Supply to Changing Demand.* World Agroforestry Centre, Nairobi, Kenya.

World Bank. 2007. *Cultivating Knowledge and Skills to Grow African Agriculture.* World Bank, Washington, DC.

—. 2007a. *World Development Report 2008: Agriculture for Development*. World Bank, Washington, DC.

—. 2008. *Agriculture for Development*. World Bank, Washington, DC.

—. 2012. *PovcalNet*. World Bank, Washington, DC. http://iresearch.worldbank.org/PovcalNet

—. 2013. *Africa Pulse. An analysis of Issues Shaping Africa's Economic Future*. Vol. 7. World Bank, Washington, DC.

—. u.d. *Agriculture and Achieving the Millennium Development Goals.* World Bank and the International Food Policy Research Institute (IFPRI). Washington, DC.

World Economic Forum (WEF). 2013. *Achieving the New Vision for Agriculture: New Models for Action.* A report by the World Economic Forum's New Vision for Agriculture Initiative Prepared in collaboration with McKinsey & Company.

# CHAPTER TWO

# CHANGING PARADIGMS, EMERGING TRENDS, AND SCENARIOS FOR SUSTAINABLE AGRICULTURAL DEVELOPMENT IN SUB-SAHARAN AFRICA

JOHAN VAN ROOYEN

## Introduction

Agriculture, including agribusiness, is projected to be a US$1 trillion industry in sub-Saharan Africa by 2030 (World Bank, 2013). It is therefore understandable that agricultural development in Africa is back on the global development and investment agenda. In September, 2009, the G20 recognised agriculture as a central activity for African development, echoing the World Bank Development Report (World Bank, 2007) and the *Our Common Interest* report (Commission for Africa, 2005). Most recently, the African Heads of State at the World Economic Forum in Cape Town called for increased agricultural investment (NEPAD, 2013). All emphasised the notion that agriculture is a key driver of economic development and growth, poverty alleviation, and food security on the African continent.

The performance of African agriculture directly and indirectly affects all eight United Nations Millennium Development Goals (UN, 2009). The recent World Bank (2013) report, *"Growing Africa—Unlocking the Potential of Agribusiness,"* recognised that agribusiness, in particular, is strategically positioned to overcome constraints and to encourage strategic "good practice" investments. At the 2013 World Economic Forum in Cape Town, the African Union chairperson confirmed this view and stated that "Africa is on the road to prosperity. African agriculture is the true driver of economic growth, but we need the private sector to kick-start this process" (NEPAD, 2013).

With investments from sources around the globe, African agribusiness started booming in the 2000s. In addition, soaring grain prices and global food price inflation spurred investor interest in African farming.

Africa has sufficient available land and space to expand farm production significantly. Due to the growing interest of the private sector, many farming projects—which are linked to the value chain through agrifood processing, food retail establishments, and service networks—are increasingly doing business on the continent. African agriculture attracted more than US$100 million in private equity investment in the first half of 2012, compared to US$50 million for all of 2011. Early in 2013, Standard Chartered Bank invested US$74 million in the grain and fertilizer trade in sub-Saharan Africa—to gain a stake there—and another US$20 million in Zimbabwe's horticulture sector.

From a development viewpoint, this "rediscovery" of the agribusiness sector provides an exceptional opportunity for planners to design and implement strategies that will support sustainable agricultural development and good agribusiness practices in Africa. These developments call into question the continued application of conventional approaches and policies, such as those influenced by the well-known and respected Johnston and Mellor paradigm of the 1960s and 1970s and the structural adjustment and Washington consensus periods that followed.

The Johnston-Mellor paradigm still seems to have validity, as it underscores the role of agricultural productivity in rural poverty reduction, demographic change, and economic development. The paradigm, which recognized the impact and the role of small-scale farming, generally directed agricultural development policy as well as strategies for development agencies and donor funding. On the other hand, the paradigms of the 1980s and 1990s underplayed the role of agriculture and drove a free market/privatisation agenda with little support for agricultural development. This left a somewhat confused state of affairs around the turn of the millennium. However, the focus has shifted again toward unlocking Africa's agricultural potential.

***Key Messages***

- Agricultural development in Africa is *back on the global development and investment agenda.*
    - *Agricultural development remains* a key driver of economic development and growth, poverty alleviation, and food security; and its strategic and societal importance is clearly illustrated by its relationship to, and impacts on, the eight UN Millennium Development Goals.
    - The African Union chairperson stated in 2013 that "African agriculture is the true driver of economic growth; but we need the private sector to kick-start this process" (NEPAD, 2013). Agribusiness, in particular large agri-food corporations, are responding with a willingness to finance agriculture in Africa through integration of primary production into business portfolios.
- Agriculture in Africa has much to offer, but realization of this potential requires a *new agricultural development paradigm.* African agriculture must be prioritised and positioned as a key growth and development sector, supported by appropriate policies and developmentally directed investment protocols. A new paradigm must encompass African agriculture in terms of the roles it fulfils, the environmental changes that affect it, and the drivers of its development.
- *Trends that will shape agriculture's roles in Africa* include meeting the growing demand for food at global, regional, city, and household levels; activating Africa's immense untapped agricultural production potential and possible production and price scenarios; directing the evolving agri-food business systems and investments in African agriculture; strengthening economic management and good governance practices; and creating human capital development.
- A new paradigm for the future development of African agriculture must accommodate three fundamental sets of current *changes affecting the agricultural development domain:* the emergence of integrated agri-food value chains and a host of technological changes; increasing resource scarcity and climate changes which impede sustainability; and the widening of agricultural development objectives and scope. The latter involves a multi-functional and multi-dimensional holistic agenda that has economic, human, social, ethical, and environmental considerations, and has good decision-making and governance mechanisms as its central components.
- The *future drivers of development in African agriculture* are global market growth for food and fibre; transformation in the agri-food business system and the emergence of agri-food value chain networks; the move towards a bio-based green economy; Africa's untapped agricultural potential; and scale factors in farm production. Scale factors concern the mixed-mode farming typology that is emerging in African agriculture. In this mode, large-scale and smallholder agriculture co-exist through coordinated, synergistic and collaborative activities, based on commercialisation and economies of scale.

Looking toward the future development of African agriculture, a few questions need to be considered:

- Is a new paradigm required to guide thinking, agricultural development opportunities, and strategic investment?
- What functions and actions should be prioritised for agriculture, and who should drive or initiate them?
- What good practice models and good governance protocols will be required to strategically direct development paths towards sustainable, equitable growth?
- What will be needed to enable African agriculture to contribute its full potential over the next decades without again turning into an exploitative scramble for Africa?
- How should initiatives be structured, directed, and implemented? How will the directions taken affect the possible scenarios that emerge for African agriculture?

Keeping these issues and questions in mind, this chapter sets out to do the following:

- Review the roles of agriculture, as defined by the classic Johnston-Mellor paradigm, in the context of the anti-agricultural biases of the 1980s and 1990s;
- Look at the renewed emphasis on African agriculture and the changed environment in which it functions and discuss whether a paradigm shift is required.
- Identify drivers and trends that will shape agriculture's roles in Africa. These include: meeting the growing demand for food at global, regional, city, and household levels, activating Africa's immense untapped potential for agricultural production and exploring possible production and price scenarios, directing the evolving agrifood business systems and investments in African agriculture, strengthening economic management and good governance practices, and creating human capital development.
- Consider a strategic framework—comprising functions, thrusts and actions—in the context of different scenarios for African agricultural development and growth.

## The Renewed Emphasis on Agriculture in Africa

In the 1960s and 1970s, agricultural growth was considered the key pillar for economic development, with farm production promoting value-addition and rapid industrialisation throughout the wider economy. It was accepted that agricultural development would initiate productivity gains and the structural transformation of the economy (Johnston and Mellor, 1961; Hayami and Ruttan, 1971; Timmer, 1986; Eicher and Staatz, 1998; Mellor, 1998). In the Johnston-Mellor paradigm, the roles of agriculture in development included:

- producing agricultural surpluses in the form of increased food supplies and exports of food and fibre products;
- enhancing demand through large and significant income multipliers for a range of products and services, obtained through market development in rural and urban environments;
- releasing or shedding farm labour for more productive employment elsewhere in the growing economy and in value-adding agrifood industries that provide positive labour linkages; and
- mobilising capital and investment in a range of activities related to agricultural production and rural development.

The Johnston-Mellor paradigm was widely advocated by development agencies and funding institutions such as the World Bank and credited with the success of the Green Revolution, in particular in Asian countries (Hayami and Ruttan, 1971, 1985; Lipton, 1989; Streeten, 1987) but also in some African countries (e.g. Zimbabwe). However, its effective implementation was subverted in the 1980s and 1990s (De Janvry, 2009). During these period influential economic development policies (structural adjustment and the Washington consensus) directed resources away from agricultural development towards industrial activities that had a strong anti-agricultural bias. They promoted, for example, local production of agricultural inputs and machinery, which employed significant numbers of local workers and sustained families and rural economies, rather than lower cost imports (Krueger et al., 1991).

Ineffective public support systems in agricultural extension, research, funding, and market support, as well as inefficient state-led bureaucratic approaches, further exacerbated the neglect of agriculture. In addition, particularly in the African context, support to smallholder farmers proved problematic because of political tensions at national and local levels, weak infrastructure (transportation and communication networks), and many missing or under developed markets for land, inputs, and agricultural produce. This all added up to restricted agribusiness initiatives (Dumont, 1962; Haggblade et al., 1988; Eicher and Staatz, 1998; Delgado, 1998; Cheru, 2002; Cheru and Bradfort, 2005; Kirsten et al., 2009). Where these initiatives were properly structured to support smallholder farming effectively, as was the case in Zimbabwe (Rukuni and Eicher, 1994) and the former South African Bantustans (Singini and Van Rooyen, 1995), good results were recorded.

Thus, the agricultural growth necessary to activate the full impact of the Johnston-Mellor paradigm was generally highly constrained and too limited to be effective. Furthermore, in contradiction to the classic agriculture-led development paradigm, the stabilisation and structural adjustment policies of the 1980s under the Washington Consensus favoured job-creating public works programmes along with direct cash transfers to the poor, compromising the role of agriculture in economic development and effectively undermining the contributions of agribusiness and the private sector (Byerlee et al., 2009).

The above economic policy scenario changed dramatically in the 2000s. First, a number of economic, financial, social, and environmental crises—related to issues such as rising food insecurity and hunger, stagnation in per capita value-added activities in sub-Saharan agriculture, increased rural poverty, increasing urban-rural disparities and tensions, serious environmental degradation, misuse of scarce agricultural resources, loss of biodiversity, and reduced environmental services—emerged in the late 1990s and early 2000s (Cheru, 2002; Binswanger, 2006; Ravallion et al., 2007; World Bank, 2009). Secondly, lucrative agribusiness investment opportunities emerged due to Africa's largely untapped agricultural resource base. Recognition of agriculture as both a contributor to these crises and as a potential instrument for solutions brought renewed attention to agriculture (FAO, 2009a; World Bank, 2005, 2007; Byerlee et al., 2009; Badiane, 2009) and activated a stream of business-driven investments (FAO, 2009a; OECD, 2010; Sanghvi et al., 2011; Anseeuw and Ducastel, 2012; Connolly et al., 2012; World Bank, 2013). In other words, although agriculture is back on the development agenda at levels similar to those of 1960–1970, this renewed interest is due to a combination of traditional and recent considerations, crises, food price increases, and considerable investment and business opportunities.

One interesting point must be noted here. Despite all the negative influences that constrained development in sub-Saharan African agriculture, the sector did show resilience and responded positively to the limited support it was granted. Agricultural production increased steadily, especially from the mid-1980s onwards, growing from around 570 kg of agricultural produce per capita to around 740 kg per capita in the mid-2000s (Badiane, 2009; Adesina, 2009; Vink, 2010). Farming, particularly large-scale operations and agribusiness, showed positive growth and upward development trajectories.

In the late 2000s, expansions—often linked to so-called land grabs by foreign countries and firms—drew world-wide attention to investment in and exploitation of African agricultural resources (Oakland Institute, 2011). Information on investment opportunities that render lucrative returns is generally available (World Bank, 2013; Connolly et al., 2012). All this is a good indication that agriculture in Africa has much to offer, but it must be prioritised and positioned as a key growth and development sector, supported by appropriate policies and developmentally directed investment protocols.

But is a new paradigm for the future development of African agriculture called for? Would the current views on agriculture's roles be sufficient to drive the process?

## Towards a New Paradigm?

An important question for policy makers, funding agencies, and strategic investors is whether the renewed interest in agriculture and, in particular, in agribusiness is sufficiently guided by a strategic development framework. Is the Johnston-Mellor paradigm still valid, or is a new paradigm required to (re)direct policy, strategy, investment, and action in the rapidly evolving African agricultural environment?

In the Elmhirst Lecture at the 2009 Conference of the International Association of Agricultural Economists in Beijing, Alain De Janvry, a leading development agricultural economist, argued for a newly designed development paradigm for agriculture in the developing world, one that moves away from the Johnston-Mellor view of agriculture as the main driver of economic development. According to De Janvry (2009), three fundamental sets of changes affecting the environment in which developing agriculture currently operates should be accommodated.

- *A widening of agricultural development objectives and scope.* Instead of focusing on growth and industrialisation through farm production per se, development is opening to a multi-functional and multi-dimensional holistic agenda that has economic, human, social, ethical, and environmental considerations, with robust decision-making and governance mechanisms as central components.
- Changes in the structural context in which agricultural growth and development take place mean there is a need for new institutions to govern occurrences such as rapidly globalising food systems, the emergence of integrated agrifood value chains and a host of technological changes.
- Increasing resource scarcity and climate change put demands on agriculture to operate in an environmentally sustainable manner and, where possible, to serve as custodian of the environment.

### New roles for African agriculture

In this new structural setting, agriculture can neither be viewed as the leading sector for production to bring about industrialisation, as was the case in 1960s and 1970s, nor as a series of sporadic activities driven only by market forces and profit motives that result in fluctuating production supplies and unstable rural environments, as has been the case since the 1980s (Delgado, 2009). Instead, agricultural development in the new millennium should contribute to several dimensions of development (De Janvry, 2009), namely:

- accelerating the growth of gross domestic product (GDP) through increased farm production and added value activities in the agrifood chain;
- providing surpluses to meet the growing demand for food and fibre, both locally and globally, through exports;
- reducing rural poverty and improving food security (De Cock, et al, 2012; Jones, Ngure, Pelto and Young, 2013 for a comprehensive definition of food security in an economic policy context to include supply, distribution, access, nutrition, stability, etc.)of poor households;
- narrowing the rural-urban income gap;
- supporting environmental sustainability;
- contributing to employment creation in the wider economy; and
- supporting economic development through domestic economic specialisation, structured urbanisation and rural growth, and regional and global integration and trade.

Most of these contributions, however, still correspond to the Johnston-Mellor paradigm, although the scope and focus have been widened (Timmer, 2010). Agriculture still clearly constitutes more than farm production expansion and yield increases. It now includes value-addition; has strong trade, social, political, cultural, and environmental impacts and roles; and must be positioned in the context of the global quest for food and fibre. By positioning the sector to support multiple development objectives related to economic, social, political, and environmental concerns, its contributions will be multifunctional, anchored in a set of sustained farm production activities and focusing on producing for local and global markets (Cheru and Bradfort, 2005; IAASTD, 2008; De Janvry, 2009).

Whereas food production will remain an important function, agricultural production will also be directed to producing energy, health and medicine, and even recreation and tourism activities such as wine tours and eco-adventures. Fantu Cheru (2002), in "African Renaissance: Roadmaps to the Challenge of Globalisation," argued for a "middle of the road" position that would go beyond state-led versus market-led scenarios to link Africa's economic development to evolving global processes.

An important, potential tipping point in favour of a new agricultural paradigm is the willingness of agribusiness, in particular large agrifood corporates, to finance agriculture in Africa "by integrating primary production into their business portfolios." For this industrial agribusiness model, Zeder, a South African investor in African agriculture, requires a stake of at least 30 percent in primary production in order to secure the required supply-chain activities. This trend implies the introduction of supply-chain management on both the input and output sides, supported by business contracts and relationships. Anseeuw and Ducastel (2012) argue that this agribusiness-driven system would generally transform family farmers and smallholders into "service providers on their own lands." In addition, the global notion of investing in agriculture only where a minimum 20 percent rate of return can be achieved (see note 1) clearly calls for new ways of thinking in the evolving agricultural development paradigm, in other words a much stronger emphasis on agribusiness economics and long-term sustainability to counter sporadic and exploitative land grabs and a new scramble for Africa's agricultural resources. This notion is supported by the recent World Bank (2013) report entitled, *"Growing Africa: Unlocking the Potential of Agribusiness."*

In addition, it will be necessary to direct decision-making, institutional innovation, and governance in the more complex, integrated, and globalised agribusiness and social/political environment within which agriculture now functions, and to address the notion of the 20 percent-plus investment paradigm for agribusiness development (De Janvry, 2009; Anseeuw and Ducastel, 2012; World Bank, 2013).

A paradigm change will accommodate:

- re-conceptualisation of complementarities and trade-offs among the multiple objectives for agricultural development;
- redefining the roles of the state, civil society, and the private or corporate sector, including farmers, and strategically supporting agribusiness and investment agencies;
- setting priorities among the many conflicting economic, social, political, and environmental objectives;
- redesigning development institutions so they can provide the necessary innovation and returns, establishing new good governance structures and experimenting with new approaches to stabilise and sustain agriculture (agricultural investment agencies and agricultural education and training are cases in point);
- obtaining the necessary commitments from all involved to support agriculture and agribusiness in the long term, as a contribution to achieving the stated development objectives.

While the Johnston-Mellor paradigm still has some fundamental value in the current situation, important adjustments are needed. A paradigm shift is required to effectively direct the renewed attention to African agriculture.

While still emphasising the Johnston-Mellor roles of agriculture, this new paradigm will also need to:

- re-interpret these roles in the context of expanded dimensions of development;
- note the agribusiness drive within the rationale of a 20 percent plus investment paradigm;
- accommodate the complex, globalised environment impacting on agricultural systems, from business, social and environmental perspectives; and
- translate all this into a new strategic framework for agricultural business development in Africa.

## Setting a New Strategic Framework

Setting a new strategic framework for agricultural development in Africa will require:

- identifying and interpreting the major future drivers of agricultural development;
- clarifying the required economic and social management processes and policies to direct these drivers; and
- defining and structuring a set of core actions and thrusts that will effect the growth and development of African agriculture over the next decades.

### Drivers of development and change in African agribusiness

African agriculture faces a unique set of circumstances and characteristics that differentiate it from the agriculture of other regions. For example, the Asian Green Revolution in the 1960s and 1970s had an immediate and positive impact on economic development and growth, economic structure, governance, and political and human capital development. Yet efforts to repeat the Asian experience in Africa failed to recognise regional differences and to consider lessons learned, so the Green Revolution in Africa often had negative consequences. Low yields and imbalances between the high costs of inputs and low output prices frequently created a price-cost squeeze. Due to global trade mechanisms, the visibility of social impacts, including social media, and internal dynamics on the continent, Africa's future will differ significantly from its past.

What are the expected occurrences that will drive opportunity and challenges in African agriculture over the coming decades? The emerging consensus among development institutions and strategic investors, referred to in the introductory paragraphs, must be appreciated as a primary set of forces driving the restructuring of agricultural development in Africa. The universally accepted Millennium Declaration, which emphasises agriculture's prominent role in all eight of the MDGs, together with declarations by African leaders on the future role of agriculture, clearly situate agriculture's contributions within the broader economic-political context of both global and local realities impacting on Africa. However, this still falls short of wide-scale implementation and continues to show underinvestment and mis-investment by many governments and international donors (De Janvry, 2009).

Recent statements and financial commitments by influential groups are encouraging. Among those expressing support of African agriculture are the World Bank and the 2008 World Development Report, the Food and Agriculture Organization of the United Nations (FAO), and several high profile international donor foundation—notably the Bill & Melinda Gates Foundation, Rockefeller Foundation, Ford Foundation and Kellogg Foundation. The launch of programmes such as the Comprehensive African Agriculture Development Programme (CAADP) of New Partnerships for Africa's Development (NEPAD), and the Alliance for a Green Revolution in Africa (AGRA) must also be noted. These have paved the way for new funding and investment instruments in African agriculture globally. There have also been commitments by, *inter alia*, local African business entities, such as the Sanlam/Kellogg AgriVie Fund, the Future Growth Fund from Old Mutual, funding from Standard Bank to support AGRA, and investments by Standard Charters and Zeder to invest in new agriculture-related ventures in many African countries.

## Driver 1: Global market growth for food and fibre

By 2050, the major growth, via income and population growth, in demand for food and fibre products is expected to come from the markets of North America, Western Europe, and China, and these trends will drive commercial food business systems (Swinnen, 2007; McCoullough et al., 2008). Demand will be influenced by population growth, per capita income trends, lifestyle aspirations, and related consumer preferences. Consumer demand in these regions has become and will continue to be more exacting, fragmented, and geared to convenience, food safety, and quality. While these consumers will seek attributes of fun, surprise, and taste sensation in their food experiences, they will also be sensitive to environmental, ethical, and social considerations (Hughes, 2007; Vermeulen et al., 2008). These trends are expected to have profound effects on the sourcing of food products and will link farm production, processing, wholesaling, and retailing. This means that the agrifood value chain will increasingly be affected by consumer preferences and values—

such as concerns about food safety, environmental and ethical valuations, certification requirements, traceability, and demand for monitoring systems—in these markets. Civil society, activist groups, and social media can be expected to monitor these factors and maintain their visibility.

In addition, exploding urbanisation in Africa, combined with its expected increases in per capita food consumption (FAO, 2009a; Vink, 2010), will directly expand local demand and provide opportunities for localised (short) value chains, directly linking farming areas to growing urban centres on the continent.

The evolving trade patterns confirm these trends. Both exports and imports are increasing, especially imports of high-value food products. African agriculture is increasingly connected to global food trade patterns with a complicated array of contracts, value chain specifications, and business opportunities.

*Household food security in urban and rural environments*. Although food security is defined in terms of food supply, access, distribution, and nutrition (DBSA, 2009), having a positive food self-sufficiency index (SSI)—the ability to produce sufficient food for the nation or region—does not necessarily lead to food security at the household level. Many African countries with positive SSIs still deal with serious hunger, food vulnerability, and malnutrition. South Africa is a case in point (Vink and Van Rooyen, 2009).

An adequate food supply can generally be secured through a combination of trade and local food production, but the problem cannot be solved solely on the basis of national agricultural production. A broader, holistic view is needed —one that focuses on farm production at national and household levels, on trade and distribution, and on aspects of food safety, household-level access, vulnerability, and nutrition education (DBSA, 2009).

Food security at the household level is generally attained through income security or the ability to purchase food stuffs as required. However, in poor environments, food security is vulnerable and highly sensitive to household income level and food price fluctuation. Thus, food security in poor societies should focus on strategies to enhance household income generation and on creating resilient food production capacities in both rural and urban environments using proven strategies such as the support of food gardens. In the quest to eliminate food vulnerability, there must be support for the view of the International Assessment of Agricultural Science and Technology for Development (IAASTD), which states that "achieving food security and sustainable livelihoods for people in chronic poverty depends on ensuring access to and control of resources by small-scale farmers" (IAASTD, 2008). In line with the IAASTD and the recommendations of international development institutions such as the Bill & Melinda Gates Foundation, the Kellogg Foundation, Oxfam, and FAO, emphasis must thus be placed on support of existing smallholder farms.

**Driver 2: Africa's untapped agricultural potential**

By 2050, global food production may need to have increased by 70 percent (over 2005–2007 levels) in order to feed the world's growing population. Cereal production will have to expand by 43 percent and meat production by 74 percent (FAO, 2009a). While Africa and South America have an abundance of untapped agricultural resources to deal with this challenge, other continents will find it increasingly complicated to expand their use of agricultural production resources (see Table 2.1).

| **Region** | **1980–2004** (percent) | **To 2050** (percent) |
|---|---|---|
| World | 21 | 15 |
| South America | 16 | 40 |
| Asia | 46 | 12 |
| Central & North America | -2 | 2 |
| Europe | 114 | -2 |
| **Africa** | **18** | **60** |

**Table 2.1. Utilisation of arable land resources (1980–2050) (percentage of land potential). Sources: FAO, 2009a and Standard Bank Agribusiness, 2012.**

Table 2.2 illustrates the current potential for food grain production on agricultural land. This clearly shows that Africa, the region with the most abundant land resources, provides by far the greatest potential increase in the supply of food and fibre through expanding land under cultivation to meet the growing global demand.

Sources of agricultural production growth. While agricultural production in Africa is often viewed as stagnant, the production increase recorded by African agriculture since 1984 tells another, very encouraging story. Per capita production increased from below 200 kg in 1984 to almost 600 kg in 2006 and from 300 million tonnes total in 1984 to almost 750 million tonnes in 2006. These increases were achieved through a combination of factors illustrated in Table 2.3, including increased area of land under cultivation, increased yields, and higher cropping intensity. These data demonstrating that it is possible to increase agricultural production markedly suggest that exceptional opportunities exist for African agriculture.

| | **1980–2004** (percent change ) | **2050** (percent change) |
|---|---|---|
| South America | 80 | 60 |
| Asia | 64 | 47 |
| Central & North America | 40 | 21 |
| Europe | 80 | 44 |
| **Africa** | **75** | **150** |

**Table 2.2. Food grain production growth potential: Actual 1980-2004 and projected 2005-2050. Source: FAO, 2009a; ABSA/Barclays Bank, 2009 as quoted by Van Rooyen, 2009.**

Climate change, variation, and instability such as variable rainfall during peak production periods constrain farm production and must be addressed through research and development (R&D) and insurance support schemes to sustain food security and reduce food vulnerability in the region.

Climate change is expected to cause shifts in production space and locations over time. However, no major changes in the potential for food grain production are projected for the sub-Saharan region over the next twenty years. The potential impact of climate change requires careful assessment, especially of its spatial effects on population and smallholder households.

**Driver 3: Transformation in the agrifood business system and the emergence of agrifood value chain networks**

The driving forces of income growth, demographic shifts, globalisation, and technical change have led to a reorganisation of the agrifood system since the 1990s, with supply chains and support networks becoming dominant features linking farm to plate. The agrifood industry has been fundamentally restructured to meet the ever expanding global demand for food and fibre, to keep up with changing consumer preferences and technological innovation, and to comply with the worldwide deregulation of agricultural business and trade (Zuurbier, 1999; Vorley et al., 2007; Hughes, 2007; Swinnen, 2007; McCoullough et al., 2008; Reardon et al., 2009; Webber and Labaste, 2009; Swinnen, 2010). Important shifts include:

- A move from spot-market-based transactions, which are largely opportunistic in nature, towards value chain management and coordination, which are largely governed by trust, long-term contracts, and relationship management;

- A shift from local sourcing of farm products towards global sourcing in wider regional and international markets;
- A change from public- or government-based standardisation and certification processes (food and quality assurance, fair labour practice, etc.) towards private sector-driven norms and standards in food safety and ethical/environmental considerations (supermarket-initiated certifications and audits).

| | Arable land expansion | | Increased yield through technology and production management | | Cropping intensity increases through multiple cropping, etc. | |
|---|---|---|---|---|---|---|
| | **1961 – 1999** (%) | **2000 – 2030** (%) | **1961 – 1999** (%) | **2000 – 2030** (%) | **1961 – 1999** (%) | **2000 – 2030** (%) |
| **Developing countries** | 23 | 21 | 71 | 67 | 6 | 12 |
| **Sub-Saharan Africa** | 35 | 27 | 34 | 61 | 31 | 12 |
| **Near East and North Africa** | 14 | 13 | 72 | 68 | 14 | 19 |
| **East Asia** | 26 | 5 | 79 | 81 | -5 | 14 |
| **South Asia** | 6 | 6 | 80 | 81 | 14 | 13 |
| **Latin America** | 46 | 33 | 55 | 46 | -1 | 21 |

**Table 2.3. Sources of growth in African agriculture: recorded for 1961–1999 and projected for 2000–2030. Source: FAO, 2009a; ABSA/BAR CLAYS BANK, 2009 as quoted by Van Rooyen, 2009.**

Interest in agriculture is now clearly stretching beyond farmland and focusing on all aspects of the value chain. As a result, sourcing of produce from farmers is increasingly directed through contracts or long-term arrangements and subjected to certification and monitoring procedures to conform to standards required by markets. Farm producers have effectively become members of corporate food business networks, operating in an integrated and coordinated manner with input suppliers, service providers, funding agencies, processors, retailers, and supermarkets—agrifood value chain networks (Reardon et al., 2009). Trust and relationship management within a particular agrifood chain network have become important drivers of sustainable agribusiness. The World Bank (2013) pointed to the importance of promoting initiatives that include small, medium, and large companies in the value chains, creating new opportunities and jointly diagnosing and relieving bottlenecks and constraints. The aforementioned all promote a range of commercially directed farming models and governance systems meant to enhance the business performance of a particular value chain network vis-à-vis other competitive agrifood chain networks.

*The location of future production sites*. Because of the vast potential of unused natural resources and land availability, Africa and, to a lesser extent, Latin America will become important future locations for farm production and sourcing of raw materials to satisfy the rapidly growing demand for food and fibre.

With a so-called Green Revolution, Africa could increase the value of its agricultural output from US$280 billion a year in 2011 to more than US$500 billion by 2020 and US$880 billion by 2030. Growth of this magnitude would increase demand for a whole range of upstream products and services such as fertilisers, seed, pesticides, machinery, market advice, and food safety services, and also for downstream activities such as grain storage and refining, food processing, and biofuels. This demand for upstream and downstream activities is projected to reach US$275 billion a year by 2030 (Sanghvi et al., 2011).

Meeting this growth in demand will require the emergence of global agrifood chains and networks that can source raw materials from African producers. In this case, cost effectiveness, innovation, traceability, food safety, and quality checks will drive the different processes in the chain. Localised agrifood chains that link farm production to rapidly growing urban areas in Africa will also feature in the food business system of the future. The evolving agrifood business system and agrifood value chains have direct and dramatic effects on the prospects for African agriculture and the structure of industries related to agriculture (Shepherd, 2008; Webber and Labaste, 2009; Reardon et al., 2009; Poulton and Lyne, 2009).

*Changing investment patterns.* The restructuring of the agrifood business system is also expected to change the nature of agribusiness investments. For example, in order to consolidate value chain actions and allow for economies of scale, investments will need to focus on wholesale market infrastructure. The singular purpose of providing farm producers with access to the next added value level (storage-processing-packaging-manufacturing-branding-retailing) in the value chain will thus shift the emphasis to the development of infrastructure and support systems to facilitate the functioning of all levels of operation in the total agrivalue chain. The new drivers of business opportunities in the food system will be investments in agribusiness mentoring, coaching, and extension for producers participating in the value chain; and on processing, retail functions, and mechanisms for support and food safety for consumers (Swinnen, 2010; Reardon, *et al.* 2009; Sanghvi et al., 2011; World Bank, 2013).

A 2010 Organisation for Economic Cooperation and Development (OECD) report on "*Private financial sector investment in farmland and agricultural infrastructure*" highlights the expectation that investment in agriculture worldwide will "double, even triple in the medium to long-term" (the estimated investment by the private sector in 2013 amounts to US$10 to 25 billion). The geographic focus of such investment activity is shifting noticeably toward South America (led by Brazil) and Africa—both areas with land resource abundance.

One major advantage for Africa is the lower land acquisition cost for large-scale farming operations in southern and central Africa, boosting the advantages of economies of scale. In Brazil, land costs range from US$1500 to $3000 per hectare, compared with US$300 to $500 per hectare in Africa. There are also transportation cost advantages on the east coast of Africa relative to the Western Hemisphere as far as exports to India and the rest of Asia are concerned (OECD, 2010). These advantages will result in the emergence of global agrifood chains and networks that will source raw materials from African producers, with cost effectiveness and innovation, traceability, food safety, and quality checks, all driving the different processes in the chain. Localised agrifood chains that link farm production to rapidly growing urban areas in Africa will also feature in the food business system. The evolving agrifood business system and agrifood value chains will directly and dramatically affect African farming prospects and typologies (Shepherd, 2008; Webber and Labaste, 2009; Reardon et al., 2009; Poulton and Lyne, 2009).

The emerging investment patterns in Africa are complex. Local capital markets are still in the early stages of development, and land tenure regulations that govern land titles and concessions are evolving. Investment schemes typically constitute a tiered corporate holding structure with subsidiaries overseeing farm production processes, often across different countries. These large-scale farmland developments do offer benefits, such as sustained, higher wage-earning opportunities, better and more stable working conditions and employment benefits, local capacity and skills development, and improved agronomic and business practices for neighbouring farmers, as well as increased farm productivity and access to markets. Where outright ownership of land is not possible, long-term concessions, which often include commitments by investors to provide support for social projects such as schooling, health care, and skills development, are generally negotiated with governments,. These operations are often the biggest and most compliant tax contributors, expanding the tax base for local communities. Governments thus have generally been favourably disposed to such private capital development, transforming of farmland, and investing in agricultural and related rural infrastructure (OECD, 2010).

The 20 percent-plus investment paradigm constitutes an increasingly attractive option for African agriculture, governments, and local entities. However, the social sustainability of such production systems is an important political and financial concern and will be discussed in the sections to follow.

As to the environmental sustainability of such farmland development, the 2010 OECD report indicates that many investors in large-scale agriculture cultivate close working relationships with multilateral organisations such as the World Bank, the International Finance Corporation (IFC), and the FAO to ensure sustainable farming practices.

Another feature of the changing investment pattern in African agriculture focuses on the improvement of logistics infrastructure. The World Bank's 2012 Logistics Performance Index (LPI) rates most African countries low, with eight of the ten lowest LPI countries being in Africa. In many respects, the key to unlocking Africa's agricultural potential will be driven by investment in logistical infrastructure.

*Smallholder agriculture and agrifood business transformation.* Because they have assured markets and can aspire to long-term prospects for a share in value-adding opportunities and more secure financial support services, farm producers who are directly linked to agrifood value chains are generally better off; they are able to offer better quality produce and get better prices as a result (Zuurbier, 1999; Vorley et al., 2007; Swinnen, 2007; Shepherd, 2008; Webber and Labaste, 2009).

Food companies and supermarkets generally prefer to source from large-scale farm operations rather than from large numbers of smallholders in order to avoid high transaction costs and problems with consistent volumes, quality, and delivery. Larger farms are also better equipped to benefit from this emerging value chain-driven food business system (Kirsten et al., 2009).

Does a commercial value chain-dominated agrifood system work against smallholder agriculture, or can smallholders access such business systems? The underlying hypothesis to this question is that the transforming agrifood business system will effectively exclude smallholder agriculture from future business opportunities. In general, it must be noted that various studies on the position of smallholder agriculture in agrifood value chain-driven business systems do not necessarily support this view (Swinnen, 2007; Shepherd, 2008; Webber and Labaste, 2009; Kirsten et al., 2009; Reardon et al., 2009).

Various cases can be quoted where raw products are sourced from smallholders—not as an act of charity or corporate social responsibility, but because their inclusion was found to be profitable, even when large producers operated in the same sector. Sugar, vegetables, milk, fruit, and meat are examples of produce sourced in this way (Nestle, 2009; Shoprite/Checkers, 2009, to name a few).

These studies also found that where smallholders dominate the agrarian structure, as in many African regions, food companies tend to purchase from those smallholders operating in localities where certain elements are in place and functioning effectively. These elements include production infrastructure and assets such as irrigation systems, farm equipment, and farm equity; collective action organisations such as farmers' associations and producer cooperatives; and access to transport and communication systems. Partnerships with government agencies could also be considered where appropriateare also sought to enable such situations. Supportive government policies and systems generally enhance the successful inclusion of smallholders in the value chain. Resource and input supply contracts between smallholders and operators serving a particular agrivalue chain are important in dealing with constraints faced by smallholders, such as the lack of credit, lack of production inputs, and limits to technology transfer and extension.

Value chains depend on sound partnership arrangements. Partnership business models such as out grower schemes, contract farming, cooperatives or business trusts, and coaching and extension schemes accommodate the aforementioned features and will enable smallholders to participate in future business prospects and to move on a development trajectory towards higher levels of commercial farming (Karaan, 2009; Van Rooyen, 2009, Mabaya et al., 2011).

Linking smallholders (and other small- and medium-scale businesses) to commercial business systems should thus be a continued major focus in situations dominated by smallholder agriculture. Policy directives, government support to business-focused strategies, programme interventions and projects, strategic financial support instruments, public-private sector partnerships, and appropriate governance systems (including long-term transactions and contracts) should be designed and introduced to support African smallholder participation in the evolving agrifood system.

A particular focus area to support sustainability in the future value chain network driven economy will be will thus be to consider under which conditions will poor people, small holder, etc. benefit from such trends. Gomez, et al, 2011, propose the introduction of a food value chain research agenda to record and analyse this "rapidly changing landscape". These include focus on opportunities in the domestic market – in particular in the rapidly urbanising environment,; noting the indirect effects of such systems; enhancing market channel efficiency; the reduction of post-harvest losses; on-farm conservation practices; and appropriate certification and check-points. The ultimate aim would be to increase the demand for and supply of foods with economically, ethically, socially and environmental attributes.

### Driver 4: Scale factors in farm production

Most African agricultural production is still done by smallholders. However, production within this system is largely stagnant, and its commercialisation—linking farm producers to commercial markets—is generally complex and difficult (Kirsten et al., 2009).

Larger-scale farming in countries such as South Africa, Zimbabwe, Zambia, and Tanzania has proven to be successful in increased production and value creation from which smallholder farmers can benefit (refer to the section on the future of smallholder farming). So have recent large or mega-scale corporate farm development initiatives that range from 30,000 ha to 1 million ha in the grain, livestock, sugar, and fruit industries in sub-Saharan Africa (Byerlee, 2010; Emergent Asset Management, Citadel Capital and Dominium Farms, cases quoted by the OECD, 2010; Connolly et al., 2012). In particular, these large-scale ventures have provided the required investment returns, as well as technical innovation, yield-increasing practices and market responses, and social community support. These business-driven systems have contributed significantly to the production increases noted in African agriculture (see previous sections). They produce large volumes cost effectively for growing markets and consistently provide food grains, vegetables, and fruit for local consumption and for export.

What is underlying this trend towards large-scale farming in Africa? Apart from a range of economies-of-scale advantages in the farm production process—in areas such as mechanisation, labour management, and so forth—large-scale farming generally lowers the transaction costs of delivering to agrifood value chains, while also providing consistency in production volumes and quality. Input-provision and food-processing companies therefore generally prefer to source from large-scale over smallholder agriculture. Large-scale farming also competes more effectively with non-agricultural business activities for investment and financial reward for its managers and workers (Reardon et al., 2009; Kirsten et al., 2009).

## Land transfers and the notion of "a land rush" in Africa

From the above discussion, it is clear that Africa has again become subject to outside interest in its natural resources, especially land and water. Dramatic changes are occurring throughout Africa, including in South Sudan, Ethiopia, Madagascar, Tanzania, Congo, Kenya, Liberia, Senegal, Zambia, and Mozambique, with leading investors coming in from Asia (China, India, and South Korea), the Netherlands, France, Canada, Italy, the UK, Sweden, and even South Africa, as well as from international funds (FAO, 2009b). Recent research by Anseeuw and Ducastel (2012) identified deals made for 948 projects encompassing 161.6 million hectares in Africa, compared with 240 projects encompassing 17.6 million hectares in Latin America. Main investors in the African deals originated from China (77 deals), Saudi Arabia (65 deals), the UK (60 deals), India (60 deals), South Africa (41 deals), the U.S. (38 deals), and the Netherlands (26 deals). Table 2.4 presents examples of recent land-related farm developments in Africa.

Africa's abundance of land and water (see note 2) allows the production of valuable agrifood commodities. With local African governments eager to capitalise these resources, the number of land deals in Africa has soared, with an unparalleled transfer of land ownership. While some view this as a form of land grab or new colonialism, others emphasize the potential to increase food security, employment, and income generation.

While agreeing that these cash-for-cropland deals might improve growth prospects and food security for all participants, many NGOs, such as Oxfam and World Watch, have raised concerns regarding the real intentions behind such deals and the potential impact of removing indigenous people from their lands, with no alternative jobs or resources provided. Some also question the shift in focus towards large-scale agribusiness and value chain-led investments (FAO, 2009b; Oakland Institute, 2011). The World Bank counters that supporting governments in regularising land tenure systems provides a much improved base to encourage investments that will support those whose rights have not been formalised. Smallholders need to be included in such developments, *inter alia*, through good practice investments and protocols (World Bank, 2013).

In their study of "land grabs" in Africa, Connolly et al. (2012) concluded: "[e]ither way, when large scale farming operations first come to a region, social disruption is inevitable, [and] outcomes are difficult to predict. Availability and pricing of both labour and resultant agricultural products are certain to change. Opportunities for entrepreneurs, either as part of a supply chain, or in ancillary goods and services, are also likely. The response of the community may also come into play, whether through cooperation or resistance."

- *Angola:* 140,000 ha mixed crops, AgriSA, South Africa; 25,000 ha rice, Lonrho, UK
- *Benin:* 350,000 ha crops and 300,000 ha oil palm, South Africa
- *Ethiopia:* 815 foreign-financed agricultural projects approved between 2007 and 2010; a 150,000 ha livestock farm; 300,000 ha farm land to an Indian company, Karuturi; 50,000 ha crops by a number of South African companies, including Richmond, New Dawn, Dinaledi, Batau
- *Congo:* 68,000 ha eucalyptus, Mag Industries, Canada; 200,000 ha mixed farming, AgriSA, South Africa
- *Cameroon:* 55,000 ha mixed crops, GP Bollere, France
- *Gabon:* Olam, Singapore, acquired 300,000 ha to produce palm oil
- *Madagascar:* Daewoo (Korea) attempted to lease 1.3 million ha for farm production; a 450,000 ha biofuel project; mixed crops on 100,000 ha by Osho of South Africa; 30,000 ha under jathropha by Neo, France
- *Mozambique:* 35,000 ha teak and 140 000 ha pine, Gsff, Sweden; 15,500 ha jatropha, Avia Spa, Italy; 20,000 ha livestock, Agriterra
- *Liberia:* A Malaysian conglomerate signed a 63-year lease for 230 000 ha to grow palm oil and rubber; an Indonesian producer signed an agreement to develop 220 000 ha for producing palm oil
- *Mali:* 540 000 ha farm project; 50,000 ha rice project by Saudi Arabia; 40,000 ha crops, Seed Rock, Canada
- *South Sudan:* 800,000 ha land lease near Darfur; United Arab Emirates acquired 750,000 ha farm lands; Saudi Arabia leased 376,000 ha to grow wheat and rice
- *Sierra Leone:* 45,000 ha mixed crops, Sierra Gold, Canada
- *North Sudan:* South Korea companies bought 700,000 ha for wheat farming
- *Ghana:* 13,000 ha under crops by Kimminic, Canada
- *Kenya:* 160,000 ha biofuels by Bedford, Canada
- *Tanzania:* 45,000 ha sorghum, Cams Group, UK; 100,000 ha trees, Norway; 100,000 ha jatropha, Diligent Energy, Netherlands; 100,000 ha crops, KRC, Rep of Korea
- *Zambia:* 200,000 ha crops, United Arab Emirates;15,000 ha sugar, Agzam, South Africa; 27,000 ha crops, DWS Galof, Germany
- *Ethiopia, Kenya, Madagascar, Senegal and Mozambique:* 80 Indian companies invested around US$2.4 billion in buying and leasing plantations.

**Table 2.4. Examples of recent African farmland developments: country, investors, and crops[a]. Sources: Connolly et al., 2012.**

[a] Note: some of these developments are intentions to invest and others are projects in implementation. These examples do, however, show the degree of interest in African agricultural investment.

From this, it can be concluded that any new paradigm and strategic framework will have to focus in particular on sustainability, as defined from various social and livelihood perspectives. A clear view on the future position of smallholder farming will therefore also be relevant.

*The future of smallholder farming:* Do these economic advantages and business preferences for large-scale farming and the related land transfers effectively leave African smallholder farming out in the cold, so to speak?

First, there is growing evidence that smallholder systems can increase farm production through yield-increasing technologies and improved access to agricultural support services—if the technologies and services are supplied to smallholders and if they are linked to agrivalue chains (Future Agricultures, 2011; World Bank, 2013).

Second, the "efficient but poor" hypothesis (Schultz, 1984) also provides evidence that although smallholder farmers can operate at higher economic efficiency levels than large-scale agriculture, they generally remain poor because of the low income-producing potential inherent in small-scale agriculture. Technological innovation may enable smallholders to increase their income levels, although this generally also requires increased capital investment, risk insurance, application of management, and technical know-how. In other words, such innovation requires operating with a higher production function. However, as increased capital, risk insurance, etc. are scarce factors in smallholder agriculture, innovation is constrained, and the "efficient but poor" hypothesis is generally confirmed.

Third, in addition to the large-scale and smallholder farming typologies, a mixed-mode farming typology is emerging in African agriculture. In this mode, large-scale and smallholder agriculture co-exist through coordinated, synergistic, and collaborative activities based on commercialisation and economies of scale. This typology is increasingly observed in labour-intensive farm activities such as 1) producing vegetables, fruit, sugar, and dairy products, which link to value-added processing and trade; 2) grain production with scaled economic advantages in off-farm storage and processing; and 3) livestock farming, which provides animal products to large abattoirs and value-added meat processing. This mixed-mode farm typology also supports opportunities for smallholders to depart from low-income smallholder agriculture and gain access to the higher income-earning opportunities of agrifood value chains and distribution networks.

The expansion of mixed-mode farming in Africa provides a sound basis for introducing institutional innovations, activating scaled economic advantages in production and support systems (both logistical and financial), and providing risk mitigation. The institution of collective actions and support systems would ensure consistent production output, quality maintenance, and food safety among suppliers, including smallholders. Achieving this outcome would require the mobilisation of community partners to implement viable interventions for optimal usage of common resources such as arable land, grazing areas, and water; institutional support would need to come from co-operatives, farmer organisations, and leadership development programmes. These institutions should also support the provision of skills and capacity-building and assure effective and fair contracting. The role of stable governance systems is also crucial (Vorley et al., 2007; Reardon et al., 2009; Masuku, 2009; Karaan, 2009).

Smallholder farming will remain an important feature of Africa's agriculture, helping secure many millions of livelihoods—mostly meagre, but nonetheless, livelihoods. Socially this typology will also remain important, because it provides a fall-back to many who lose their employment in the non-farming environment. Uma Lele (1984) referred to this as a residual function of smallholder agriculture that provides at least a degree of household food security to members and a refuge for the elderly and children.

Support to smallholder farming will require particular mechanisms—including public sector support and, where possible, the integration of smallholder farms into supply chain-driven agrifood systems. In addition, any agricultural development framework with a focus on smallholders must recognise the need to empower African women and youth in particular, as these are the groups mostly served by smallholder production.

Lessons from other regions (Connolly et al., 2012) exposed to commercial farming forces should also be analysed. Two cases come to mind, namely the so-called miracle of the Brazilian *cerrados*, where poor soil was improved enabling highly productive farms, and India's support to millions of small-scale farmers, which offered biotechnology tools to non-food crops, small packets of technology including seeds, fertiliser, herbicides, pesticides and mechanisation for farmers with less than 1 ha of land, along with training and extension. The Chinese approach—focusing on the mobilisation of smallholder production through the provision of modern inputs and farming innovation—also provides lessons and direction, although the notion of central planning has to be considered. The African situation will require a unique set of solutions to fit the many diverse local situations.

## Driver 5: Towards a bio-based green economy

The idea of a bio-based green economy encapsulates the vision of a future where the world is no longer wholly dependent on fossil fuels or industrial raw materials for energy (CGIAR, 2009a). Agriculture could stand central to the bio-based economy, providing source materials for agricultural processes—liquids, fuels, chemicals, and production inputs. At the same time, agriculture will continue to provide food and animal feed that is more environmentally sustainable, healthy, and safe.

A bio-based economy will require agriculture to expand well beyond its current core function of food production. Positioning agriculture as a supplier of energy could be controversial; because the added demands on the sector could potentially lead to increased food prices, some view this as a threat to food supply and food security. However, this scenario also offers exciting opportunities. Biomass production can provide the energy needed in production processes, plus it is a renewable resource – input/output balances however need to be considered. Thus, African agriculture should explore the future possibilities of biomass with emphasis on the next generation of biofuels, and the production of bio products (CGIAR, 2009a).

*Pressures, preferences and trends.* Consumers, private agencies in food retail and trade, and civil society organisations will continue to pressure the agrifood sector to focus on environmental integrity and sustainable production. Environmental regulation and certification, biotechnology and bio-based items such as liquid fuels, agrichemicals, and animal feed will all be important features of the evolving commercial agribusiness environment. The establishment of environmentally sustainable and energy-saving crop and livestock systems will thus be a priority for African agrifood R&D systems. Livestock systems, which constitute the largest land-use activity on earth with high energy-to-food ratios, will be particularly challenged in this context (Swanepoel et al., 2009).

*Research links, networks, and protocols.* CGIAR initiatives to establish and link global research networks are encouraging and must be supported in the strategic framework for African agriculture (CGIAR, 2009a). These initiatives promote agricultural systems striving to establish a bio-based economy that reduces water, land, nutrient, and chemical use in order to achieve the envisaged doubling of food production by 2050. Linking research networks could provide incentives to encourage and direct the new trend of commercial African agriculture to move towards greener practices. Implementing these greener practices will require establishing new science policies, good governance practices, and interdisciplinary collaboration among interested groups such as governments, scientists, civil society organisations, consumer groups, food producing businesses, farm agencies, and farmers. For these efforts to succeed, the various groups must be able to draw from a strong pool of human resources; there is an urgent need for trained personnel having the appropriate qualifications for success.

## Economic Management and Governance

The above drivers of agricultural change will affect Africa's complex social, political, economic and natural environments, which are dealing with complementarities, trade-offs, redefined roles, changing institutions, and long-term commitments. To achieve faster agriculture-based growth rates and equitable participation processes, favourable macro-economic and trade policies must be in place, along with good governance practices, good infrastructure, and access to credit, land, knowledge, and markets. These will establish an enabling environment that will incentivise both small- and large-scale farmers to adopt new and sustainable technologies and diversify production into higher value crop systems that can support agribusiness systems and value chains. Agro-industrialisation, as is expected in the African environment, is generally regarded as ushering in periods of individual and collective stress, discontinuous change, and economic disorder. However, the positive, negative, and neutral implications of such processes must be noted.

For the aforementioned enabling environment to be established, sound economic management systems will be required. The following section introduces the types of systems needed.

*Africa-directed trade development.* The reduction of trade barriers, especially for higher value-added products, would encourage agribusiness-driven value chains to explore lucrative opportunities in regional and global environments. Supporting African agricultural trade will require a range of actions, including the reduction of domestic price support programmes and the eventual removal of border protection by OECD countries; development of African production and regulatory capacity to meet the exacting standards of importation in global markets; reduction of inter-Africa trade barriers and red tape, which suppress the competition required for African industry to gain sustainable advantages in evolving and lucrative global markets; improved transportation systems; and the addressing of impediments to efficient African rural-urban market linkages for food and fibre. It also calls for simplification of overly complex intra-African trade agreements.

*Improved public-private partnerships (PPPs).* Collaboration between government and the private sector is called for as an important strategy to increase financial, technical, human, and social capacity. PPPs need to focus on a range of activities including training, extension, and skills development for farming and agribusiness activities; human capacity development through internships, mentorships, and skills transfers; development of agricultural research and technology transfers; the development of information and communication systems; expansion of rural infrastructure; the establishment of agro-industrial clusters and production schemes; technical support for land reform and transfer schemes; the identification and resolution of value chain constraints; and the opening of opportunities as business system tipping points (Mayers and Van Rooyen, 2011). The provision of access to financial resources and risk mitigation instruments geared to selected agricultural development activity also needs to be structured.

*Good governance practices and investment protocols* must be viewed as key components of a new paradigm and strategic framework to direct agricultural development and ensure that economic and business management processes are implemented in an accountable, transparent, and responsible manner. This will require institutional innovation to deal with the complexities of competition and value distribution along value chains as well as other matters confronting the African agricultural environment (IFPRI, 2006, Swinnen, 2007; Webber and Labaste, 2009; Poulton and Lyne, 2009; Gabre-Madhin, 2009; World Bank, 2013). Good practice investments that agitate against corruption and exploitative deals are also called for. Rural environments in particular will require strong new institutions and governance structures to, for example, enable farmers and agribusiness groups to negotiate fair deals in agrifood value chains dominated by supermarkets or branded manufacturers. New governance structures will also be required to provide a fair voice to civil society and in particular to women in agriculture and youth. These institutions and structures must direct the sustainable use and exploitation of natural resources and attempt to resolve conflicts such as private versus public, individual versus community, and business versus social interests.

The World Bank (2013) views the implementation of economic management and good governance as a "daunting challenge for governments with weak capacity and little experience in engaging the big players in the private sector." Good governance in African agriculture can be built through the development of social capital and collective action, where participation in such networks increases the availability of information, helps to enforce property rights regimes, and reduces opportunistic behaviour concerning the use of common resources such as land, water, and information (Ostrom, 2009). International donor agencies and private businesses should collaborate with public sector institutions in processes aimed at establishing and maintaining the required good governance practices.

The notion of setting good practice investment protocols to direct development patterns towards sustainable growth paths must be afforded priority. The engagement of private investors and agrifood corporations chasing the 20 percent-plus baseline will be of vital importance. South Africa's Broad Based Black Economic Empowerment (BBBEE) codes and scorecards can be viewed as a potential model in this context. BBBEE allocates points for transformational protocols, such as preferential procurement from historically disadvantaged groups and individuals, human capital development, equity opportunities and ownership, and corporate social investments. These South African protocols are currently in operation and provide for a range of business development models such as out-grower and equity share schemes, contracting opportunities and social services (Mabaya et al., 2011).

Reforming the legal systems to, *inter alia*, regularise land tenure systems will support good governance and help those whose rights have not been formalised, in particular women and smallholders.

*The role of civil society, women and youth in African agriculture.* According to the African Women in Agricultural Research and Development (AWARD) programme, women are currently the backbone of African agriculture (CGIAR, 2009b). CGIAR (2009b) finds that women produce, process, and market most of the food Africa depends on. Yet, women farmers receive only 5 percent of agricultural extension training and less than 10 percent of rural credit. The CGIAR also documented that women represent less than a quarter of the agricultural researchers, and fewer than one in seven of the leadership. This reality needs to be addressed as a priority, if smallholder farm production is to be increased and commercialised. The CGIAR view that "we cannot defeat hunger and poverty in Africa unless women have a strong voice" is echoed by most prominent institutions in their efforts to support the renewal of African agriculture.

It is also generally observed that young people often want to escape rural farming life. This is not surprising, as smallholder farming renders few incentives to remain on such small holdings. The concept of a thriving and ever modernising farming system, linked to technical innovation, value chains, markets, and business deals requires intellectual capacity and skills that are generally not available to the aging rural population and to farm producers. Finding a way to mobilise the youth to take up farming as a scientific and business-driven career remains a big challenge.

Support programmes and funding schemes giving preference to the young and to female farmers must receive priority in development strategies.

## Towards a Framework Matrix for African Agricultural Development

### Core functions and strategic thrusts

Future growth in African agriculture is expected to be achieved largely through increased farm-level production. Mechanisms for growth include:

- supporting land expansion, technological innovation, and increased yield;
- commercialising agriculture at smallholder and large-scale farming levels;

- building linkages to agrifood chains that will activate intensification and value-adding opportunities for African agriculture; and
- applying good governance practice through institutions that direct the design and management of complex, globalising agrifood business systems.

A framework matrix can be created through the interaction of a set of core functions and development thrusts.

*Core functions.* An integrated development strategy needs to be designed to activate four core functions:

- income generation, including foreign exchange, through the provision of food and fibre at primary production and value-adding levels to supply growing and differentiating global markets;
- employment creation in the full agrivalue chain and in the wider economic environment;
- household level food security through local supply and production;
- provision of environmentally sustainable products and services including bio-based energy and fuels.

*Cross-cutting strategic development thrusts.* Focusing these core actions will require five strategic thrusts for agricultural development in the sub-Saharan African environment, each cross-cutting to the four core actions:

- stimulate market-pull opportunities: support the exploitation of prospects and opportunities opening up to African agriculture—both large and small scale—in order to gain access to the growing agrifood business system and to ensure farmers' commercially sustained inclusion (empowerment of women farmers and support to young farmers will be important in achieving this);
- improve household level livelihoods and food security: promote the support of food security and deal with food vulnerability that is a concern of many African households in rural and urban environments;
- promote environmental stability: design environmentally sustainable agricultural practices to support, *inter alia,* innovation towards the bio-based economy and development of appropriate systems as incentives for farming communities and agribusinesses to introduce and maintain such practices in their production systems;
- improve economic management and governance: instil leadership qualities, support good governance practices and codes—accountability, transparency, predictability, and participation—and give agribusiness, farm producers, and women-based groups each a voice in public affairs and in the design and implementation of development plans;

- develop human capacity: emphasise and support human capital development through an interactive and responsive AET system, providing the human capacities, skills, and agents required to drive African agricultural development.

## A scenario matrix for African agricultural development

The nature of the interaction between the four core activities and five cross-cutting development thrusts creates three possible scenarios (see Table 2.5).

- *Hit and Run Development (H&R).* Agricultural development is driven by short-run market linkages and exploitative social and environmental practices, while governance is driven by short-run considerations, accompanied by corrupt deals to gain rapid access to natural resource potential. Development investment is segmented and only focused on the short-term prospects. Civil unrest and destructive actions increasingly jeopardise a sustainable development future. Human capital development is focused mostly on the development of short-term skills or on expatriate management and subject matter or professional expertise.
- *Equitable and Sustained Growth (E&SG).* This scenario is the opposite of H&R. It establishes the environment and building blocks to empower African society to participate fully in the governance and management of agricultural development and to benefit from its participation in the creation of its own future. The E&SG scenario is driven by market linkages, good governance and economic management, social and environmental directed codes and protocols, and sustained and integrated human capital development at all levels—school, higher education, and life-long education and training systems that identify and build needed skills.
- *Stop and Go (S&G).* This falls between the H&R and E&SG development paths, with sporadic, opportunistic, uneven, and inconsistent investments, business practices, and systems. There is less emphasis on good governance and human capital development. Short-term vibrant growth and initiatives may be experienced, but there is failure to achieve a sustained and equitable growth path for the agricultural sector, which is required in order to benefit African society at large.

| THRUSTS: | H&R | S&G | E&SG |
|---|---|---|---|
| 1. Equitable market linkages | segmented | opportunistic | interactive |
| 1. Social/livelihood | exploitative | uneven | equitable |
| 2. Environmental practice | exploitative | opportunistic | sustainable |
| 3. Governance systems | opportunistic | inconsistent | accountable |
| 4. Human capital use | exploitative | inconsistent | empowering |

**Table 2.5. Development scenarios for African agriculture**

Note 1: H&R = Hit & Run; S&G = Stop & Go; E&SG = Equitable and Sustained Growth

Note 2: Historically, investment opportunities with returns of around 6–10 percent above inflation were viewed to be sufficient for the agriculture sector.

Note 3: "With land comes the right to withdraw the water linked to it, in most countries essentially a freebie that increasingly could be the most valuable part of the deal" (Brabeck-Lemanthe, 2009).

## References

Adesina, A. 2009. *Institutional Changes in African Agriculture.* Panel discussion, IAAE Conference. Beijing, China.

Anseeuw, W. and A. Ducastel. 2012. *New Agricultural Investment Models and Agrarian Change in South Africa.* Agricultural Economic Association of South Africa Conference. Bloemfontein, South Africa.

Badiane, O. 2009. *Promoting the Competitiveness of African Agriculture.* The 47th AEASA Conference. Durban, South Africa.

Binswanger, H. 2006. *Empowering Rural People for their own Development.* Elmhirst Lecture, IAAE Conference. Gold Coast, Australia.

Brabeck-Lemanthe, P. 2009. The Next Big Thing: H20. *Foreign Policy Journal.* http://www.foreignpolicy.com/articles/2009/04/15/the_next_big_thing_h20

Byerlee, D., A. de Janvry, and E. Sadoulet. 2009. Agriculture for Development: Toward a New Paradigm. *Annual Review of Resource Economics* 1:1, 1–20.

Byerlee, D. 2010. *Globalisation and the Rise of Corporate Farms in Land-Abundant Developing Countries.* Paper presented at the 54th Annual Conference of the Australian Agricultural and Resource Economics Society. Adelaide, Australia.

Cheru, F. 2002. *African Renaissance: Roadmaps to the Challenge of Globalisation.* New Africa Books, Palgrave, St Martin's Press: New York.

Cheru, F. and C. Bradford. 2005. *The Millennium Development Goals: Raising the Resources to Tackle World Poverty.* Zed Books: New York.

CGIAR. 2009a. *Agriculture Beyond Food: Science for a Bio-based Economy.* Science Forum, 2009.

—. 2009b. *African Women in Agricultural Research and Development.* CGIAR Gender and Diversity Programme. Nairobi, Kenya.

Commission for Africa. 2005. *Our Common Interest: Report of the Commission for Africa. London: Commission for Africa.* http://www.commissionforafrica.org/english/report/thereport/ cfafullreport_1.pdf

Connolly, A., K. Connolly, and M. Lyons. 2012. A Seismic Change: Land Control in Africa. *International Food and Agribusiness Management Review* 15: 2, 171–177.

DeCock, N. D'Haese, M. Vink, N. Van Rooyen, C.J. Staelens, L. Schonfeldt, H & D'Haese L. 2012. Food security in rural areas of Limpopo province, South Africa. Food Security, DOI 10.1007/s12571-013-0247-y. Springer.

Delgado, C. 1998. *Agricultural Growth Linkages in sub-Saharan Agriculture.* IFPRI, Washington D.C.

—. 2009. *Global Food Crisis: From Urgent Short-term Responses to Viable Longer-Term Solutions.* Agriculture and Rural Development, World Bank, Washington D.C.

Development Bank of Southern Africa (DBSA). 2009. *Food Security Project Report.* Midrand, South Africa.

De Janvry, A. 2009. *Agriculture for Development: New Paradigm and Options for Success.* Elmhirst Lecture, IAAE Conference. Beijing, China.

Dumont, R. 1962. *L'Afrique noire est mal partie.* Editions du Seuil: Paris.

Eicher, C. and J. Staatz. 1998, *International Agricultural Development.* The John Hopkins University Press: London.

FAO. 2009a. *1.02 billion People Hungry: One Sixth of Humanity Undernourished—More than ever before.* FAO, Rome.

—. 2009b. *Land Grab or Development Opportunity? Agricultural Investment and International Land Deals in Africa.* FAO, Rome.

Future Agricultures. 2011. *Small Farm Commercialisation in Africa: Reviewing the Issues.* Research Paper 023/April. *www.future-agricultures.org*

Gabre-Madhin, E. 2009. *Exchange, Contracts, and Enforcement.* In Kirsten, J., A. Dorward, C. Poulton, and N. Vink, Eds. Institutional Economic Perspectives on African Agricultural Development. IFPRI, Washington D.C.

Gomez, M.I., Barrett, C.B., Buck, L.E., De Groote, H., Ferris, S., Gao, H.O., McCollough, E., Miller, D.D., Outhred, H., Pell, A.N., Reardon, T., Retnanestri, M., Ruben, R., Struebi, P., Swinnen, J., Touesnard, M.A.,

Weinberger, K., Keatinge, J.D.H., Milstein, M.B. and Yang, R.Y. 2011 Science, vol 332, June, 2011

Haggblade, S., P. Hazell, and J. Brown. 1988. *Farm-nonfarm Linkages in Rural sub-Saharan Africa*. World Bank, Washington D.C.

Hayami, Y and V. Ruttan. 1971. *Agricultural Development: An International Perspective*. First Edition. John Hopkins University Press: Baltimore, Maryland.

—. 1985. *Agricultural Development: An International Perspective*. Second Edition. John Hopkins University Press: Baltimore, Maryland.

Hughes, D. 2007. *Super Markets in the Global Food Business Systems*. Keynote paper, IAMA World Forum. Monterey, USA.

IAASTD. 2008. *International Assessment of Agricultural Science and Technology for Development.* Synthesis Report of the IAASTD. Johannesburg, South Africa.

International Food Policy Research Institute (IFPRI). 2006. *Agriculture and Achieving the Millennium Development Goals.* Series nu.32729-GLB. Washington, DC.

Johnston, B. and J. Mellor. 1961. The Role of Agriculture in Economic Development. *American Economic Review* 87: 2, 566–93.

Jones A.D., F.M. Ngure, G.H. Pelto, and S.L. Young. 2013. What are we assessing when we measure food security? A compendium and review of current metrics. *Advances in Nutrition* 4: 481-505.

Karaan, A. 2009. A Transaction Cost Approach to Enterprise Modelling and Coordination between Small Growers and a Large Firm. In Kirsten, J., A. Dorward, C. Poulton, and

N. Vink, Ed. *Institutional Economics Perspectives on African Agricultural Development.* IFPRI, Washington D.C.

Kirsten, J., A. Dorward, C. Poulton, and N. Vink. 2009. *Institutional Economic Perspectives on African Agricultural Development*. IFPRI, Washington D.C.

Krueger, A., M. Schiff and A. Valdés. 1991. *The Political Economy of Agricultural Pricing Policy*. John Hopkins University Press: Baltimore, Maryland.

Lele, U. 1984. *The Design of Rural Development: Lessons for Africa.* Books on Demand.

Lipton, M. 1989. *New Seeds and Poor People* (with Richard Longhurst). Unwin Hyman: London.

Mabaya, E., K. Tihanyi, A. Karaan, and C. Van Rooyen. 2011. *Case Studies of Emerging Farmers and Agribusiness in South Africa.* Sun Press, Stellenbosch.

Masuku, M. 2009. The Role of Trust in Contract Enforcement: Analyses of Smallholder Farmers and Sugar Millers in Swaziland. In Kirsten, J., A. Dorward, C. Poulton, and N. Vink, Eds. *Institutional Economic Perspectives on African Agricultural Development.* IFPRI, Washington D.C.

Mayers, W. and C. Van Rooyen. 2011. *Creating "tipping points" in the Ethiopian Agribusiness Environment.* Report to the Government of the Netherlands.

McCoullough, E., P. Pingali, and K. Stamoulis. 2008. *The Transformation of AgriFood Systems: Globalisation, Supply Chains and Smallholder Farmers*. FAO, Rome.

Mellor, J. 1998. Agriculture on the Road to Industrialisation. In Eicher, C. and J. Staatz, Eds. *International Agricultural Development.* John Hopkins University Press: Baltimore, Maryland.

NEPAD. 2013. *Report on the World Economic Forum.* Cape Town. May 2013. *www.nepad.org*

Nestle. 2009. Personal interview with Dr. Hans Johr, International Agriculture, on linking smallholders to their food supply chains. Montreau, Switzerland.

Oakland Institute. 2011. *Understanding Land Investment Deals in Africa.* http://www.oaklandinstitute.org/files/OI_Mali_Land_Investment_report.pdf-agricultural-development

OECD. 2010. *Private Financial Sector Investment in Farmland and Agricultural Infrastructure.* Food, Agriculture and Fisheries Working Paper No. 33, OECD Publishing.

Ostrom, E. 2009. *Managing Common Resources.* Plenary Address. International Association of Agricultural Economists Conference. Beijing, China.

Poulton, C. and M. Lyne. 2009. Coordination for Market Development. In Kirsten, J., A. Dorward, C. Poulton, and N. Vink, Eds. *Institutional Economics Perspectives on African Agricultural Development.* IFPRI, Washington D.C.

Ravallion, M., S. Chen and P. Sangraula. 2007. *New Evidence on the Urbanisation of Global Poverty.* World Bank Policy Research Working Paper No. 4199. World Bank, Washington, DC.

Reardon, T., C. Barrett, J. Berdegué and J. Swinnen. 2009. Agrifood Industry Transformation and Small Farmers in Developing Countries. *World Development*; 37 (11): 1717-1727.

Rukuni, M. and C. Eicher. 1994. *Zimbabwe's Agricultural Revolution.* University of Zimbabwe Publications Office.

Sanghvi, S., R. Simons, and R. Uchoa. 2011. *Four Lessons for Transforming African Agriculture.* McKinsey & Company.

Schultz, T. 1984. *Transforming Traditional Agriculture.* New Haven Publishers: New Haven, CT.
Shepherd, A. 2008. *Approaches to Linking Producers to Markets.* FAO, Rome.
Shoprite/Checkers. 2009. Personal interview with Mr. E. Nel, Director, on the linking of smallholders in to their food supply chain in African countries. Stellenbosch, South Africa.
Singini, R. and J. Van Rooyen. 1995. Serving Small-Scale Farmers. Development Bank of Southern Africa, Midrand, South Africa.
Streeten. 1987. *What Price Food*? St Martin's Press: New York.
Swanepoel, F., A. Stroebel, and S. Moyo. 2009. *The Role of Livestock in Developing Communities: Enhancing Multifunctionality.* WCAP Satellite Symposium. Cape Town, South Africa.
Swinnen, J. 2007. *Global Supply Chains, Standards and the Poor.* CABI: Massachusetts.
—. 2010. *Value Chains in Senegal.* Keynote Paper Presented at the joint American Association for Agricultural Education and Agricultural Economic Association of South Africa Conference. Cape Town, South Africa.
Timmer, P. 1986. *The Agricultural Transformation Issue.* 223 Development Discussion Paper, Harvard Institute for International Development, Harvard University.
—. 2010. *Lessons from Food Crises Past.* Centre for Global Development, Washington, DC.
United Nations. 2009. *The Millennium Development Goals Report 2009.* UN, New York.
Van Rooyen, C. 2009. *The Future of Agribusiness in Africa.* Final Panel Session of the IFAMA World Forum. Budapest, Hungary.
Vermeulen, S., J. Woodhill, F. Proctor, and R. Delnoye. 2008. *Chain-wide Learning for Inclusive AgriFood Market Development.* International Institute for Environment and Development, London, UK.
Vink, N. and C. Van Rooyen. 2009. *Agriculture and Food Security in South Africa.* Paper Prepared for the Development Bank of Southern Africa Food Security Report. Midrand, South Africa.
—. 2010. *Agribusiness Policy and Strategy in the Future.* 2010. Conference of the Agricultural Business Chamber (ABC), Lord Charles Hotel. Somerset West, South Africa.
Vorley, W., A. Fearne, and D. Ray. 2007. *Regoverning Markets: A Place for Small-Scale Producers in Modern Agrifood Chains.* Gower Publishing: Aldershot, UK.

Webber, M. and P. Labaste. 2009. *Building Competitiveness in Africa's Agriculture: A Guide to Value Chain Concepts and Applications*. World Bank, Washington, DC.

World Bank. 2005. *World Development Report 2006: Equity and Development*. World Bank: Washington D.C.

—. 2007. *World Development Report 2008: Agriculture for Development*. World Bank: Washington, DC.

—. 2009. *World Bank Commodity Markets Review*. DECPG, World Bank: Washington, DC.

—. 2013. *Growing Africa: Unlocking the Potential of Agribusiness*. AFTFP/AFTAI. World Bank: Washington, DC.

Zuurbier, P. 1999. *Supply Chain Management*. Waginingen University, the Netherlands.

# Chapter Three

# Agricultural Education and Training for Development: Lessons from sub-Saharan Africa

Irene Annor-Frempong and Monty Jones

## The Context for Agricultural Education and Training Reforms

Africa is moving into an agricultural transformation era – a transformation predicated on the change in economic structure as Africa's agriculture sector moves from "producing mainly for its own consumption" to "producing mainly for the market." With the progression of such economic development in African countries, the underlying innovation patterns for strengthening forward and backward linkages of primary agriculture offer an expanding role for agricultural input industries and services, and for agricultural innovation in both the public and private sectors (Anderson et al., 2013).

Innovation depends on the development of individual and organisational capabilities and strengthened institutions (Babu et al., 2011; Lynam et al., 2011). This underpinning paradigm shift shapes the dynamic development path for agricultural education and research systems in Africa. The key transformation required in Africa is to connect the missions and visions of advanced learning institutions with new local and global contexts (Juma, 2011). This means that educational approaches and learning philosophies applied to agricultural education and training (AET) in Africa need to cover a wide list of options that can accommodate different types of individuals and supporting institutions. If education is to "work for development," AET must be better integrated with other actors in the innovation system, including national, sub-regional and regional knowledge institutions, to allow all links to work together towards the needed impacts at the smallholder level.

AET is the teaching of agriculture, food science, natural resource management, land management, and related social sciences using hands-on experience and guidance to prepare students for entry-level jobs or for advanced agricultural jobs. Certification levels span vocational, craftsmanship, and technicianship to university graduate degrees.

At the national level the AET system, as illustrated in Figure 3.1, encompasses the vertical and horizontal networks and linkages within educational institutions (Spielman et al., 2012). AET provides and prepares skilled professionals who can support agriculture through education, extension, information, research, and entrepreneurship. In addition, non-formal (extension) education is needed for training farming families as well as for promoting indigenous knowledge and farmer-to-farmer sharing of information, as these are essential for achieving food security.

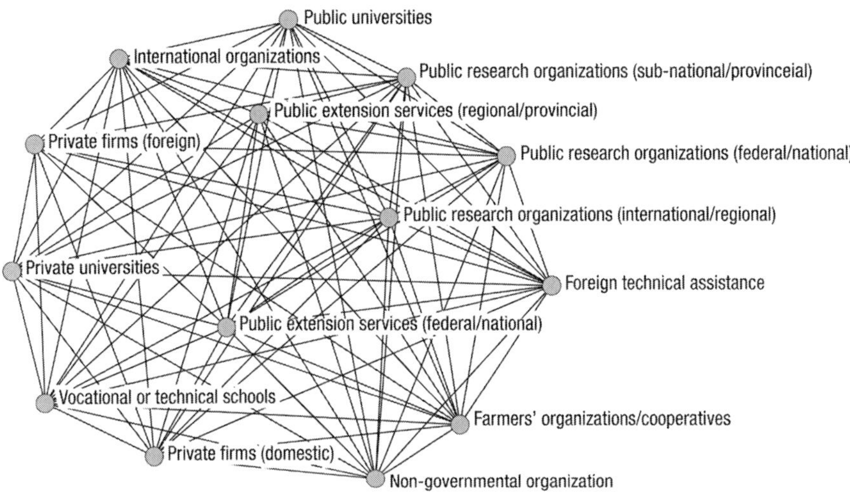


Figure 3.1. Theoretical model of AET (Spielman et al., 2012)

Since independence, many African countries in sub-Saharan Africa and other developing countries have received significant technical cooperation resources, many of which have been aimed at supporting institution-building or human capital development. While some important skills and institutions have been supported, there have been challenges in the form of lack of sustainability, fragmentation of resources, untargeted interventions, low domestic investment, and uncoordinated external support.

***Key Messages***

- The economic structure of the agricultural sector is changing from producing mainly for direct consumption to producing mainly for the market. This is resulting in a *paradigm shift from research towards innovation,* which in turn requires AET to develop both individual and organizational capacities.
- The Framework for African Agricultural Productivity (FAAP) articulates five *key dimensions of capacity development* that require intervention in order to transform AET in Africa: the overall context of education and training; management of education institutions; agricultural education networks and their linkages among organisations; public sector environment; and investments.
- The current systems approach to agriculture research and development derives from changes in agricultural innovation policies and strategies over the past two decades: from a National Agricultural Research Systems (NARS) (1980s), to agricultural knowledge and information systems (1990s), and more recently an *agricultural innovation systems (AIS)* perspective (2000s).
- The AIS perspective involves *a broader set of actors* than the traditional education, research, and extension institutions and places more emphasis on the role of markets in the innovation process. This is particularly relevant for AET in view of its multi-dimensional nature and the variety of actors who engage in it. In this perspective, AET extends beyond the traditional institutions to involve strengthening networks, policies, and institutional conditions from which innovations emanate. AET must integrate with other actors in the innovation system—including national, sub-regional, and regional knowledge institutions—to assure impact at the smallholder level.
- In order to apply systems approaches to AET, scientific and technical information has to be integrated across different disciplines, and practitioners need to be equipped with a range of system skills.
- *Successful professional networks* have the ability to keep researchers scientifically active—in Africa—and focused on making measurable contributions to the broader AIS. These networks are unimpeded by institutional and national boundaries, span both public and private sectors, and involve postgraduate training and collaborations that strengthen institutions.
  - The Strengthening Capacity of Agricultural Research for Development *(SCAR-DA)* programme is one such network. It includes research and AET institutions and is focused on embedding change management processes in the capacity strengthening efforts of participating institutions and countries.
- To support agricultural transformation, *AET reform requires* planning that aligns strategies with national Agriculture and Food Security Investment Plans; integration of capacities across AIS actors; holistic AET strengthening to encourage organizational and institutional excellence; preparation of a new, high quality, gender-balanced generation of agricultural professionals; responsive AET agendas that meet the needs of different user communities; forward-thinking African-led initiatives and institutions that ensure ownership and sustainability; appropriate utilisation of ICTs for AET development; an appropriate policy environment; and instruments for monitoring and evaluating AET performance.

Post-secondary AET in Africa developed considerably between 1960 and 1990. The output of graduate agricultural researchers quadrupled during this time period (Global Education Digest, 2006). However, in the 1990s, domestic investment in AET collapsed in the wake of structural adjustment, resulting in a negative impact on higher education, research and extension services. At the same time, the reforms associated with structural adjustment led to the liberalisation of agricultural markets, further reshaping the demand for a new set of agriculturalists. The period since 1990 has also seen the rapid growth of private AET provision, which is now offered by more than one hundred institutions in Africa. Africa has about three hundred universities with Faculties of Agriculture (Global Education Digest, 2006). For example, a study by Spielman et al. (2012) reports that Ethiopia has eight institutions of higher learning in the field of agriculture and over twenty-five Agricultural Technical and Vocational Education and Training (ATVET) colleges.

However, these agricultural education institutions tend to provide training and conduct research on traditional agricultural issues such as plant breeding, animal health, and agricultural economics, with limited or no specialty courses in emerging issues such as climate change, biotechnology, bio-energy or food systems. Given the Agricultural Innovation System (AIS) perspective, a lot is needed to bring the capacity of African countries to the required level to spur agricultural transformation. In many developing countries, higher education in agriculture seems to be at a crossroads. Financial constraints are severe, yet at the same time demand for high quality education has never been greater. There is a need for greater educational relevance and higher quality graduates, as well as a need to enrol more women and to produce students who are adequately prepared for positions of leadership.

The Framework for African Agricultural Productivity (FAAP) calls for a radical transformation of education and training in Africa and espouses fourteen key principles for achieving this (FARA, 2006). These cover dimensions of capacity that should be the major areas of analysis and the key levels of intervention, namely: 1) overall context of education and training; 2) educational institutions and their management; 3) agricultural education networks and their linkages among organisations; 4) public sector environment; and 5) investments.

The central argument of this chapter is that the AET system should have strong horizontal linkages with other organisations and actors within the innovation system, including those in research, extension, policy, and public and private business. The chapter therefore brings out the contextual issues and lessons surrounding AET reforms and focuses on situating AET in the innovation system with the goal of increasing agricultural productivity in Africa.

## Changes in the underlying Paradigm: Agricultural Innovation Systems

The current agricultural research and development (R&D) paradigm concerning systems perspective derives from shifts in agricultural innovation policies and strategies that have occurred in the past two decades. These range from a National Agricultural Research Systems (NARS) perspective (1980s) to an Agricultural Knowledge and Information Systems (AKIS) perspective (1990s) and, more recently, the Agricultural Innovation Systems (AIS) perspective (2000s). The latter involves a broader set of actors than the traditional agricultural research, extension, and education agencies and places more emphasis on the role of markets and market actors in the innovation process (Anderson et al., 2013).

This innovation systems perspective extends AET capacity-building options beyond a focus on AET, research, and extension organisations. It involves strengthening interactions among networks, policy, and institutional conditions, from which innovations emanate (World Bank, 2011). An assessment of agricultural innovation experiences in smallholder farming in African countries such as Benin indicated that formal research did not initiate or play a leading role in many innovation cases (Triomphe et al., 2012). Rather, the ideas and initiatives came from different sources, including farmers. Mbabu and Hall (2012) present a wide range of capacity-building and education programme options that emanate from the systems perspective. These options include:

- developing entrepreneurial activity;
- building value chains that connect farmers to new markets;
- establishing innovation platforms to connect ideas to opportunities;
- establishing specialist innovation broker agencies to build links among partners and negotiate policy change;
- establishing sector coordination bodies to build coherence across member organisations;
- supporting public-private partnerships;
- establishing decentralised design and manufacturing arrangements for agricultural machinery;
- identifying new modes of competitive funding for consortium development;
- funding existing innovation trajectories that hold the promise of success;
- investing in participatory planning, scenario planning, and foresight exercises; and

- recognising seed price policy, intellectual property rights, finance and tax trade regimes, and a range of policy measures aimed at improving the environment for innovation.

According to Spielman et al. (2012), "Innovation depends on the development of organizational cultures and behaviors that transform AET systems into conduits for transmitting new applications of existing information, new products and processes, and new organizational cultures and behaviors." This suggests that innovation will spawn from AET professionals' active participation in networks, partnerships, and other interactions that link a wide range of stakeholders in an AIS.

From an innovation systems perspective, new levers for transformation on which future restructuring of agricultural higher education in Africa should depend have been put forward. These include:

- populist movements tackling long-standing problems of inequities and exclusion;
- the reorganisation of knowledge systems to accommodate emerging complex fields, such as climate change, that demand overcoming disciplinary barriers to problem formulation and problem-solving, and require renewed appreciation of indigenous bodies of knowledge;
- the growing importance of the private sector and value chains compelling the incorporation of a business schools into research and training; and
- the effect of globalisation on the reduction of time and space influences relationships among institutions, knowledge production and other agents of the agricultural innovation system (Moock, 2011).

The previous sections show that for AET to respond to developmental needs, it should be perceived within a national innovation system – a web that comprises five interlaced elements: i) knowledge-producing organisations in the education and training system, ii) a proper macro-economic and regulatory framework, iii) innovative private enterprises and Non-governmental Organisations (NGOs), iv) adequate communications infrastructure, and v) other factors such as access to global knowledge resources and market conditions that support innovation. Higher education institutions are important for such systems, as developers of high-level skills but also as providers of information hubs (Spielman *et al.*, 2012).

## Changes in the Public Sector Environment and the Role of Government in the AET System

The NEPAD Planning and Coordinating Agency (NPCA) has specifically targeted agricultural development as a central component of its overall development framework. NEPAD created the Comprehensive Africa Agriculture Development Programme (CAADP), an African-led vehicle, to facilitate this work (NEPAD/CAADP, 2010). CAADP is an integrated programme commissioned to address poverty and food security through agricultural growth, rural development, and increased trade. It affords nations and regions the space to develop policies and programmes for strategic investment based on locally determined priorities while, at the same time, building the institutional infrastructure to contribute to the broader NEPAD vision of increased continent-level integration and global participation. However, cost of making CAADP a reality is yet to be assessed. A bold step is required to fix the capacity deficit and give Africa a real chance for sustained development. This does not imply merely undertaking one big initiative but rather assessing needs, harmonising efforts, planning investments, and mobilising resources at the right scale and quality.

In 2003, African heads of state meeting in Maputo pledged to allocate 10 percent of their annual national budgets to agriculture and rural development by 2008. By 2007, eight countries had reached or surpassed the 10 percent spending target (Burkina Faso, Ethiopia, Mali, Malawi, Ghana, Niger, Senegal, and Zimbabwe). An additional sixteen countries are spending between 5 and 10 percent of their budgets on agriculture.

Implementing agencies of CAADP at regional, sub-regional, and national levels provide support to policies that ensure education reforms under the CAADP framework; similarly, partnerships supported by CAADP can position the public sector to enable private sector input for AET reforms.

At the national level, the role of the private sector is increasing in African countries where it plays a significant role in social progress and economic growth. However, governments are ultimately responsible for creating a framework for the private sector to optimise its contribution to development and to AET specifically. In the past decade, the private sector has responded well to the very large demand for higher education in most African countries. Development thinkers and practitioners have reached a broad consensus that a "capable government"—able to perform key functions and create effective national frameworks for different systems and institutions—is a precondition for development.

Consequently, most capacity-building analyses and strategies, and indeed much donor support for capacity development, have focused on the public sector. However, with this focus comes the need for increased public-sector capacity to identify problems and to formulate appropriate policies and programmes to respond to these problems as well as to rapidly changing international conditions. The need to foster broad participation in debate and decision-making is a driving force behind CAADP's current emphasis on developing capacity in countries with Agriculture and Food Security Investment Plans (AFSIPs) in place. Transparency and accountability are required for the public sector to function effectively, meaning that capable governments are prerequisites for a strong and effective education and training system.

A public sector reform programme would enable a shift in the allocation of public expenditures, releasing resources to pay adequate compensation to public sector employees and improve operation and maintenance budgets. This would benefit the AET system as it would improve the retention and productivity of public-sector AET staff. External resources may also be available if donors are willing to provide targeted budgetary support so that public-sector salaries can be adjusted upwards in a rational manner over an agreed-upon period. Without such reforms and the accompanying changes in donor practice, interventions aimed at capacity development are unlikely to succeed.

The World Bank has broadened its support of development to include strengthening public institutions and fostering demand for public service improvements. It has added new diagnostic tools to assess a country's capacity to manage its public financial resources and has increased the range of lending instruments that support increased country ownership of these activities. The Bank has also expanded corporate and regional programmes that directly support capacity-building. These changes are relevant because they recognise that capacity-building is a long-term process requiring a systems approach, demand for improved public-sector performance, and a supply of well-structured organisations and skilled personnel.

## Networks and Linkages among AET Organisations

The success of the AET system requires the coordinated activities of a range of organisations. The role of networks is to reinforce and concentrate on problems that require collective action and to pool talents to reach the critical mass and synergy necessary for realising creative solutions. These networks are becoming an increasingly popular way of building a strong human capital development infrastructure and harnessing gains from innovation in the research process. Moock (2011) characterised successful professional networks as having the ability to keep researchers in Africa, keep them scientifically active, and focus them on making measurable contributions to the broader system of innovation in the agriculture sector. How these networks function, and the nature of formal and informal interactions among them, are important aspects of the AET system's organisational performance.

AET networks span both public and private sectors and include postgraduate training and collaborations that strengthen institutions. Unimpeded by geography, such a collection of agricultural scientists can capitalise on greatly improved mobility and telecommunications to transcend institutional and national boundaries. A number of such agricultural networks now exist on the continent. Notable among them are the Regional Universities Forum for Capacity Building in Agriculture (RUFORUM), African Network for Agro-Forestry Education (ANAFE), Collaborative Masters of Agricultural and Applied Economics (CMAAE), Education for African Crop Improvement (EACI), Biosciences Eastern and Central Africa (BECA), Programme for Emerging Agricultural Research Leaders (PEARL), and *Réseau pour l'Excellence de l'Enseignement Supérieur en Afrique de l'ouest* (REESAO).

Moock (2011) analysed five strategic networks involved in university-based training and research and enhancement of agricultural productivity, namely RUFORUM, CMAAE, EACI, BECA and PEARL. The analysis presents ten principles and challenges of network innovation including:

1. aligning vision and mandate with national aspiration;
2. determining core competencies and comparative advantage;
3. ensuring that new approaches in academia can be mainstreamed within the university system;
4. increasing the participation and voice of women;
5. investing in appropriate information technologies;
6. identifying the full range of clientele;
7. designing strategies for cost recovery and growth on a manageable scale;
8. contributing to enhancement of the policy environment;
9. building strong management and governance; and
10. fostering sustainability through better evaluation and risk management.

## Investing in AET: International and Regional Initiatives

Unless there is sustained human resource development through investment in education, the goal of sustainable economic and social development will not be realised. During the 1990s, development assistance to Africa declined along with government funding of AET. Over this period, donor support for the three components of the agricultural knowledge triangle (education, research, and extension) became highly imbalanced, with higher education receiving very little donor support. The Organisation for Economic Cooperation and Development (OECD) evaluation of World Bank support of capacity building in Africa highlighted the limited impact of its support in the face of the still weak public sector in many countries (OECD, 2011). If education is to make a significant contribution to social and economic advancement, it has to be perceived as a long-term investment.

The World Bank's agricultural strategy calls for recommitment to AET. This is echoed by the Bank's Africa strategy and supported by the International Development Association (IDA)-funded African Centres of Excellence (ACE) Programme (World Bank, 2012). The new strategy emphasises the importance of agriculture to African development and recommends strengthening AET, with a particular focus on educating women farmers. Similar sentiments can be found in a range of regional documents linked to the NEPAD process and championed under the Framework for African Agricultural Productivity (FAAP). To support the FAAP and as its main vehicle for carrying forward African research and development including AET strategies, the World Bank has put forward an African Agriculture Productivity Programme (AAPP). This includes the West Africa Agricultural Productivity Programme (WAAPP), the East Africa Productivity Programme (EAPP) and a similar operation for Southern Africa. The UK Department for International Development (DFID), in its systematic review of the continental programmes on Strengthening Capacity of Agriculture Research and Development in Africa (SCARDA), shows intentions to support AET holistically within the wider capacity-building perspective (Posthumus et al., 2012).

## Advances and Challenges in AET in Africa

The agriculture industry is short of skilled technicians, entrepreneurs, and change agents who can drive innovation. With the resulting weak capacity for agricultural innovation, Africa has been unable to take advantage of the current knowledge economy. With the calls for reform in macroeconomic policies, trade regimes, and marketing policies, globalisation opens more export opportunity for private traders and agro-processors. Globalisation has allowed for increased focus on the value chain and high-value products driven by changes in consumer demand, increased regional markets for traditional food crops, and governments giving priority to agriculture.

The connection between science and practice should capture prime space in all efforts to increase productivity and reduce poverty and food insecurity. New advances in science and technology are influencing the subject matter and types of courses students require if they are to understand today's agriculture. Food processing and post-harvest technologies, biotechnology, agribusiness management, and farming systems development are some of the subject areas that need to be incorporated into curricula. In addition to new scientific knowledge, the most important source of knowledge for agricultural development is the rural people themselves and the time-tested systems of production that embody their knowledge. Therefore, an understanding of rural people and their production systems should be an integral part of agricultural education. This requires that agricultural education institutions refocus – not only on academic aspects, but on engaging local communities and providing them with the relevant local knowledge. Understanding the contributions that local people can make to solve their own problems is the key to sustainable rural development and should underpin the essence of AET for development (AET4D), as illustrated in Figure 3.2.

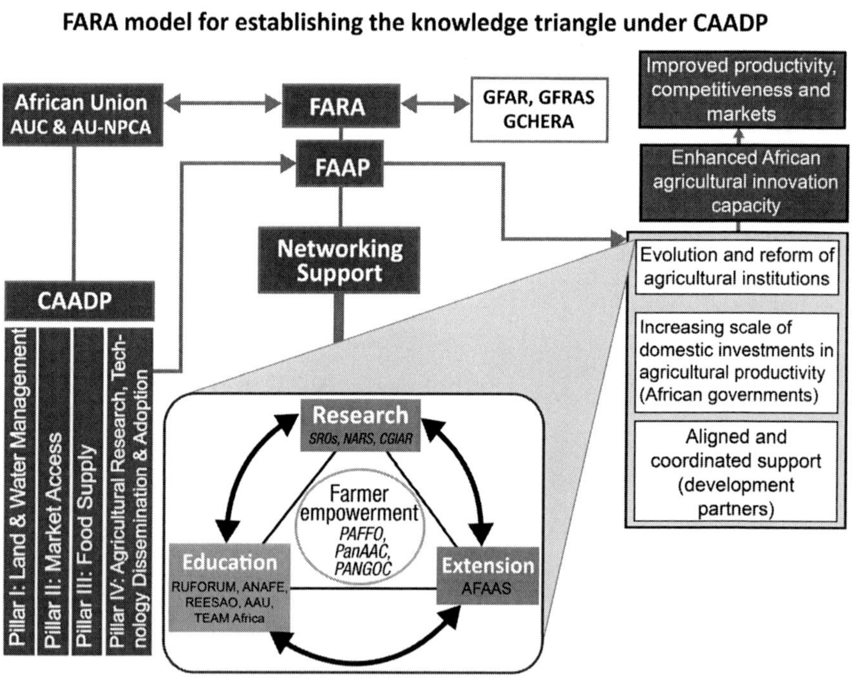


Figure 3.2. Establishing the knowledge triangle within the innovation system (Annor-Frempong, 2012)

## The AET situation in sub-Saharan Africa

Agriculture education in Africa has made strides in terms of increased enrolments in schools and universities (Table 3.1). However, it has also been confronted with a myriad of challenges, such as

- mismatch between students' needs and courses offered, and irrelevance of courses;
- erosion of staff and skewed age distribution;
- weak linkages between higher education (the supply side) and R&D institutions (the demand side);
- poor organisation and management of research and existing capacity and lack of systematic reinforcement (Swanepoel and Stroebel, 2012; World Bank, 2007).

Effective performance and high productivity require a well-educated and well-trained workforce that has both the "know-why" and the "know-how," respectively. Developing the human resource base at all levels of the human capacity pyramid (Figure 3.3) will require specialised training and professional education, including in-service training for role-specific activities.

| **Region** | **Higher education enrolment** | | | |
|---|---|---|---|---|
| | **1991–1996** | **1996–1999** | **1999–2004** | **1991–2004** |
| **sub-Saharan Africa** | 4.5 | 9.0 | 8.9 | 7.2 |

**Table 3.1. Average annual growth in higher education enrolment. Source: Global Education Digest (2006)**

There is a need for well trained professionals and technicians, in both the formal and informal sectors, in all aspects of any industry and its related services. Thus, Africa will require a "whole-institute approach" to develop its human resource base across the entire value-chain and to accommodate the range of academic, professional, and technical capacities required to strengthen the continent's ability to build capacity. This approach for restoring the balance to the national human resource base underscores the integrated capacity development of the Forum for Agricultural Research in Africa (FARA).

In terms of human and institutional capacity, a number of interrelated constraints—all largely the result of uneven funding patterns—have affected AET in Africa. The major factors include decreased funding and increased demand. Rapid expansion can lead to:

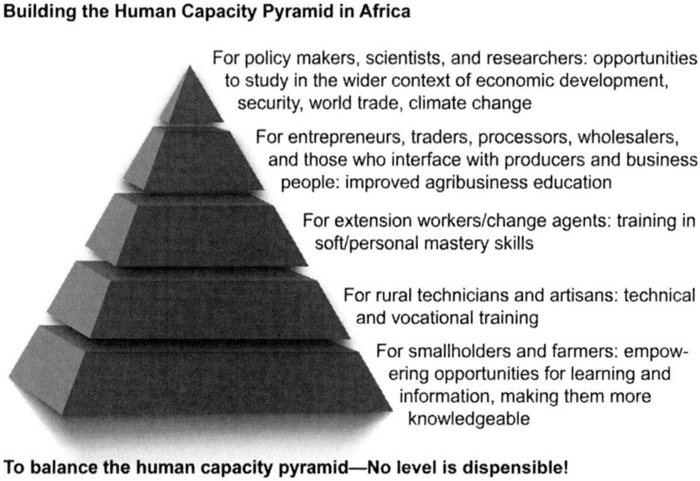


Figure 3.3. Theoretical model of the human capacity pyramid (Von Kaufmann and Annor-Frempong, 2011)

- sacrificing quality for quantity in higher education;
- out-of-context curricula;
- outdated teaching methods and inadequate teaching materials, resulting in high levels of unemployed and under-employed graduates;
- brain drain;
- restricted recruitment under structural adjustment programmes;
- uneven distribution across tertiary agriculture sectors;
- low technical training capacity;
- low math and science enrolment; and
- outdated curricula and pedagogy.

African leadership has reaffirmed the role of higher education in its knowledge-based development strategy under the broad framework of the African Union's Second Decade of Education for Africa (2006, 15), which identifies strengthening higher education and building the research capacity of universities as key goals (African Union, 2006). The framework also:

- emphasises technical and vocational education and training systems as part of a broader strategy to increase the role of education in economic development;
- advocates for more partnerships and networks among African institutions and organisations, as well as between institutions in the South and North; and
- calls for quality assurance, regional harmonisation, and the on-going support of the Pan-African University it has helped establish.

Current efforts are being directed at ensuring that transformative processes lead to improved quality of education systems and to their integration into the wider innovation system. Key networks and partners including RUFORUM, ANAFE, and Tertiary Education for Agriculture Mechanism in Africa (TEAM-Africa) are working with FARA and NPCA to set the path for restoration. Reforming the Technical and Vocational Education and Training (TVET) sector is principal to rural and agricultural development and would require a greater insertion of agriculture-related and rural development skills into TVET and higher education policies. The promotion of the ATVET initiative under the NEPAD Planning and Coordination Agency (NPCA) indicates that the TVET provision can be transformational in contributing to rural economic regeneration and should also concentrate on agricultural education. A number of development agencies are supporting approaches that promote inclusion of TVET programmes and development of specific AET strategies such as rural skills interventions in poverty reduction and rural development projects. Country investment plans should make special reference to TVET as a critical component of capacity building.

## Shifts in the locus of labour demand and curricular content

For agricultural education to keep pace with scientific progress, curricula must continually be updated and competitive programmes designed to elicit student interest and ensure graduate employability in a wide choice of agricultural careers. Scientific knowledge is changing very quickly as modern communication technologies facilitate global sharing of information among scientists and educators. The lead times between "new" and "old" knowledge become shorter each day, making it essential that students develop the skills and attitudes that will allow them to continue learning and developing throughout their professional lives.

Improving employment opportunities for graduates requires that curricula give equal attention to specific technical knowledge—which will quickly become less applicable—and to the processes of thinking and problem-solving, which will enable students to respond to societal needs. Students should learn skills and abilities that are transferable to a wide range of occupations.

This section draws mainly on a study conducted under the aegis of SCARDA, in all regions of sub-Saharan Africa. The study targeted employers of various levels of agricultural graduates but could not systematically trace unemployed agricultural graduates in all countries. For example, in Ghana, the highest numbers of unemployed graduates were in the crops and livestock disciplines, while agricultural engineering and forestry had the highest numbers employed. As shown in Figure 3.4, the study found that the second largest employer other than the public sector for agriculture graduates was NGOs (accounting for 4 percent in Mali, 7 percent in Gambia, and 14 percent in Ghana).

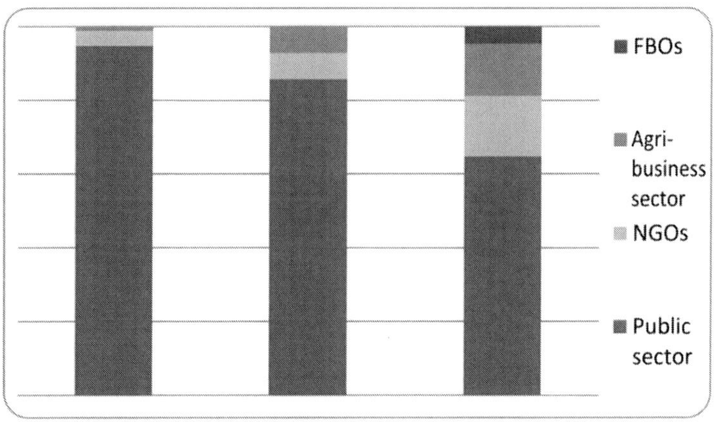


Figure 3.4. Graduate absorption sectors. *Source: Sanyang et al., 2011; Blackie et al., 2011; Macala, 2011. Reports commissioned by SCARDA.*

The agribusiness sector employed 1.2 percent in Mali, 7 percent in Gambia, and 12 percent in Ghana. It was only in Ghana that farmer organisations reported employing agricultural graduates, and these accounted for 4 percent of the total. About 55 percent of employed students in Mali were employed in crops and agricultural engineering, with another 20 percent in forestry. In Ghana, about 50 percent of the employed graduates were in economics and crops, with animal science graduates representing 17 percent. Although Congo reported 40 percent female agricultural graduates, Gambia indicated 5 percent at most. Mali reported 13.7 percent female agricultural graduates in the public service, while Ghana indicated about a 5 percent female unemployment rate compared to its male unemployment rate of 11 percent. In Mali, Ghana, and Gambia, employers reported that 4 to 14 percent of the graduates they hired were women (Blackie et al., 2011).

The locus of labour demand for agricultural skills and competencies is shifting (Figure 3.5).Tschirley and Démbele (2011) reported shifts from the traditional courses to those relating to marketing and distribution as well as to processing, packaging and regulation.

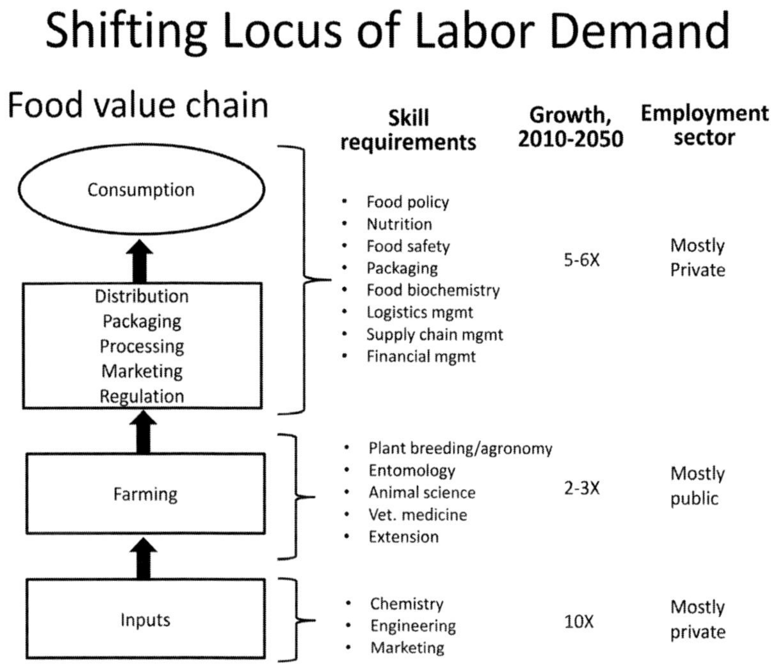


Figure 3.5. The shifting locus of labour demand (Tschirley and Démbele, 2011).

An emphasis on systems approaches is particularly relevant for AET and ARD in view of their multi-dimensional natures and the variety of actors who engage in them. In order to apply systems approaches, scientific and technical information has to be integrated across different disciplines including AET, and practitioners need to be equipped with a range of "system skills" in areas such as entrepreneurship, facilitation, leadership, and communication. This enables them to engage successfully in effective multi-actor partnerships, plan activities, solve problems, and monitor progress. Competence in these domains can best be acquired by learning in real-life situations and in the context of group interaction. However, many education and research organisations in developing countries continue to rely on outdated curricula and traditional modes of teaching and research with a strong disciplinary focus. This means that students are poorly prepared for the challenges of the workplace, and employers are not able to recruit graduates with suitable skills (Ludemann et al., 2012; ANAFE, 2013).

## Governance and management of AET institutions

A project of the Pan African Institute for Development (PAID) identified five broad governance challenges confronting AET institutions:

1. efficiency: how to develop the universities' governance practices and make them real agents of development;
2. technical challenges: how to help universities rationalise their means and realise their missions;
3. strategic challenges: pooling tools and exchanging good practices;
4. community: how to realise an African community of higher education that transcends language and political barriers; and
5. political: maintaining a network of partners and experts.

The project, set up by the *Agence Universitaire de la Francophonie* (AUF) and the Association of Commonwealth Universities (ACU) in support of the Ministry of Higher Education of Cameroon, provides a unique support structure for improved practices in the administration of higher education in Africa. The project's mission is to assist in modernising the governance of higher education through implementation of expertise, modules of training, seminars and workshops, and especially tools of management, analysis, and evaluation. The SCARDA programme launched by FARA and partners (FARA, 2010)[5] included research and AET institutions and focused on embedding change management processes in capacity-strengthening efforts for all participating institutions and countries. The programme provided good insights into establishing organisational change and management.

A systematic review of National Agricultural Research Systems' (NARS) capacity-strengthening experiences, commissioned by DFID as part of its support for SCARDA, identified critical components of successful capacity development (Posthumus et al., 2012). A strong commitment on the part of senior managers and staff to support capacity-strengthening interventions is

[5] During phase II, SCARDA has taken a decentralised approach, with follow-up activities at both regional and sub-regional levels. ASARECA and CORAF have made this approach to capacity building a top priority.

one component. This commitment is often part of a process that requires new ways of thinking and behaving, adoption of new systems or structures, and establishment of adequate management frameworks to capture the benefits and share good practices. Babu et al. (2011) compared the SCARDA approach to the PhD programme in Plant Breeding[6] of the African Centre for Crop Improvement (ACCI) at the University of KwaZulu-Natal, South Africa, and to the Collaborative Masters in Agricultural and Applied Economics (CMAAE) programme in Eastern, Central, and Southern Africa.[7] They concluded that the capacity development approach needs to combine human and implementation capacity in order to advance the implementation of CAADP at country level. The SCARDA programme has emerged as a formidable approach to strengthening both individual human and institutional capacities and advancing organisational change in Africa.

The Universities, Business and Research in Agricultural Innovation (UniBRAIN) Programme, which establishes incubator consortia by linking universities to private sector business and research at governance levels, is proving a unique way of ensuring strong engagement of university curricula and actions with business (FARA, 2013; ANAFE, 2013).

## Concluding Comments

### Strategic directions for AET

The Comprehensive Africa Agriculture Development Programme provides a framework for coordinated programming from local to continental levels based on stock-taking, roundtable negotiations, and country and regional compacts involving governments and development partners. While Agricultural Food Security Plans have identified capacity deficits, the corrective actions proposed are not rooted in addressing the origins of weaknesses in the agricultural education institutions, as agriculture was the mainline ministry engaged with the CAADP stock-taking processes. Thus, the deficits will inevitably reappear in the next generation of AFSIPs and similar agricultural development initiatives. Reform of AET systems is imperative, and problems need to be addressed vigorously at the root and within the systems perspective. A nine-point strategic direction is offered through the CAADP framework to set AET on the path to agricultural transformation in Africa. It must be emphasised that different actors (communities, national governments, the

[6] This section draws on ACCI reports and publications, along with consultations with several people associated with the program, including its director, Mark Laing.

[7] This section is primarily based on reports of the CMAAE programme and on personal conversations with those involved in the program.

private sector, and sub-regional and regional entities) all have roles to play in strengthening the AET system to support Africa's agricultural transformation. The recommendations provided below should be seen in this context.

### Align AET strategies to national AFSIPs

At the national level, AET systems need to move into closer productive relationships with other actors in the innovation system to build on the comparative advantages of different actors and institutions to achieve economies of scale and scope, reduce transaction costs, exploit complementarities, and realise synergies in the process of innovation. This can be achieved by aligning AET systems with national AFSIPs. It is a fact that CAADP has improved national planning processes and the policy environment for agricultural transformation. However, in many countries, the AET system has not featured in these planning and alignment processes (FARA, 2011). It is critical to ensure that strategies, visions, and mandates of AET organisations align with national development aspirations by promoting new types of educational programmes. The TEAM Africa initiative, facilitated by FARA and implemented by RUFORUM and Agroforestry and Natural Resources Education, is a major effort towards articulating a shared strategic vision for Tertiary Agricultural Education (TAE) capacity-building in Africa. The vision includes developing an African strategy for TAE, establishing a CAADP-TAE African working group, and providing information for priority-setting in TAE investments to all interested parties, but especially to Development Partners and African TAE institutions. This can be done by:

- providing guidance on best approaches to investment in TAE;
- strengthening existing and promising new TAE networks; and
- fostering alignment and coordination of donor support to TAE.

It is currently envisaged that a larger multi-donor trust fund (MDTF) will be established for long-term support of TAE in Africa. This MDTF should be aligned with the longer-term (usually five-year) strategic plan developed by TEAM-Africa, which by this point would be widely assessed and officially endorsed by TAE stakeholders within the framework of CAADP.

### Establish equilibrium of the capacity pyramid at national level by integrating capacities across actors in the innovation system

Efforts at capacity building in general and at strengthening AET systems in particular have had low impact. Africa still lacks the critical mass of human resources needed for implementing programmes aimed at improving agricultural productivity. This is because capacity development efforts have been fragmented, heavily supply-driven, and not targeted to local needs. At the national level, AET has over-emphasised a few levels of the capacity pyramid and neglected others, especially the vocational and technician levels. Investment in education and training is small in relation to need, with unacceptable duplication of efforts. Reforms in AET are required across the board for all levels, disciplines, and skills because weaknesses in one area impede the effectiveness of workers in other aspects of the industry. For example, irrigation schemes need engineers and pump mechanics as well as agronomists and business managers. This means that the focus now has to be on restoring the essential equilibrium of the "human capacity pyramid." This will require policies and programmes to focus not only on improving formal AET organisations, but also on expanding technical and vocational training institutes, in-service and on-the-job programmes, distance education, and other modalities specifically adapted to the needs of diverse actors in the innovation system. Policies and programmes should also focus on private sector sources of AET as a necessary complement to the formal, public sector AET system.

### Promote holistic AET strengthening to induce organisational and institutional change and excellence

Policies and programmes should also focus on strengthening individual and organisational incentives to develop and retain capacity on a national scale, and on introducing organisational and managerial innovations into the AET system itself (Spielman et al., 2012). Strengthening professional capacity needs to be complemented by institutional and organisational capacity to strengthen agricultural research, technology development and dissemination, and strategic policy analysis (Timmer, 2011; Tamboli and Nene, 2011) in order to generate sustainable capacity for implementation of agricultural development programmes (Posthumus et al., 2012; Annor-Frempong et al., 2012)

Sustainable changes will be driven by new demands emanating from the smallholder sector, high-value crop production systems, domestic agro-industries, and other parts of the agriculture sector including incentives within AET organisations and among AET professionals. Faculties of agriculture are fast becoming recognised as important contributors to national research capacity and often have strong human resources, while research institutes with reasonable research infrastructure tend to have poor human resources. Integrating national research capacity (upstream research, education and community research work) across university faculties and research institutes could provide the critical mass needed for agricultural development. This would reduce the cost of duplicating resources, avoid the unnecessary competition that sometimes occurs, and auger for excellence at the national level. Policies and programmes need to encourage organisational change by promoting linkages with a wide array of other stakeholders who can effectively serve both AET organisations and the country's innovation system. AET organisations need to show greater openness to collaborating with private firms, informal AET service providers, civil society organisations, and traditional/indigenous knowledge institutions. In addition, these organisations should conduct problem-solving, demand-driven research and training in order to encourage private-sector partnerships that can help to build business and entrepreneurial capacity.

**Make a dedicated plan to prepare the new generation for agricultural professions and ensure gender equity**

The new generation of scientists and managers will have a difficult transformation to bring about, as farms in rural areas adapt to meet the requirements of an emerging market. The great challenge of the farming sector for the years to come will be preparing its youth, including young women, to lead this process of farm transformation. Faced with the extent of the training needs of rural stakeholders, the agricultural education system is clearly ill-prepared. The current system trains few producers, trains more engineers than technicians, and no longer meets the requirements of a profoundly changed farming profession.

The implementation of the training strategy as part of a national strategy should involve:

- re-designing the educational system to offer two clear streams: a path to pursue either farming/agribusiness or research—a model that has been successful in developed countries such as the Netherlands and New Zealand;
- redefining educational content to take into account farming and its ever-changing environment;
- re-balancing flows of graduates based on the new sectoral needs;
- regionalising the system to ease its integration into the framework of decentralisation measures; and
- securing sustainable long-term funding for these reforms.

Innovations have historically marginalised women. The adoption of technology must therefore aim for gender neutrality (Meinzen-Dick et al., 2010). Because Africa's agriculture sector has multiplier effects in many sectors of the economy, the modernisation of the sector across its value chain offers different kinds of opportunities for women. There is a need to create openings for young people to take full part in society and the economy. Agriculture, the largest sector in most African economies, is a natural place to increase youth employment. New ideas and technologies can enable youth to channel their creativity; transfer knowledge, information and ideas; and also stay connected. With entrepreneurship, these ideas can create tangible value in the agriculture sector. The FARA-led Universities' Business Research in Agricultural Innovation (UniBRAIN) initiative aims to support the establishment of agribusiness innovation incubators with emphasis on youth across the agricultural value chain (FARA, 2011). Other innovative programmes, such as the Junior Farmer Field and Life Schools (JFFLS), offer not only holistic training in agriculture techniques, but also life and business skills.

### Develop responsive AET agendas for meeting needs of different user communities to nurture high quality human capital

In the medium term, efforts to improve AET in sub-Saharan Africa should focus on the provision of learning opportunities geared to the specific needs of different actors in the innovation system, rather than on traditional benchmarks set by standards of public service or academia (Spielman et al., 2012). Many of the programmes that target the individual level are limited because they are not linked to the overall development of the institutions—students' study programmes are not necessarily aligned with the priorities of research and education organisations in developing countries (Ludemann et al., 2012).

FARA's integrated capacity development approach is geared towards strategic human capital formation at the country level—through linking the demand and supply of capacities to ensure the AET agenda is demand-driven (FARA, 2013). The essence is to link AET and the research agenda more closely to the needs of different user communities such as smallholder farmers, rural traders, agro-processors, and consumers, in order to foster stronger linkages between formal AET organisations and national extension systems (public/private), and to explore new ways of leveraging expertise and resources from the international community.

### Emphasise African-led initiatives and institutions and adopt a forward-thinking outlook to ensure ownership and sustainable agricultural development

A key lesson from the SCARDA programme is that African ownership is critical for the sustainability of initiatives (FARA, 2010). The DFID 2008–

2013 strategy argues that public policies work best when they are designed and implemented by local actors (DFID, 2008). CAADP has also shown that local ownership and buy-in is critical to successful development interventions and programme sustainability. Without locally generated data and analysis, well-intentioned programmes often do not respond to realities on the ground. Capacity strengthening is about change, i.e. changes in the practices and cultures of both formal and non-formal AET as well as internalisation of new skills in applied problem-solving and critical thinking. "Agri-preneurship" will take a generation to become common practice; thus it is critical to strengthen African foresight capabilities, which are strategically linked to the AET and research systems. FARA's effort to establish an African chapter of the Foresight Academy holds great promise for AET for development. The academy will:

- operationalise African foresight by stimulating, catalysing, encouraging, and enabling foresight studies on issues of importance to Africa, led and executed by African teams;
- provide a platform that acknowledges young African professionals' expertise and allows them to participate in policymaking processes at the highest level; and
- support the development of African foresight capabilities at individual and organisational levels in all countries and sub-regions of the African continent (FARA, 2013).

**Advance ICT to strengthen AET systems**

Use of Information and Communication Technology (ICT) in agriculture and AET can make an enormous difference in providing quality education to meet the current and increasing demand for higher education through distance learning. AET systems and institutions have to mainstream and embed ICTs and use them extensively to foster agricultural innovation. This requires changes in agricultural curricula in universities that teach ICT applications, including training in how to use ICT tools in research, education, and extension, and fostering further research in ICT use and application in agricultural science, innovation, and technology generation.

Effective and efficient use of advances in information and communication sciences and ICTs requires the following:

- improved communications infrastructure and bandwidth, to be achieved by investing in lower-cost hardware, software, and applications that connect science right along the development chain;

- increased and improved formal education and training in information and communication sciences that contribute to innovation in the use of new ICTs in agriculture;
- extended generation and dissemination of data and information content as a public good that is widely accessible and licensed to be easily re-used and applied;
- support for applications that integrate data and information or foster the inter-operability of applications and information systems, allowing safe and ethical access while protecting necessary rights;
- effective uptake and use of data, information, and knowledge, particularly focusing on the capacity-building dimensions necessary for the outputs of science to have impact; and
- support for innovation in the workflows, processes, and tools used to create, share, publish, visualise, and connect the outputs of agricultural science and the people engaged in it.

It is encouraging that, under the transformative agenda of CAADP, ICT education and training constitute two out of its four strategic thrusts.

### Create an appropriate policy environment

The long-term transformation of AET systems in sub-Saharan Africa will require an appropriate policy environment and policy makers with the knowledge and will to facilitate the transition process. Appropriate policies are fundamental to encouraging investment in AET. A number of regional, sub-regional, and national organisations are currently working in this area and providing advocacy for creating enabling policies. Organisations and initiatives including the Association for the Development of Education in Africa (ADEA), the Food, Agriculture and Natural Resources Policy Analysis Network (FANRPAN), FARA, and AFAPP need to be promoted and encouraged to form relevant collaborations to support national policy dialogues and platforms that target AET.

### Promote instruments for monitoring and evaluating AET performance and for tracking graduates

Capacity building in general should be considered a core objective for development programmes in Africa. However, most capacity-building activities lack standard quality assurance processes at the design stage, and they are not routinely tracked, monitored, or evaluated. Continuous evaluation of AET organisations and the AET system as a whole is necessary. These evaluations can play a critical role in helping organisations redefine their mandates and goals relative to changes in the country's AET system, the wider AIS, and the availability of resources. Evaluations would also assist in the efficient reallocation of AET resources across teaching, research, extension, and private enterprise. Tracking of graduates and trainees provides a basis upon which more sophisticated, ongoing measurement and monitoring protocols can be designed (Annor-Frempong et al., 2012). Special effort should be directed towards empowering specific organisations to collect, maintain, and analyse relevant data so as to provide important baselines for AET systems.

## References

African Union. 2006. *Second Decade of Education for Africa*. Addis Ababa, Ethiopia.

ANAFE. 2013. *Linking Training of Agriculture to Agribusiness Development.* A Tracer Study of Universities involved in UniBRAIN.

Anderson, J., H. Roseboom, and Weidemann Associates, Inc. 2013. *Draft Issues Paper for USAID's Roundtable Discussion on NARS*. This publication was produced for review by USAID. Unpublished Report.

Annor-Frempong, I, J. Roseboom, and N. Ojijo. 2012. A Pilot Study on Institutional and Organisational Changes in Selected National Agricultural Research and Education Institutes in sub-Saharan Africa. FARA, Accra, Ghana. http://www.fara-africa.org/publications/

Annor-Frempong, I. 2012. *FARA's Mission and Relevance of TAE in Africa.* Proceedings CCCEdu Meeting, March 17, Burkina Faso.

Babu, S., I. Annor-Frempong, and K. Asenso-Okyere. 2011. Enhancing Capacity for African Agricultural Research: Selected Models and Lessons ASTI/IFPRI-FARA Conference, Working Paper 1.

Blackie, M., I. Annor-Frempong, S. Sanyang, J. Macala, and A. Ekwamu. 2011. An Analysis of the Demand for Agricultural Graduates in the CORAF, WECARD, ASARECA, and SADC Sub-Regions of Africa. Unpublished report.

Blackie, M., M. Mutema, and A. Ward. 2011. "ASARECA/RUFORUM THRUST 4: A study of Agricultural Graduates in Eastern, Central, and Southern Africa: Demand, Quality, and Job Performance Issues." Unpublished report commissioned by SCARDA.

DFID. 2008. Capacity Building Strategy 2008–2013 Working Paper Series. London.

FARA. 2006. *Framework for African Agricultural Productivity.* Accra, Ghana.

—. 2010. *Strengthening Capacity for Agricultural Research and Development in Africa.* End of Pilot-Phase Report. SCARDA, Accra, Ghana.

—. 2011. *Universities, Business, Research in Agricultural Innovation (UniBRAIN)*. Implementation Document. Accra, Ghana.

—. 2013. *Strategic Plan and Medium Term Operational Plan.* In Press. Accra, Ghana.

Global Education Digest. 2006. *Comparing Education Statistics across the World*. UNESCO Institute for Statistics, Montreal.

Juma, C. 2011. *The New Harvest: Agricultural Innovation in Africa*. Oxford University Press: New York.

Ludemann, R., M. van Dorp, A. Groot-Kormelinck, and T. Chancellor. 2012. *Capacity Development in Agricultural Research for Development*. GIZ-EIARD Report.

Lynam, J., N. Beintema, and I. Annor-Frempong. 2011. *Agricultural R&D: Investing in Africa's Future—Analyzing Trends, Challenges, and Opportunities.* Synthesis paper. Unpublished.

Macala, J. 2011. Synthesis Report: Tracer Study on Effectiveness of Agricultural Training Programmes in Botswana, Lesotho and Zambia. Unpublished report commissioned by SCARDA.

Mbabu, A. and A. Hall, Eds. 2012. *Capacity Building for Agricultural Research for Development: Lessons from Practice in Papua, New Guinea.* United Nations University–Maastricht Economic and Social Research Institute on Innovation and Technology (UNU-MERIT), Maastricht, the Netherlands.

Meinzen-Dick, R., A. Quisumbing, J. Behrman, P. Biermayr-Jenzano, V. Wilde, M. Noordeloos, C. Ragasa, and N. Beintema. 2010. *Engendering Agricultural Research*. IFPRI Discussion Paper 00973. International Food Policy Research Institute, Washington, DC.

Moock, J. 2011. *Network Innovations: Building the Next Generation of Agricultural Scientists in Africa*. ASTI/IFPRI-FARA Conference Working Paper 1.

NEPAD–CAADP. 2010. *New Partnership for Africa's Development: Comprehensive Africa Agriculture Development Programme*. Pillar 4: Agricultural Research. http://ww.nepad-caadp.net/pillar-4.php

OECD. 2011. *Aid Effectiveness 2005–2010: Progress in Implementing the Paris Declaration.* OECD Publishing: Paris.

Posthumus, H., A. Martin, and T. Chancellor. 2012. *A Systematic Review on the Impacts of Capacity Strengthening of Agricultural Research Systems for Development and the Conditions of Success.* EPPI-Centre, Social Science Research Unit, Institute of Education, University of
London, London.

Sanyang, S., S. Ly, S. Ennis, L. Jobe, L. Diarra, and P. Bantaba. 2011. Employment of agricultural graduates: Who are we training for? Unpublished report commissioned by SCARDA.

Spielman, D., K. Davis, E. Zerfu, J. Ekboir, and C. Ochieng. 2012. An Innovation Systems' Perspectives on Tertiary-Level Agricultural Education in sub-Saharan Africa: Evidence from Ethiopia. *Ethiopian Journal of Education and Science* 7: 2, 15–32.

Swanepoel, F. and A. Stroebel. 2012. *Capacity Building for African Agricultural Development: the Need for Transformation in Tertiary Agricultural Education.* Keynote address at the Third TAE-Partnership Workshop on Tertiary Agricultural Education in Africa. November 13–15, Stellenbosch, South Africa.

Tamboli, P. and Y. Nene. 2011. *Revitalizing Higher Agricultural Education in India: Journey towards Excellence.* Agri-History Foundation: Hyderabad, India.

Timmer, P. 2011. *Managing Food Price Volatility: Approaches at the Global, National and Household Levels.*
http://foodsecurty.stanford.edu/events/managing_food_price_volatility_approaches_at_the_global_national_and_household_levels/

Triomphe, B., A. Waters-Bayer, A. Floquet, G. Kamau, B. Letty, S. Vodouhe, T. Ng'ang'a, J. Steens, J. van den Berg, N. Selemna, B. Bridier, T. Crane, C. Almekinders, N. Oudwater, and H. Hocdé. 2012. *Joint Learning to Enhance Innovation Systems in African Agriculture.* A report of a Conference on International Research on Food Security, Natural Resource Management, and Rural Development, organised by Georg-August Universität Göttingen and University of Kassel-Witzenhausen. Tropentag, September 19–21, Göttingen, Germany.

Tschirley, D. and N. Dέmbele. 2011. *On Managing the New Food Price Environment in Countries with Food Insecure Populations*. Michigan State University.

Von Kaufmann, R. and I. Annor-Frempong. 2011. *Rebuilding the Capacity Pyramid.* Conference on Higher Education and Research (GCHERA).

World Bank. 2007. *Cultivating Knowledge and Skills to Grow African Agriculture: A Synthesis of an Institutional, Regional and International Review.* World Bank, Washington, DC.

—. 2011. *Agricultural Innovation Systems*. World Bank, Washington, DC.

—. 2012. *Project Assistance Document on African Centres of Excellence (ACE) Programme*. World Bank, Washington, DC.

# PART TWO

# INTERNATIONAL INSIGHTS

# CHAPTER FOUR

# TRANSFORMING AGRICULTURAL EDUCATION AND TECHNOLOGICAL IMPROVEMENT IN ASIA: A VISION FOR STRENGTHENING ASIA-AFRICA LINKAGES

CANAGASABY DEVENDRA

Agricultural productivity and economic growth have had an overwhelming impact in Asia, contributing to an unprecedented rate of rural growth, improved livelihoods, and increased prosperity. In the early 1960s, Asia's agricultural productivity was low and stagnant—a poor source of income at a time when population growth was rapid. The transformation since then has been spectacular, bringing with it rapid growth in the rural, non-farm, and urban economies as well as cost-effectiveness in the agro-industry sector.

The engine for this transformation was the Green Revolution, which spanned the 1960s to the 1980s, providing farmers with improved crop varieties to increase yields—notably rice, wheat, and maize—along with new technologies in the form of fertilizers and irrigation techniques. Effective application of technological packages significantly improved productivity across small and large farms, and also distributed the benefits of growth across variable income groups. The drivers responsible for this transformation, and in particular for reducing poverty, were market-oriented and open trade policies, aggressive investments in technological change and infrastructure, and improvements in human capital. The Green Revolution also attracted and encouraged other organisations to establish partnerships in rural development, most notably the international agricultural research centres (IARCs).

Two strategies were involved in this process. One was rural economic growth, which greatly benefitted the economic status of, for example, poor farmers and the landless in Punjab and Haryana states in India. The other significant effect was the impact of education on human resource development. The strong agricultural growth brought with it knowledge that had positive effects in many other Asian countries, providing information on ways to lower production costs, increase productivity, and address poverty. In effect, an economic transformation in the rural areas fuelled further development in areas such as diversification and commercialisation.

Associated with the above, it is appropriate to keep the main features of Asian agriculture in perspective. Asian agriculture is characterised by mixed-farming activities that form the backbone of farming systems. The diversity of mainly small-farm, mixed-farming, or crop-animal systems, usually operated by poor farmers (Devendra, 2010), is widespread in all agro-ecological zones and regions and includes annual and perennial crops, ruminants (buffaloes, cattle, goats, and sheep), and non-ruminants (poultry, pigs, and ducks).With rapid population increase, there is compelling pressure to accelerate food production for the future, presenting major challenges for farming systems to stimulate agricultural and economic growth in the Asian region.

Of the world's total of 470 million farms of less than two hectares, 87 percent, or 348 million farms, are found in Asia (Nagayets, 2005). These farms are home to about 2 million resource-poor people and form the backbone of Asian agriculture. Many are models of efficiency in natural resource management and low resource-input systems, producing food for subsistence and occasionally for income. The contribution to food production from these small farms is significant and maintains an estimated 51 percent of the human population. The farmers' lives, however, are often characterized by deprivation, illiteracy, subsistence-level survival, and vulnerability, and fraught with a never-ending syndrome of poverty, fragility, lack of hope, and low life expectancy (Devendra, 2010). Prioritizing development for this population is justified by the urgent need for increased agricultural productivity, food security, and poverty alleviation. At the same time, sustained pro-poor investment can provide an important means to maximize the potential productivity of these small farms.

The Asian Green Revolution successfully illustrated what improved technology can do to dramatically increase production. The strong productivity growth was based on effective input use of technology packages and aggressive public investments in infrastructure and education. Investments in research were particularly significant and demonstrated high rates of return and profitability from the outlays.

***Key Messages***

- The *Asian Green Revolution* illustrated how technology can improve production, but also brought problems such as bias towards resource-rich regions and wealthy farmers; technologies that required large amounts of fertiliser and irrigation; increased dependence on expensive imported inputs; and increased disparities between rich farmers and the poor and landless.
- Among the lessons learned is that *the development of human capital at all levels is instrumental in making the application of new knowledge more productive and progressive.*
- Efforts to *transform AET should be viewed from a systems perspective,* recognising that together with farming systems research it is fundamental to improving natural resource management and addressing poverty, food insecurity, and the environment. Systems approaches are also needed to link the work of researchers, extension agents, and farmers.
- AET needs to be more dynamic and sensitive to emerging problems. Formal AET undertaken in Western environments is often steeped in sectoral and disciplinary interests. Yet the *complex issues facing small-scale farmers require interdisciplinary perspectives relevant to the farmers' aspirations.*
- *Empowerment should be at the heart of all education and training.* This means enabling people to have control of their own resources and to set their own agendas. AET should therefore deliver well-rounded graduates who understand farmers and the rationale for technology development and use, and who can participate in shared impact-oriented progress assessments that promote self-reliance.
- The *traditional model of research-extension-farmer technology delivery should be replaced by participatory methods* for diffusing technology to spur change and development. Much depends on the change agents, their ability to work well with farmers, and the process of participation.
- A good *AET strategy will include a focus on formal degree specialisation with curricula that build on good science education at school and combine a strong disciplinary orientation with systems perspectives and systems methodologies.* Non-formal education should be intensified at different levels, with appropriate training of trainers as change agents.
- *Linkages between Africa and Asia should be strengthened,* promoting networking, the concerted sharing of information and innovations, and the exchange of scientists, technicians, extension agents, and farmers.

However, the revolution and its socio-economic benefits were not without problems:

- the technologies developed were biased toward more favourable environments, resource-rich regions, and wealthy farmers;
- the technologies used required large amounts of fertiliser and irrigation;
- there was an increased need for credit and dependence on expensive imported inputs such as fertilisers;
- there was increased disparity between the rich farmers on the one hand and the poor and landless on the other.

Several lessons were learned from the outcomes of the Green Revolution. Without being exhaustive, the following are five of the most important:

- For small farmers and the landless, inputs can be very expensive and thus beyond their reach. This indicates the need to keep costs of production low through cost-reducing measures such as maximising the use of locally available resources.

- Appropriate technology application and resultant economic growth is an important means to reduce poverty. The key is to ensure that the technologies available are pro-poor, economical, and socially acceptable.
- Investment in infrastructure is important because it encourages farmers to improve production and move forward in commercialising the value chain.
- Development of human capital at all levels is instrumental in making the application of new knowledge more productive and progressive.
- In farming systems, community-based participatory efforts increase the involvement of the poor and the landless. This is an important determinant of growth in productivity. Such participation helps to push agricultural growth and revitalise the rural economy.

It is evident that natural resources, market-friendly policies, public investments in infrastructure, and agricultural education and training were all important components for improved productivity in Asian agriculture. At the same time, diversification of farm activities also contributes to rapid growth in the non-farm sector. The complexity of the interactions necessitates two important considerations: 1) the need to look at the system as a whole, recognise the implications of the interactions, and identify and prioritise the constraints; and 2) the need to recognise that this approach enables application of systems' perspectives that, in the example of Asian agriculture, led toward efficiency in the use of available resources. In both cases, formal and non-formal education were central to the strategies. This chapter discusses both these components to provide a more holistic view of education and training and the links with agriculture in Asia. For the purposes of this chapter, education and training refer to teaching and instruction.

*Education.* Education is an important determinant of productivity growth in agriculture. It underpins investments in many ways, including the development of human capital. Research and development capacity are most essential in influencing the use of new knowledge and encouraging progress in productivity. In turn, rapid growth in agriculture and the rural economies is noticeable, and these spur an additional demand for increased investment in development of the human capital required to ensure that rural-urban migrants who must diversify into non-farm activities have the relevant and appropriate skills to make use of these new opportunities.

Agricultural education in Asia and the developing world is focused on development of human capital and technology-induced progress. It underpins all aspects of agriculture, and has benefited greatly from the significant technical and economic advances made in developed countries. These advances have come from a combination of factors such as concerted research, improved methodologies, and education, extension, and efficient delivery systems for the adoption of new technologies. Systematic research, investment, and enlightened foresight have resulted in three significant achievements:

- development and use of improved technologies that drive increased productivity;
- education and training that have produced large numbers of competent agricultural scientists, technicians, and extension agents in various disciplines;
- the diffusion of technologies to farmers and the private sector to promote change.

A historical sweep of the major agricultural advances in industrialised countries over the past three decades suggests that, together with rapid social and economic development, research, education, and extension have made it possible for farmers to benefit greatly from new approaches and technology applications. The situation in Asia today is no different. While progress is being made, Asia faces numerous complex constraints and challenging issues that affect agricultural development and the rate of progress. For example, Pakistan and India face a huge problem due to salinisation, which occurs when evapotranspiration exceeds rainfall, causing a concentration of salts on the soil surface. These constraints and challenges include threats to the environment and natural resources (crops, animals, land, water, and production inputs), inadequate research capacity, several million affected farmers of whom many are landless, poor services related to markets and value chains, and poor access to government services and credits.

Effective transfer of technical knowledge is an important process in development. Encouraging farmers and agricultural advisors to adopt technological change that promotes agricultural production requires full understanding of the various issues, the local environment, social and economic benefits, and pathways to significant impacts. New knowledge and technologies require improved understanding and more skills to promote adoption by farmers. In addition, large-scale adoption and impact are best achieved through community-based participation of farmers, where farmers are seen as major partners in research and development (R&D). Education in agricultural systems and application of farming systems perspectives can make a significant contribution to agricultural productivity and growth. This is a complex and vast subject and raises several questions:

- What are the best strategies and technological options for individual agro-ecological zones?
- What innovative methodologies are required to ensure adoption and impacts at the farm level?
- Is the traditional model of diffusing technology through research-extension linkages inadequate and outdated?
- What changes are necessary at the formal level to promote understanding and resolution of the prevailing issues, and at the non-formal level to promote economical and socially beneficial acceptance of improved technologies and innovation?
- Are prevailing education and training programmes equipped to cope with the threats and effects of climate change on agriculture?
- Is there a need to fortify policy on the teaching and application of systems perspectives?

The answers to many of these questions remain weak or unknown, which underscores the need for ways and means to improve the situation. One tentative conclusion—and a distinct means for improvement—calls for the public sector to actively support investments in education and the development of technology. Strengthened policy definition can be helpful in this regard. This chapter considers these and other issues concerned with the changing nature and relevance of agriculture and the implications for agricultural education and training.

## Waning Agricultural Development in Asia

In recent years, agriculture has been neglected and agricultural development has waned in Asia. Recognising its vital importance in food security, good health, and improved livelihoods, urgent efforts are targeting the revitalisation and transformation of this sector in many countries. Its importance in economic and rural growth is widely recognised, and consequently there is new impetus for stimulating agricultural development.

According to the Economic and Social Commission for Asia-Pacific (ESCAP, 2008), Asia has some 1100 million hectares of surface area, about 16 percent of which is agricultural land. The rain-fed regions of Southeast and South Asia comprise about 99 and 116 million hectares respectively. For example, in India this ecosystem occupies 68 percent of the total cultivated area, supports 40 percent of the human and 65 percent of the livestock populations, and also produces 4 percent of the area's food requirements. In the Asia-Pacific region, the ecosystem provides jobs for 60 percent of the working population and generates a quarter of the region's gross national product (GNP).

In recent years, changing consumer tastes and dietary preferences have brought with them a commercialisation of agriculture. At the same time, agriculture's share of gross domestic product (GDP) has declined significantly as low productivity has resulted in slow growth (ESCAP, 2008). In essence, low product prices and high input prices have made agriculture less attractive. In South Asia, ESCAP (2008) reported that growth in agricultural output dropped from 3.6 percent in the 1980s to 3.0 percent in 2000–2003. In East Asia and the Pacific, it dropped from 3.0 percent in the 1980s to a mere 0.1 percent in 2000–2003. Nevertheless, the sector still makes a significant contribution to the economies of many countries in Asia and has shown some apparent improvement. In Southeast Asia, the share of agriculture gross value-added to the GDP during 1990–2010 was 21.2 percent (ADB, 2012).

Table 4.1 summarises the household characteristics, biophysical data and patterns of farming systems of the five rain-fed locations in China, Indonesia, the Philippines, Thailand, and Vietnam that were chosen for the Improving the Crop-Livestock Production Systems in Rain-fed Areas of Southeast Asia Project (CASREN). The general objective of the project was to conduct collaborative, multidisciplinary research to generate technology and policy options for increased productivity of smallholder crop-livestock systems in Southeast Asia. The average annual rainfall was between 1500–2300 mm. Rice-based cropping systems were common, but other annual crops and tree crops, as well as ruminant and non-ruminant livestock, were also supported. Animals contributed 10–25 percent of total farm income (Devendra, 2005).

The presence of both animal and crop diversity provided a variety of crop-animal interactions (Devendra and Thomas, 2001) that affected productivity, livelihoods, and sustainable agriculture, providing major opportunities for research and development activities. The major constraint was the annual five to seven month dry period.

[a] BC = Beef Cattle, DC = Dairy Cattle, Bu = Buffaloes, S= Sheep, G = Goats, Pi = Pigs, Po = Poultry, Fi = Fish
[b] TLU = Tropical livestock unit, equivalent to a ruminant animal of 250 kg body weight. Only ruminant species have been considered for its estimation, according to the following equivalencies: cattle and buffalo = 1.0, sheep and goats = 0.01; in all species, mature male = 1.0; mature female = 0.75, growing animal = 0.5; pre-weaned animal = 0.25

Waning agricultural development and decreasing productivity have major impacts on poverty. ESCAP (2008) reported that although poverty was declining, the rate of that decline has slowed since the 1980s. The study noted that agriculture alone could lift an estimated 641 million people out of poverty and that a 1 percent increase in agricultural productivity would lead to a 0.37 percent drop in poverty in the Asia-Pacific region. Using Thailand as the benchmark for productivity gains, it has been further calculated that increased agricultural productivity can free about 218 million people in the region from poverty.

The majority of these poor people are either landless or small scale farmers. Assisting them to increase productivity from agriculture will require investments in human capital—especially in rural informal education to empower participation in technology-driven activities. Such activities will stimulate wider economic growth in the non-farm rural economy and contribute to overcoming food insecurity and alleviating poverty. Irrespective of where the smallholder farms are found, a few key descriptive words can be applied to the livelihoods of all smallholder farmers. Up to the mid-1990s, these were: deprivation, subsistence, illiteracy, and survival. Now, due to globalisation and its many related externalities, the additional key word is "vulnerability." Education can ensure survival and serve as a catalyst for demonstrable impact in subsistence agriculture. At the heart of these issues, human dignity is at stake (Devendra, 2010).

| Item | Vietnam | Thailand | Indonesia | Philippines | China |
|---|---|---|---|---|---|
| **Household attributes** | | | | | |
| Average farm size (ha) | 2.59+_ 1.97 | 5.24+_2.74 | 0.55+_0.73 | 1.26+_0.45 | 0.27+_0.12 |
| Household size (persons) | 4.94+_1.74 | 4.79+_2.08 | 4.46+_1.54 | 5.52+_2.33 | 4.73+_1.29 |
| percent of farms managed by women | 28.4 | 31.6 | 4.2 | 30.0 | 26.0 |
| **Item** | **Vietnam** | **Thailand** | **Indonesia** | **Philippines** | **China** |
| **Biophysical attributes** | | | | | |
| Location | Dong Tam, Bin Phuoc | Amphur Muang Maharasakham | Dangiang, Cilawu, Garut | Don Montano, Umingan, Pangasinan | Bixi Xiang, Nanjian, Yunnan |
| Mean annual rainfall (mm) | 2170 | 1500 | 2200 | 2300 | 760 |
| Dry season (months) | 6 | 6–7 | 5–6 | 6 | 7 |
| Predominant crop-animal systems | Rice-based beef cattle production | Rice-based dairy cattle production | Rice-based cattle fattening and sheep raising | Rice-based beef cattle and goat production | Wheat/maize-based beef cattle and goat production |
| Predominant animal species 1 | BCa, Pi, Po | DC,Pi, Po | BC, Bu, S, G, Fi | BC, Bu, G, Pi, Po | BC, Bu, G, Pi, Po |
| Number of animals (TLU'sb) 2 | 0.67 +_1.69 | 7.34+_4.32 | 0.70+_0.78 | 2.26+_1.97 | 1.07+_0.73 |

| Item | Vietnam | Thailand | Indonesia | Philippines | China |
|---|---|---|---|---|---|
| Main crop-animal interactions | Crop residues as feeds, use of draft animals | Crop residues as feed, manure as fertilizer, use of draft animal | Crop residues as feed, manure as fertilizer | Crop residues as feed, manure as fertilizer | Crop residues as feed, manure as fertilizers, use of draft animals |
| Contribution of livestock to total income (%) | 13 | 10-20 | 10-15 | 15-20 | 20-25 |
| Mean annual rainfall (mm) | 2170 | 1500 | 2200 | 2300 | 760 |
| Dry season (months) | 6 | 6–7 | 5–6 | 6 | 7 |
| **Farming systems** | | | | | |
| Predominant crop-animal systems | Rice-based beef cattle production | Rice-based dairy cattle production | Rice-based cattle fattening and sheep raising | Rice-based beef cattle and goat production | Wheat/maize-based beef cattle and goat production |
| Predominant animal species[a] | Bca.Pi, Po | DC,Pi, Po | BC, Bu, S, G, Fi | BC, Bu, G, Pi, Po | BC, Bu, G, Pi, Po |
| Number of animals (TLU'sb)[b] | 0.67 +_1.69 | 7.34+_4.32 | 0.70+_0.78 | 2.26+_1.97 | 1.07+_0.73 |
| Main crop-animal interactions | Crop residues as feeds, use of draft animals | Crop residues as feed, manure as fertilizer, use of draft animal | Crop residues as feed, manure as fertilizer | Crop residues as feed, manure as fertilizer | Crop residues as feed, manure as fertilizers, use of draft animals |
| Contribution of livestock to total income (%) | 13 | 10–20 | 10–15 | 15–20 | 20–25 |

**Table 4.1. Some household attributes and biophysical characteristics in rain-fed farming systems in five countries in Southeast and East Asia**

## Agricultural Education and Human Capital Development

Agricultural education can make important and robust contributions toward improving nutrition and health, increasing food security and income, reducing poverty, and empowering scientists, extension officers, and farmers. Its effect on the poverty, fragility, lack of hope, and vulnerability of small farmers and the landless is largely driven by the extent of the technology application and community-based participation of farmers.

Several studies confirm that better educated farmers benefit from higher rates of return on land (World Bank, 1990) and reduced poverty (Wodon, 1999). Fan and Chan-Kan (2004) demonstrated that the returns to investment in education are considerably higher in the less-favoured than in the better-favoured areas of Uganda, China, and India. Yet in absolute terms, returns on investment in education are substantially lower than investments in roads or agricultural research and development. These and other studies emphasise that for agriculture in general, and smallholder farming in particular, education is vital for promoting growth and food security and reducing poverty.

### Formal education and training

Throughout Asia, formal agricultural education and training is usually steeped in sector and disciplinary interests. The latter are so entrenched that integrated efforts are negligible. The subject matter is essentially production-oriented, without much evidence of agendas that link research and technology transfer processes for farmers. Thus, examples of subject matter areas include biological diversity, type of soil and soil fertility, patterns of land use, crops and cropping patterns, integrated pest management, post-harvest losses, animal breeding, animal nutrition, meat and egg production, post-production and processing, markets and value chains, and information communication technology.

Mahadevan (1984) distinguished between two levels of education in agriculture: formal degree education at universities, and intermediate level training in agricultural colleges. University graduates are mainly first degree holders who are seldom effective for advisory or other extension work. Their education is usually concerned largely with the development of technology at university or research centres. Interestingly, there is a great need for a significant number of such graduates to be involved with the transfer and adoption of technology. The intermediate level training produces graduates who are primarily responsible for helping farmers understand how to introduce changes in practice.

Like commodity and animal science research—which emphasises individual crops, animals, and animal products such as meat, milk, and eggs—formal education in technology development focuses on disciplines. This practice perpetuates challenges that are prevalent in agricultural research, as illustrated by Devendra (1996) with respect to animal science research (See Box 1).

**Box 4.1. Challenges in Animal Sciences Research in Asia**

- Disciplinary research is dominant in areas where animal nutrition and breeding are prominent.
- Fundamental and applied research are most common.
- Technology development and availability is far in excess of the utilisation of research results.
- Duplication of research is quite evident, especially in animal nutrition.
- Research priorities are often neither needs-based nor targeted to resolve on-farm constraints, using systems methodologies and community-based efforts
- Technology delivery mechanisms were often not clear and measures of impact systems approaches were rarely used in the research agenda, nor were participatory efforts with farmers.
- Research to link and serve development is generally weak and neglected.

Source: Devendra, 1996

Systems perspectives and methodologies that deal holistically with natural resource management, rural communities, and the environment were scarcely considered in the formal curricula of most countries, although Indonesia, the Philippines, and Thailand may be exceptions. Although problem identification and priority setting are important tools, they were inadequately pursued. Of equal concern is the fact that the looming threat of climate change and its implications were rarely considered as subjects for study.

## Non-formal education and training

Non-formal education and training, widespread in agriculture, are conducted at several levels in colleges, universities, and government institutions as well as in rural areas at provincial, district, sub-district, and farm levels.

The rationale and justification for such efforts may be determined by prevailing circumstances, constraints, or particular needs. Non-formal training initiatives are often driven by the ad hoc formulation of new projects or limitations within an ongoing project, and can also be a regular affair to ensure success.

At research station and university levels, topics include project formulation, participatory methodologies, team-building, understanding systems perspectives, developing and using questionnaires, taking field measurements, building research-extension-farmer linkages, and coping with constraints. The farm level covers topics such as developing confidence, collecting and storing data, dealing with feedback, resolving problems, developing and promoting wider adoption of cooperatives and self-help groups.

Careful priority setting and monitoring are needed to ensure progress and impact, especially where the courses are linked to provide cumulative outcomes. Local training is usually undertaken by government and local groups, which vary from country to country (Box 2).

**Box 4.2. Groups responsible for non-formal education and training in various Asian countries**

- China – provincial and district offices
- India – farmer training and knowledge centres (*krishivigyankendras* [KVKs])
- Indonesia – district and subdistrict *(kechematan)*
- Philippines – local government units (LGUs) and focus groups
- Korea – provincial rural development administration (PRDAs)
- Malaysia – farmers' organisations
- Vietnam – provincial, district and hamlet offices

## Training of trainers

Training of trainers as agents of change is an important component of both formal and informal education. Training and retraining are essential, as is the need to remain updated with current knowledge and trends and to understand participatory ways of working with farmers. The essential ingredients include understanding the process of technological change, maintaining close links with research scientists and subject matter specialists, building skills in effective communication of new technologies, and understanding the approaches that work successfully with farmers.

## Empowerment and self-reliance

Empowerment, at the heart of all education and training, means enabling people to have control and use of their own resources and to set their own agendas. For this, they require access to information and services as well as a developed capacity to determine their own futures. In the long term, such empowerment will enhance their self-reliance, that is, their ability to be resourceful to the extent possible with minimum dependence on external inputs. For example, the education of women has powerful beneficial effects on agricultural development because it empowers them to make decisions about food and nutritional security, health, productivity, and stability of farm households.

Women make a very significant contribution to food security and nutrition. Millions of women work as farmers and farm workers throughout the developing world. They account for 70–80 percent of household food production in sub-Saharan Africa. In terms food security, women process, purchase, and prepare food and are responsible for 65 percent of family food security in Asia, and 45 percent in Latin America and the Caribbean (Brown et al., 2001).

An International Food Policy Research Institute (IFPRI) study found that improvements in women`s education and status together accounted for about 55 percent of the reduction in child mortality and child malnutrition. In Asia, concerning the implementation of farming systems projects, the IFPRI found consistently that the implementation of the project and its eventual success was largely dependent on a few women farmers or participants in the project.

Farmers' willingness to assume the responsibility of managing and using their own resources is a direct result of empowerment and self-reliance. Box 3 introduces the Anand model of "Operation Flood" in India, which involves some 13 million farming families who process about 90 million kg of milk per year. In this model, farmers are shareholders of the whole chain of marketing and processing milk, and their livelihoods are improved accordingly.

## Research-Extension-Farmer Linkages: The Asian Experience

The so-called research-extension-farmer linkage is synonymous with the traditional technology transfer model. The efficiency of public sector extension services varies from one country to another, as does programme focus. Today, with agriculture waning, concerns have emerged about scope and effectiveness, the problems of globalisation and the need for more innovative strategies to deal with the changing environment, natural resource management, and food insecurity issues. These and other matters are of grave concern to the productivity of small farms and farmers' livelihoods.

The term "extension" and the systems and structures that deal with it have a diversity of meanings throughout Asia.

- The basic connotation is one of partnership and a unified public sector service that is consistent with technology transfer. Talking and persuasion are key words.
- The participatory extension approach, which engages in conversation with farmers and rural communities, helps address and resolve problems on the farm.
- Variations to the approaches exist in systems research, particularly in implementation. In Indonesia, for example, seventeen Assessment Institutes for Agricultural Technology (AIAT) have been created since 1995, involving researchers, extension agents, farmers, and local government officials. These AIATs have enabled regionalisation and decentralisation of research on individual commodities, development and transfer of location-specific technologies, and provision of technical services.
- Extension can have many approaches, ranging from those aimed at helping farmers increase production to those supporting marketing arrangements. In recent years, extension orientation has also expanded to include innovative structural, funding, and managerial arrangements (Rivera and Sulaiman, 2009).
- Sentiment is increasing in Asia that public extension services are no longer adequate to respond to the needs of farmers in many developing countries (Fleischer et al., 2004).

Additionally, discussions of potential successes of extension services in achieving change must factor in the technical capacity, commitment, and skills of extension agents, as well as understanding of the various constraints to development and the distribution methods that were used.

The demands on, and orientation of, extension are therefore many and far-reaching. Of particular concern is the fact that there are increasing and variable demands on available time, as extension appears to be shifting away from its primary function of diffusing improved technology in participatory ways to spur development. There is increasing risk that the primary task of technology transfer to meet the needs of farmers is weak and neglected.

**Box 4.3. Farmers' milk cooperatives in India become instruments of social change**

Disturbed by unfavourable price and market conditions, the producers of large supplies of buffalo milk from the rural areas of the Kaira District of Mumbai (formerly Bombay) met in January 1946 and established milk producers' societies in each village of the district. The two-tier Kaira Milk Cooperative has a District Milk Producers Cooperative at the central level, with more than 850 village Milk Producers' Cooperative Societies at the village level. Its establishment enabled farmers to negotiate from a position of strength for higher milk prices in the strong Bombay market and marginalised the middlemen who exploited the marketing system.

Each village-level Cooperative Society maintains a Milk Collection Centre with trained staff. Milk is received morning and evening, tested for quality, and payment is made for the milk delivered at the previous collection. Today, the Kaira District Cooperative Milk Producers Union is a confederation, supplying the dairy plant owned by the farmers with milk for processing and for the various products: butter, cheese, ghee, milk powder, baby foods, and chocolate. AMUL, the trade name under which the products are marketed, is well known throughout India. Headquartered in Anand, Gujaret, AMUL is the acronym for Anand Milk Union Limited but the word "*amul*" also means "priceless" or "precious" in Sanskrit.

A comparison of incomes from buffalo milk and cow's milk found that incomes were respectively 51 percent and 62 percent higher in villages with cooperatives than in villages without (Srivastra, 1970)

The AMUL complex continues to demonstrate the benefits of integrated education, research, extension and training activities as well as the importance of cooperatives. The Cooperative Dairy Development in Anand has brought about profound social and economic impacts. Not only have milk supplies increased, but the whole fabric of rural life has been enhanced, with improved nutritional well-being, higher incomes, household stability, village cohesiveness, increased security, and increased employment opportunities. In the words of the widely recognised architect of the programme, Dr V.R. Kurien, "dairying is an instrument of social and economic change."

This trend is reflected in several questions that relate to the limitations and potential for improved productivity through farming systems research:

- What are the priorities of extension to assist farmers?
- How does extension discern between the needs of different farmers and communities?
- Can extension services cope with farmers' activities and aim towards national development goals?
- Is the traditional technology transfer approach adequate to deal with the more diverse, risk-prone, rain-fed environments and their considerable biophysical variations?
- Is the extension arm effective in dealing with new and emerging issues such as mitigation and adaptation options in response to climate change and the dynamics of agriculture?
- Are extension agents sufficiently trained in systems perspectives to integrate their efforts with formal agricultural research systems, technology transfer processes, and related policy institutions?
- Are mechanisms in place to retrain extension agents to keep up with changing times?

## Developing the R&D Framework

Developing a sound research framework is a useful prerequisite for application of farming systems research methodologies. This is achieved initially by posing a few research questions that are relevant to the situation and provide some indicators of whether the research has the capacity to develop solutions to complex interacting elements such as agro-ecological zones (AEZs), natural resource management, changing environments, and the people within these systems. These questions are as follows.

- What is the researchable issue? Does it have the potential to generate new information, and is it worth the resources that are needed for this purpose?
- Recognising that R&D activities are dynamic and subject to constant critique, are the results economically beneficial and socially acceptable?
- Do the resulting improved technologies have the potential for wide replication and scaling up?
- Will they improve productivity and livelihoods?

## Application of Farming System Research Methodology and Systems Perspectives

Farming systems are the practices of agriculture. The numerous existing systems differ across locations and countries, and vary according to the types of agro-ecological zones, biophysical environment, eco-regions, and the extent and quality of the natural resources available. Farming systems are open systems with inputs from, and outputs to, their environments. The application of the methodology for farming system research follows six distinct and sequential steps:

1. Site selection;
2. Site description and characterisation (diagnosis);
3. Planning of on-farm research;
4. On-farm testing and validation of alternatives;
5. Diffusion of results;
6. Impact assessment.

"Systems" is a relatively new discipline, with systems perspectives and approaches based on a concept of organised and essential steps that deal with several prerequisites:

- recognition of the importance of interdisciplinary participatory approaches;
- formulation of research programmes that have community-based participation to set a common agenda and create ownership, involving both production and post-production systems;
- establishment of programmes that are needs-led, and have institutional and structural commitment;
- establishment of effective participatory planning, inter-institutional coordination and collaboration, research management, dissemination of information, and resolution of feedback issues; and
- long-term commitment to achieving impacts and education and training in agricultural systems and systems methodologies at various levels.

Figure 4.1 illustrates the steps involved in farming systems research.

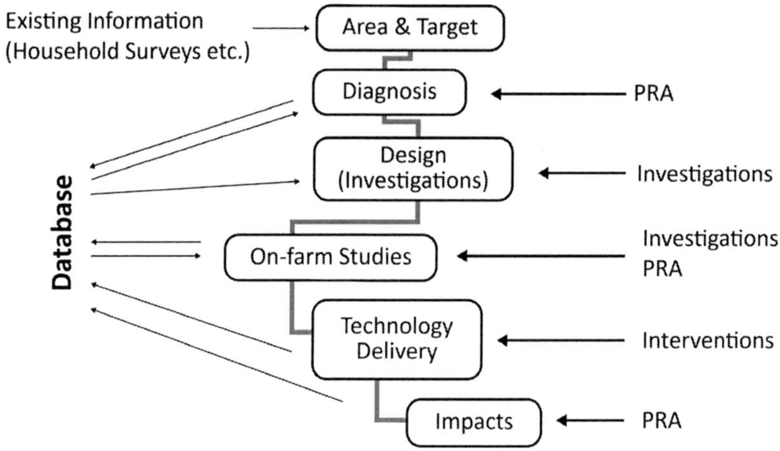


Figure 4.1. Methodology for farming systems research

The methodology of farming systems research is well known and widely applied in many countries in the developing world (Ruiz, 1995). Of the six steps identified above, the first three are fairly standard procedures for use in any location, while the last three—on-farm studies, technology delivery or diffusion of results, and impact assessment—are very location-specific and adaptive. They merit discussion and sharing of the experiences.

## Participatory rural appraisal

In order to comprehend the complexities of farming systems and their constraints in individual locations, participatory rural appraisal (PRA) was developed from the rapid rural appraisal (RRA) technique, which was found to be inadequate in terms of farmer participation. Other terms, such as participatory technology development (PTD), have been used more recently (Conroy, 2005).

Participatory rural appraisal has been effective in promoting understanding of production systems research across commodities throughout the developing countries of Asia, Africa, and Latin America. It has been used to improve the contribution of livestock in crop-animal systems in rain-fed areas of Southeast Asia, and in several countries of Latin America (Ruiz, 1995) and Africa. In Zimbabwe, participating farmers identified many problems limiting dairy farming such as late age of calving, alleged repeat breeding, long calving intervals, and low productivity due to inadequate availability and poor quality of feeds (Francis and Sibanda, 2001).

## Listening to farmers

Talking to farmers about their problems and aspirations is imperative in farming systems research. Farmers are recognised as lead participants in the research, without whom the credibility of the results would be in doubt, as would any formulation of activities to address the constraints and provide solutions. As the farmers are the ultimate beneficiaries of farming systems R&D, approaches should include farmers in the core team at every step from programme definition, formulation, and implementation to the evaluation of impact. It was therefore important to invest time and resources to address this issue.

Inasmuch as research is meaningless if it is does not drive development, the formulation of R&D programmes is futile if it does not involve farmers as partners in developing strategies to improve their livelihoods. The major advantages of listening to farmers are as follows (Devendra, 2006).

- Farmers know the biophysical environment.
- Farmers understand prevailing farming systems.
- Farmers have full understanding of their local constraints and problems.
- Farmers are the target beneficiaries of improved technologies.
- Farmers are the targets for empowerment.
- Farmers have deep knowledge of local customs, indigenous and traditional systems.
- Farmers are the agents of change and promotion of wide adoption.
- Farmers have all the local contacts.
- Farmers know current patterns of marketing.
- Farmers are the target beneficiaries of sustainable livelihoods and sustainable development programmes.
- Farmers provide an entry point for understanding of farming communities including their perceptions, preferences, and aspirations for household stability, improved livelihoods, and a better tomorrow.

The Three-Strata Forage System Project in Bali, Indonesia, introduced below, illustrates the importance of systems pathways, defining the problems, and setting the priorities and methodologies for R&D to ensure good results and long-term impact (Devendra, 2005).

## The three-strata forage system in Indonesia

Many productivity-enhancing technological options have been used to improve animal production in many parts of Asia and elsewhere. These technologies have provided a technical basis for development and are appropriate to systems combining animals with annual and perennial cropping.

The three-strata forage system (TSFS) was developed for Bali's dry, low rainfall environment (600–900 mm annual rainfall and four to eight months dry season) in Bali. The system integrates cash cropping and ruminant production (mainly cattle and goats) in a sustainable crop-animal system and enhances efficient use of natural resources, especially for small farms. The system was developed over nine-and-a-half years of R&D, with a focus on scaling out to other areas. As a result, it now has the potential for application in semi-arid areas of sub-Saharan Africa (Devendra, 2010).

The TSFS integrates the planting and harvesting of forages to ensure that a source of feed is available throughout the year. The plots are designed with a core area for growing maize, soya bean and cassava. The peripheral area is planted in three strata: 1) grasses and legumes for use during the wet season, 2) shrub legumes for use during the middle of the dry season, and 3) fodder trees for producing feeds for the late dry season.

## Plantings: grasses, legumes and trees

### Stratum 1

- Grasses: Buffel (*Cenchrusciliaris*) and Green Panic (*Panicum maximum*)
- Legumes: Stylo (*Stylosanthes gracilis*), Centrosema (*Centrosemapubescens*), and Caribbean Stylo (*Stylosantheshamata*)

### Stratum 2

- Shrubs: Gliricidia (*Gliricidiasepium*), Leucaena (*Leucaenaleucocephala*)

### Stratum 3

- Fodder trees: Ficus (*Ficuspoacellie*), Hibiscus (*Hibiscustilleacius*)
- Lane (*Lanneacorromandilica*)

### Outcomes of the TSFS include the following (Nitis et al., 1990):

- Forage production increased, enabling higher stocking rates (3.2 animal units/ha) and leading to annual live weight gains of 375 kg/ha in the TSFS compared to 2.1 animal units and 122 kg/ha/year in the non-TSFS.
- Cattle in the TSFS gained 90 percent more live weight and reached market-weigh 13 percent faster than those not in the TSFS.
- Farm income increased 31 percent.
- Soil erosion reduced by 57 percent in TSFS compared to the non-TSFS, and soil fertility increased, due to introduction of forage legumes.
- The 200 shrubs and 112 trees logged twice a year produced 1.5 tonnes/year of firewood and met 64 percent of the farmers' annual firewood requirements.
- Integration of goats in addition to cattle into the system further increased the farmers' incomes above the 31 percent noted earlier.
- The concept and technology of TSFS has become institutionalised.

## Diffusion of results: Networks and networking

Establishing networks and networking are major components of strengthened research capacity and R&D efforts. More importantly, they are essential for diffusion of research results. Both issues are very important mechanisms for enhancing R&D processes such as

- posting and exchanging of information;
- enhancing speed of information flow;
- disseminating of information;
- strengthening research capacity;
- setting up consultations and research partnerships;
- organising electronic conferences;
- enhancing personal contact nationally, regionally and internationally;
- strengthening empowerment and research capacity.

A number of R&D networks formed over time have significantly contributed to improving agricultural production in Asia. These include:

- Asian Farming Systems Research Network (AFSRN)
- Forage Network
- Small Ruminant Production Systems Network (SRUPNA)
- Crop-Animal Systems Research Network (CASREN)

The ARFSN, established by the International Development Research Centre (IDRC) in the 1970s, started with a systems approach to research mainly on rice, based at the International Rice Research Institute (IRRI). In 1993, an animal component was added to the network to give emphasis to crop-animal systems and address the totality of farming systems. Some seventy-two R&D sites in nine countries participated in the network.

During the sixteen-year lifespan of this project, it was very effective in promoting systems perspectives and systems research and methodologies relevant to farming systems research. Increased research capacity was apparent in fourteen countries in Asia, resulting in institutionalisation of crop-animal systems research. Research institutes were formed in the Philippines and Thailand, and farming systems offices in Bangladesh, Indonesia, Nepal, and Pakistan. In India, for example, a Cropping Systems Directorate was established involving sixteen universities.

## Impact assessment

The assessment of impacts is an important final step in the farming systems research process. It provides a measure of the extent to which the project objectives and the envisaged outcomes and impacts have been achieved, for whom and how, including whether any unexpected positive or negative impacts have been observed. Different kinds of outcomes and impacts can be involved, such as increased crop yields, increased profits, improved human or community capabilities, decreased poverty and improved livelihoods.

## Climate change

The discussion on the overriding importance of agricultural development and productivity is not complete without a brief defence of the relatively new science of climate change. Looming climate change—its threats and effects in agriculture—is serious. The interactions are complex, interdependent, and interrelated and present major R&D opportunities for mitigation and adaptation in the context of soil-crop-animal-people-biophysical interactions, pointing to the urgent need for interdisciplinary approaches.

Table 4.2 summarises the situation with reference to key issues, e.g., heat stress and feed resources, climate change impacts, and opportunities for R&D. The information that is presented is not exhaustive, but rather tries to capture the range of issues involved, the priorities, and the need for more R&D investments.

## Shared Vision for Strengthening Asia-Africa Linkages

The opportunities for a shared vision—one with its roots in partnerships at various levels that can be significantly strengthened—are limitless. Strong links already exist between Asia and Africa in the field of formal agricultural education at the college and university levels. This is reflected in the sizeable and increasing numbers of African students going to Asia for various studies across disciplines in such countries as India, Malaysia, Indonesia, and the Philippines. The reasons for this interest are associated with cultural compatibility, lower education and maintenance costs, comparable climates, complementary food habits, geographical proximity, and easy adaptation to the Asian environment.

On the other hand, the extent of Asia-Africa linkages in terms of sharing information and technologies, inter-university and professional contact, and networking in various aspects of agricultural development is less clear and is probably weak. Good opportunities for improvement exist; there are major pathways available to promote and strengthen existing linkages through concerted and more formal arrangements. These include:

- application of appropriate and proven productivity-increasing technologies in similar eco-regions and locations with similar climates; for example humid Southeast Asia and West Africa have much in common in the pattern of crop-animal systems (Devendra and Pezo, 2004);
- convening regular reciprocal meetings on priority issues, e.g. leguminous forages, marketing of agricultural products, and climate change;
- exchange of information, experiences, and lessons learned;
- exchange of teachers, scientists, extension agents, technicians, and enlightened farmers;
- intensive networking and increased personal contact;
- promotion of regular joint meetings to share results from case studies and to discuss common problems; and
- promotion of reciprocal visits by farmers.

Another useful opportunity is the promotion of reciprocal attendance of scientists at technical and network meetings. This can be easily organised as a matter of policy at the institutional level. The Consultative Group for International Agricultural Research (CGIAR) is well placed to execute this opportunity, which could be in many ways a very useful and beneficial investment with the potential for creating an Asian-African network to provide for stronger bonds between the two continents.

| Major issue | Potential climate change impacts | Opportunities for R&D |
|---|---|---|
| **1. Heat stress** | • Physiology<br>• Metabolism<br>• Reduced feed intake<br>• Reduced reproduction<br>• Increased mortality<br>• Low productivity<br>• Unsustainable prod. systems<br>• Reduced multifunctionality | • Adaptation<br>• Feed efficiency<br>• Measures to increase intake<br>• Supplementation<br>• Improved management |
| **2. Feed resources** (Forages, crop esidues, AIBP and NCFR) * | • Reduced quantities<br>• Poorer nutritional quality<br>• More fibrous<br>• Decreased palatability | • Use of heat tolerant plants<br>• Food-feed systems<br>• Use of multipurpose tree legumes<br>• Conservation |
| **3. Land use systems** | • Shift to dry land agriculture<br>• Droughts<br>• Water scarcity<br>• Diversification of agriculture<br>• Sustainability | • Heat tolerant plants and animals<br>• Emphasis on rain-fed agriculture<br>• Maximising feed supply |
| **4. Animal species and breeds** | • Adaptation<br>• Possible reduction in size<br>• Loss of biodiversity<br>• Migratory systems<br>• Uncontrolled and overstocking can cause environmental degradation | • Dynamics of nomadic and transhumant systems<br>• Ensuring choice for AEZ<br>• Understanding interactions with the environment<br>• Improving Vulnerability and survival of the poor and their animals |
| **5. GHG emissions:** Enteric fermentation and manure, producing global warming | • Reduced crop growth and animal productivity<br>• Poor C sequestration | • Improved use of grasses, legumes and agronomic practices<br>• Use of dietary nitrates to reduce $CH_4$<br>• Intensification |
| **6. Integrated NRM and holistic systems*** | • R and D capacity | • Interdisciplinarity<br>• Use of systems perspectives<br>• Advantage of shade in<br>• Plantations |
| **7. Semi-arid and arid AEZs including rangelands** | • Reduced feeds<br>• Overstocking<br>• Environmental damage | • Control of numbers<br>• Use of multipurpose leguminous trees<br>• Improved management. |

**Table 4.2. Major issues in animal production that will be affected by climate change impacts (Devendra, 2012)**

*AIBP: agro-industrial by-products; NCFR: non-conventional feed resources; NRM: Natural Resource Management.

The rationale for forging closer and stronger linkages between Asia and Africa is limitless, considering the many similarities in problem orientation and resolution and ensuring progress, as shown in Table 4.3. In order to be realistic and more meaningful, only the major issues (evolution, types of crops and animals and mixed farms, crop-animal interactions) are considered across regions and within subregions. The farming systems in West Africa (Fernandes-Rivera et al., 2004) and in Asia (Devendra and Pezo, 2004) have been well described. Some aspects worth noting that Asia has in common with West Africa include:

- high rural populations, high population density, small farm systems and small size of arable land per hectare, and increasing urbanization;
- crop-animal systems involving both ruminants and non-ruminants are evolving;
- feeding and nutrition are the major constraints to animal production, and the situation is more acute in rain-fed areas;
- an increasing trend to incorporate leguminous forages into cropping systems;
- rapid growth of industrial systems involving poultry and pigs,
- ruminant production systems are less well developed and provide good opportunities for increased productivity and contribution to improved livelihoods of the poor; and
- development of market-oriented dairying in peri-urban areas.

Transforming agricultural education in both Asia and Africa will need more understanding and entail use of systems perspectives and systems approaches that are fundamental to R&D efforts and deal with natural resource management, poverty, food insecurity, and the environment, as well as interactions among these elements.

Strong disciplinary training is essential, but a more rounded view is necessary to expose the formally trained degree holders to the scope of issues they may encounter in holistic systems. This is especially true with Asian farming systems, where the diversity of crops and animals, traditional methods of farming, multifunctional contributions of animals, multiple crop-animal interactions, and numerous challenges faced by farmers present a complex plethora of issues that cannot be readily resolved. Specific problems cannot be fixed by a single discipline or in a disciplinary context alone. While this has been a major weakness of many research programmes involved with natural resource management in the past, systems perspectives and pathways can ensure effective results.

| Issue | West Africa | Asia |
|---|---|---|
| Type of farming system | • Predominantly pastoral, but shifting to mixed systems | • Predominantly crop-animal systems |
| Evolution | • Shifts from pastoralism and extensive crop production to mixed farming systems<br>• Decreasing fallow period and nutrient cycling<br>• Low level of intensification | • Integration of crops and animals<br>• Systems generally stable, but have different levels of intensification<br>• Increasing integration of forage legumes to develop food–feed systems<br>• Influence of the Asian Rice Farming Systems Network (ARFSN) and more recently, the Crop-animal Systems Research Network (CASREN) |
| Types of crops and animals in mixed farms | • Mostly annual crops<br>• Cattle and small ruminants are dominant, and small ruminants pigs and poultry are secondary | • Annual and perennial crops<br>• Cattle, buffaloes small ruminants, poultry and pigs are common |
| Crop-animal interactions | • Involve mainly cattle and small ruminants (crop residues, draught and animal manure) | Many and varied and mostly beneficial, e.g. crop residues, draught, manure |
| Opportunities | • Stability and sustainability<br>• Integration of animals with perennial crops<br>• Integration of forage legumes<br>• Intensification<br>• Increased focus on semi-arid areas | • Intensification<br>• Nutrient recycling mechanisms<br>• Development of low input and sustainable systems<br>• Development of year round feeding systems<br>• Development of food-feed systems |

**Table 4.3. Similarities between Asia and West Africa of major issues and opportunities for improving crop-animal systems**

A five-pronged strategy is proposed in order to facilitate the transformation of agricultural education into a systems approach that has a holistic R&D view:

- Develop appropriate curricula for formal agricultural education that builds good science training into school systems and combines strong disciplinary orientation with systems perspectives and systems methodologies.
- Offer formal degree education and specialisation at university level that reflects strong training in systems perspectives, agricultural systems, and the environment. Introduce formal degree education in advanced countries in Asia where such courses are available.
- Intensify non-formal educational and training needs at different levels, including the training of trainers.
- Develop educational programmes that provide sensitivity to similar problems, opportunities to share experiences through inter-country and inter-regional cooperation, accelerated information exchanges and networking to introduce technology applications that support more rapid agricultural growth.
- Promote vigorous technology development and adoption that can significantly increase concerted information sharing, networking and major windows of opportunities for strengthening linkages between Asia and Africa.

There is no doubt that education and training are essential prerequisites for empowering scientists, extension workers, and farmers to promote progress in agricultural development. This process begins with good curricula and science training in school systems and must continue into undergraduate and finally postgraduate programmes that can effectively meet the challenges of the future, revitalising the potential of rural economies, enhancing agricultural productivity, and thus empowering communities to eventually achieve self-reliance.

## Concluding Comments

Agricultural education and technological change are essential determinants of sustained productivity and prosperity for the people of Asia. Together, education and technology enable efficiency, build the capacity to manage and conserve the natural resource base, and promote maximum technology-induced productivity gains that are beneficial to human welfare. The focus of this development must be directly concerned with the preponderance of small farm systems that provide jobs for 60 percent of the working population and generate a quarter of the region's gross national product (GNP). These farms are responsible for 50–90 percent of Asia's total food production. However, poverty is rampant, with the poor described in terms of *deprivation, subsist-*

*ence, illiteracy, survival, and vulnerability*, and as having lives fraught with a never-ending syndrome of *poverty, adaptation, fragility, lack of hope, and low life expectancy*. Thirty years after the Green Revolution, poverty remains persistent.

Over the past two decades, agriculture has been neglected, and agricultural development has been waning in Asia. However, its vital importance is widely recognised and concerted efforts now target the revitalising and transformation of this sector. The recognition of agriculture's importance to economic and rural growth has led to increased investments and a new impetus for agricultural development. The enormous impact of the Green Revolution on Asian agriculture is well known. There were problems, such as the negative effects of using high-cost inputs such as fertilisers and irrigation, and social inequalities such as the increased disparity between the rich and poor farmers and the landless. But important lessons were learned, such as the importance of using locally available resources to the extent possible; the importance of education and training for empowerment and the development of human capital; the importance of farming systems research for problem definition and technology development and delivery; and the importance of community-based participation and information dissemination.

Agricultural education is crucial for all aspects of agricultural development. It can make robust contributions toward improving nutrition and health; increasing food security and income; reducing poverty; and empowering scientists, extension officers, and farmers. Its effect on the *poverty, fragility, lack of hope, and vulnerability* of small farmers and the landless is largely driven by the extent of the research and development employed, the application of new technology, and the community-based participation of farmers. Several studies confirm that better educated farmers reap benefits in the form of reduced poverty and higher rates of return on land, and that the returns on investment in education are considerably higher in the less favoured, rain-fed areas than in the more favoured, irrigated areas.

Agricultural education needs to be viewed as an integral part of a concerted strategy of institutional restructuring based on a set of operational guidelines that provide for improved resource management, contribute to sustainable development, and support the waning agricultural sector. This comprises all levels of formal and non-formal learning systems, from local schools to colleges and universities, as well as all formal agricultural systems of research, technology transfer, and extension—including policy institutions. It means that educating agricultural specialists and their advisors calls for an understanding of needs-based technology development and of the processes leading to technology transfer, adoption and impact.

Both formal (i.e. colleges and universities) and non-formal (i.e. multi-level intermediate training) systems of education are important. Formal agricultural education is generally steeped in disciplinary interests. It is essentially production-oriented, with little evidence of ensuring links between research and technology transfer processes. For example, formal education focuses on areas such as biological diversity, soil and soil fertility, land use, crops and cropping patterns, integrated pest management, post-harvest losses, animal breeding, animal nutrition, meat and egg production, post-production and processing, markets and value chains, and information communication technology. While much of the formal training is undertaken in Western colleges and universities, graduates are seldom effective for advisory or extension work in their home countries. The intermediate level of training, on the other hand, produces persons whose primary responsibility is to introduce change to local farmers. Empowerment, at the heart of all education and training, means enabling people to have control and use of their own resources and to set their own agendas. The Anand model of Operation Flood in India, based on cooperative development and used as an instrument for social change, is cited as a case study for empowerment. Empowerment of women is especially important as women make the most significant contributions to food security; they are responsible for generating 65 percent of family food security in Asia and 45 percent in Latin America and the Caribbean.

Asian farming systems, essentially based on mixed farming, have to cope with a diversity of crops and animals, traditional methods of farming, low-input systems, multifunctional contributions of animals, and multiple crop-animal interactions, as well as numerous problems of farmers. To this list must now be added the challenge of climate change. Recognising that specific problems cannot be resolved by a single discipline or in a disciplinary context alone, efforts toward transformation of agricultural education and training now give emphasis to systems perspectives. Farming systems research, with its well-defined approach and methodology, provides an important solution to these issues. The five-pronged plan described in this chapter can further the importance and potential contribution of systems research.

Associated with the above is the challenge of research-extension-farmer linkages. Public sector extension services vary across countries in terms of both efficiency and programme focus. The challenge stems from concerns about extension services' scope, efficiency, and effectiveness in following through the process from discovery to delivery. The old model of "top-down" extension functions or specialised disciplines that depend mainly on talking and persuasion is no longer acceptable. The proposed changes necessitate community-based participation and interaction in which the overarching issues are the forging of interdisciplinary perspectives and a deep and holistic understanding of farmers' aspirations, problems, and other attendant issues of importance. Improved education and training provides for effective and dynamic researcher-farmer-extension partnerships to resolve constraints and promote progress.

The final part of this chapter discusses opportunities for forging a shared vision of stronger Asia-Africa linkages. The opportunities are limitless as the two regions share similar problems and potential solutions. Areas for possible collaboration discussed include types of crops and animals in mixed farms, crop-animal interactions, and opportunities for parallel R&D. Commonalties between Asia and West Africa include high rural populations, high population density, small farm systems and size of arable land, and increasing urbanisation, as well as crop-animal systems involving both ruminants and non-ruminants. These situations are constantly evolving and intensifying.

The opportunities for collaborative effort include application of appropriate and proven technologies, exchanges of information, experiences, and lessons learned; and exchanges among professional personnel.

Finally, a distinct opportunity that can be formulated as a matter of policy concerns reciprocal attendance of scientists at all technical and network meetings—a potentially useful and beneficial investment that CGIAR is well placed to formalise. An attendant issue is the potential opportunity to create an Asian-African network in agriculture to provide for stronger bonding between the two regions. Much remains to be done towards a technologically-driven transformation that can enhance agricultural productivity and self-reliance in both Asia and Africa. The transformation and revitalisation of AET can be an instrument of change in that direction. One product of that change, systems-oriented, improved R&D, is crucial for increasing productivity and food security and for alleviating poverty at this critical time. Together with policy change, multinational engagement, and new investments, R&D can contribute significantly to growth in agricultural productivity and improved livelihoods in the future.

## References

Asian Development Bank (ADB). 2012. Food Security and Poverty in Asia and the Pacific: Key Challenges and Policy Issues. Asian Development Bank, Manila, Philippines.

Brown, L., H. Feldstein, L. Haddad, C. Pena and A. Quisumbing. 2001. Generating Food Security in the Year 2010: Producers, Gatekeepers and Shock Absorbers. In The Unfinished Agenda, International Food Policy Research Institute, Washington D.C.

Conroy, C. 2005. Participatory Livestock Research: A Guide. ITDG Publishing: Bourton, Dunsmore, Great Britain.

Devendra, C. 1996. Present Status of Research and Development and Institutional Considerations for Animal Production Systems in the Asia-Pacific Region. Proceedings of the Eighth Asian-Australian Animal Science Congress, October 13–18, Seoul, Korea.

—. 2005. Improvement of Crop-Animal Systems in Rainfed Agriculture in South East Asia. In Rawlinson, P., C. Wachirapakorn, P. Pakdee and M. Wanapat, Eds. Proceedings of the International Conference on Livestock-Crop Systems to Meet the Challenges of Globalisation, November 14–18, Khon Kaen, Thailand. Vol. 1, 220–231.

—. 2006. Listening to Farmers: Participatory Approaches for Developing Goat Production. Proceedings of the Sixth International Symposium on Agriculture, Biotechnology and Environment, September 23, Daegu, Korea.

—. 2010. Small Farms in Asia: Revitalising Agricultural Production, Food Security and Rural Prosperity. Academy of Sciences: Kuala Lumpur, Malaysia.

—. 2012. Climate Change Threats and Effects: Challenges for Agriculture and Food Security. Academy Science Malaysia Series on Climate Change, Kuala Lumpur, Malaysia.

Devendra, C. and D. Pezo. 2004. Crop-Animal Systems in Asia and Latin America: Characteristics, Trends and Emerging Challenges. Comparison with West Africa. In Williams, T., S. Tarawali, P. Hiernaux and S. Fernandez-Rivera, Eds. Proceedings of the International Conference on Sustainable Crop-Livestock Production for Improved Livelihoods and Natural Resource Management, November 19–22. International Livestock Research Institute (ILRI): Nairobi.

Devendra, C. and D. Thomas. 2001. Crop-Animal Interactions in Mixed Farming Systems in Asia. Agricultural Systems 71: 27–40.

Economic and Social Survey of Asia and the Pacific (ESCAP). 2008. Sustaining Growth and Sharing Prosperity. Bangkok, Thailand.

Fan, S. and C. Chan-Kan. 2004. Returns to Investments in Less-Favoured Countries: A Synthesis of Evidence and Implications for Africa. Food Policy 29, 432–444.

Fernandes-Rivera, S., I. Elokke, V. Manyong, T. Williams, R. Kruska and S. Tarawali. 2004. Classification and Description of the Major Farming Systems Incorporating Ruminant Livestock in West Africa. In Williams, T., S. Tarawali, P. Hiernaux and S. Fernandez- Rivera, Eds. Proceedings of the International Conference on Sustainable Crop-Livestock Production for Improved Livelihoods and Natural Resource Management, 19–22 November 2002. International Livestock Research Institute (ILRI), Nairobi, Kenya.

Fleischer, G., H. Waibel, and G. Walter-Echols. 2004. Egypt: How much does it Cost to Introduce Participatory Extension Approaches in Public Extension Services? In Rivera, W. and G. Alex, Eds. Demand-Driven Approaches to Agriculture Extension. Case Studies of International Initiatives. Agriculture and Rural Development Paper 10, Extension Reform for Rural Development. World Bank: Washinton D.C.

Francis, J. and S. Sibanda. 2001. Participatory Action Research: Experiences in Smallholder Dairy Farming in Zimbabwe. Livestock Research for Rural Development 13, 1–15.

Mahadevan, P. 1984. Education of Livestock Producers and Agents of Change in the Use of New Technologies: Development of Animal Production Systems. Elsevier: Amsterdam, Netherlands.

Nagayets, O. 2005. Small Farms: Current Status and Key Trends. Proceedings: The Future of Small Farms. International Food Policy Research Institute: Washington, D.C.

Nitis, I., K. Lana, W. Sukanten, M. Suarna, and S. Putra. 1990. The Concept and Development of the Three-Strata Forage System. In Devendra, C., Ed. Shrubs and Tree Fodders for Farm Animals. International Development Research Centre, IDRC-276e, Ottawa, Canada.

Rivera, W. and V. Sulaiman. 2009. Extension: Object of Reform, Engine for Innovation. Outlook in Agriculture 38, 267–274.

Ruiz, M. 1995. RISPAL: The Latin American Research Network for Animal Production Systems. IDRC-IICA Global Workshop. San Jose, Costa Rica.

Srivastra, R. 1970. Importance of Animal Production in Rural Economy. Indian Dairyman 25, 223–243.

Wodon, Q. 1999. Microdeterminants of Consumption, Poverty, Growth and Inequality in Bangladesh. Policy Research Working Paper No. 2076: Washington, D.C.

World Bank. 1990. Poverty: World Development Indicators. World Development Report, 1990. Oxford Universty Press: New York.

# CHAPTER FIVE

# AGRICULTURAL HIGHER EDUCATION IN SUB-SAHARAN AFRICA: PARTNERSHIPS AND THE LAND-GRANT MODEL

TESHOME ALEMNEH

## Higher Education in sub-Saharan Africa

Higher education, in particular agricultural education and training (AET), is one of the most effective mechanisms available for addressing the development challenges facing sub-Saharan Africa. Such education is imperative for the continuous generation, adaptation, and application of technologies that support sustainable development. The opportunity education affords for development of human resources can enhance socio-economic transformation, promote innovation and technological progress, and provide opportunities for advancement and overall competitiveness in the global economy (Kapur and Crowley, 2008; World Bank, 2008). Given the innovation-driven, knowledge-intensive nature of this economy, a development strategy that prioritizes the training of human capital through quality higher education is not optional—it is the only path for Africa to take out of poverty towards sustainable development.

Agriculture remains dominant in sub-Saharan Africa, and improving agricultural productivity has become a critical component of the continent's growth and development. Agricultural education and training institutions—both university and non-university—are foundational to changing lives and transforming the agriculture sector. One of the most critical factors ensuring Africa's renaissance in the twenty-first century will be how effectively it cultivates its human resource and utilises its diaspora capital (Alemneh and Cornick, 2012; Yizengaw, 2008). While higher education in Africa has long been considered less critical for national development and poverty reduction than basic services such as primary education and health care, there is now a wealth of information, based on research and studies, supporting and advocating the importance of higher education for accelerated development and transformation in Africa (African Union, 2006; Kellogg et al., 2008; Parker, 2012; UNESCO/ World Bank, 2000; World Bank, 2008).

In addition to building the highly critical human resources, AET has a direct impact on agricultural productivity through research, innovation, and community outreach programmes (Davis et al., 2007; Kingslow, 2007). As part of the overall higher education system in Africa and a major source of an educated work force, AET supports knowledge generation, adaptation, and dissemination for improving local capacities and fosters the innovation and creativity that drive economic growth and social transformation. Nonetheless, while it is generally accepted that AET institutions in sub-Saharan Africa prepare agricultural professionals who improve practices and productivity and contribute to poverty reduction, their contributions to the generation of knowledge and dissemination of technologies and innovation to farmers leaves a lot to be desired.

According to Johanson and Saint (2007), agricultural education and training, though vital, has been a much neglected component of agricultural strategies and action plans in many sub-Saharan Africa countries. In many of these countries, the contributions of agriculture to growth and poverty reduction have been further constrained by neglect of long-term institutional and human capacity development. On the other hand, AET development was an integral part of national strategy in countries such as India, Brazil, Malaysia, Chile, and the Philippines that successfully developed their agricultural sectors (Staatz and Dembele, 2008). These countries have achieved notable success in establishing AET systems.

Higher education in sub-Saharan Africa faces several challenges, including limited access to and poor quality of programmes and curricula; lack of relevance and focus on problem-solving; weak governance and financial constraints; poor infrastructure and facilities; weak research and innovation capacities, and an aging faculty facing a critical shortfall in numbers. As a result of these challenges, human and institutional capacity that is too weak to generate and apply appropriate solutions to local problems has become a key problem in Africa's agricultural and economic development. Consequently, smallholder farmers have limited access to knowledge, technologies, and improved practices. This limitation hinders productivity and transformation of livelihoods and communities.

Agricultural colleges and universities in sub-Saharan Africa are poorly staffed, mainly due to losing faculty through attrition, retirement, and internal and external brain drain. The erosion of the human capital base and the

***Key Messages***

- Through research, innovation and community outreach, *AET builds highly critical human resources and has a direct impact on agricultural productivity and transformation of livelihoods* and communities in Sub-Saharan Africa.
- Higher education in Sub-Saharan Africa—including AET—faces several challenges. Poor institutional capacity limits smallholder farmers' access to knowledge, technologies, and improved practices. The lack of a close link between AET and research and extension systems means that curricula, teaching methods, and outreach programmes are not appropriately contextualised; possibilities for producing adaptable graduates are not identified; and the effectiveness of AET research and extension is limited. *AET institutions can address these challenges through policy reform.*
- *Partnerships provide more than aid and development and technical support.* Higher education partnerships are collaboration mechanisms that enable institutions working toward a common goal to enhance shared learning and mutual benefits, with collaborative contributions made by the partnering institutions.
- Regional associations, state agricultural universities, the land-grant model, and the Higher Education for Development (HED) model are some of the most *common approaches to higher education partnerships* in developing countries. This chapter details specific examples of Africa-to-Africa, Africa-Europe, and Africa-U.S. collaborations that utilise these models.
- The Regional Universities Forum for Capacity Building in Agriculture (RU-FORUM) is a regional association that manages *Africa-to-Africa* partnerships. It oversees agricultural training and networks at 29 universities in Eastern, Central, and Southern Africa.
- In terms of *Africa-Europe* partnerships, the Swedish International Development Cooperation Agency (SIDA) has helped create successful bilateral research cooperation programmes with institutions in Burkina Faso, Ethiopia, Mozambique, Rwanda, Tanzania, and Uganda.
- Ethiopia, Nigeria, Kenya, Malawi and Cameroon all have universities involved in successful *Africa-U.S. partnerships* based on the land-grant model. Examples of HED model Africa-U.S. partnerships include universities in Ethiopia, Kenya, South Sudan, Uganda, Malawi, Namibia and Burundi.
- If adapted and effectively implemented, the U.S. land-grant model—which embodies the three core functions of education, research, and extension—can improve the role of AET institutions in sub-Saharan Africa. However, the challenge is to ensure that *the three core functions are not dispersed among three public entities* (i.e. universities for education, national institutes for research, and agriculture ministries for extension). This requires a formal relationship between ministries and national research bodies to provide the space and resources for AET institutions to promote research and extension.
- AET institutions urgently need *visionary institutional leadership and structural policy changes* to fully realise their potential to establish successful partnerships for agricultural development.

weakly developed programmes of these institutions have given rise to the question of who will train the next generation of agriculturalists and agricultural scientists.

Although undergraduate programmes have expanded in many countries (Staatz and Dembele, 2008), university programmes are weakly organised, particularly at the graduate level. Of even greater concern, however, is the fact that only a fraction of Africa's higher education enrolment increases have been in science and technology. In fact, enrolment in agriculture, science, engineering, and technology, which are greatly needed for innovation, knowledge generation, adaptation, and overall national competitiveness, accounts for less than 30 percent of total enrolment in sub-Saharan Africa. This limits research, knowledge generation, and adaptation as well as development of innovation capacities for most AET institutions. Furthermore, the ability of institutions to produce graduates who can solve societal problems and bring about meaningful change in livelihoods and overall national development is being significantly influenced by a mismatch between curricula/programmes of study and labour market needs (African Union, 2006; Johanson and Saint, 2007).

Agricultural education and training in sub-Saharan Africa also lacks a close linkage to research and extension systems that might help AET adapt its curricula and teaching methods, reorient its outreach programmes, and identify possibilities for producing adaptable, qualified graduates. This, in turn, has limited the effectiveness of AET institutions in research and extension and has weakened contributions to improving lives in farming communities.

In order to address these challenges, AET institutions need to develop their human and institutional capacities and reorient their overall missions through policy reform. Collaboration with higher education institutions that have greater experience and abilities, are supported by ministries of agriculture, and include research organisations is critical. For example, promotion and investment in international collaboration and cooperation has proven successful in furthering human and institutional development capacity in institutions of higher education in sub-Saharan Africa. Such partnerships will help institutions contribute to better adaptation of experiences, technologies, and lessons that have worked elsewhere.

## Higher Education Partnerships

Partnerships provide more than development assistance and aid, and they consist of more than technical assistance and exchanges (Samoff and Bidemi, 2002). Higher education partnerships are mechanisms of collaboration that enable institutions working toward a common goal to enhance shared learning and mutual benefits, while collaborative contributions are made by the partnering institutions.

Within the context of higher education development cooperation, Wanni et al. (2010) defined partnership between educational institutions as a dynamic collaborative process that brings mutual, though not necessarily symmetrical, benefits to the partners. Partners share ownership, and their relationship is based on respect, trust, transparency, and reciprocity. Partners understand each other's cultural and working environments, decisions are taken jointly, and all partners are open and clear about what they bring to and what they expect from the partnership.

In typical higher education partnerships, entities that have established relationships come together to address societal challenges and develop and build institutional and human capacities of beneficiary, host-country higher education institutions in a cost-effective and sustainable manner. Higher education partnerships allow individual institutions to combine their particular strengths, each party giving as much as receiving. Genuine partnerships are based on mutual respect, understanding, and recognition of each other's capacities and limitations. When properly designed and executed, and based on a common purpose and long-term engagement, partnerships help ignite and rekindle innovation and the problem-solving potential of Africa's higher education community and institutions.

Partnerships are distinguished by their characteristics of long-term, shared responsibility, reciprocity, equality, mutuality and balance of power, country ownership accountability, joint decision-making, respect, trust, transparency, and sustainability (Bailey and Dolan, 2011; Gutierrez, 2008; King, 2008; Wanni et al., 2010) as well as enduring institutional support and the right institutional infrastructure (Association of Commonwealth Universities, 2011). It is important that educational partnerships be demand-driven and country owned. Without institutional commitment and ownership, partnerships risk being yet another instance of "unhelpful help" (Wanni et al., 2010).

However, partnerships have also been criticised for being a one-way flow of capacity from the North to the South and for the absence of genuine sharing. Ishengoma (2011) concluded that extensive partnerships between universities in Tanzania and institutions in the North have not made meaningful and significant contributions to the strengthening of higher education. The paper highlighted non-reciprocal and non-equal links, the perpetuation of dependence on funding from the North, and the indirect nature of funding as major drawbacks of the partnerships.

The following section of this chapter discusses some of the most common models of higher education partnership. It provides a brief description of collaboration among African institutions, and between African institutions and universities in the U.S. and Europe. It also provides more detailed description and analysis of partnerships between African and American institutions, with some emphasis on the land-grant model.

## Africa-to-Africa Higher Education Partnerships

Africa-to-Africa partnership refers to collaborations among two or more African higher education institutions working on common goals with principles of African ownership, mutual benefit, problem-solving, and transparency. For example, the Regional Universities Forum for Capacity Building in Agriculture (RUFORUM) and the Association for Strengthening Agricultural Research in Eastern and Central Africa (ASARECA) have demonstrated successful partnerships among AET institutions in Africa. ASARECA initiated a Master's degree programme in agricultural information and communication management at the Universities of Haramaya (Ethiopia), Makerere (Uganda), Sokoine (Tanzania) and Nairobi (Kenya)—an excellent example of regional cooperation. The Centre for Agricultural Research Management and Policy Learning for Eastern Africa at Haramaya University, initiated by the International Food Policy Research Institute (IFPRI), is another laudable initiative for improving regional cooperation toward developing human capital in sub-Saharan Africa.

RUFORUM, an association of twenty-five universities in Eastern, Central and Southern Africa, oversees graduate training and networks specialised in agriculture (RUFORUM, 2010). An Africa-owned and home-grown initiative, RUFORUM guides innovation that is relevant to African farmers and informs national and regional development practice and policies. This forms a community of learning with active interaction and sharing of knowledge and skills among partners to enhance the integration of education, research, and outreach functions. It also facilitates partners' increased ability to achieve strategic goals, e.g. improved teaching practices, and focuses on impact-oriented applied research in rural innovative processes—which are necessary to prepare professionals with the competencies needed to effect meaningful change in agricultural production systems. Students complete their studies well prepared for careers nationally and regionally because both the coursework and the associated field studies are grounded in local policies, agro-ecologies, and the challenges of institutions and farming systems. RUFORUM's research paradigm puts farmers and their problems at the forefront and, in turn, prepares its graduates to conduct relevant, development-oriented research geared towards problem-solving.

## Africa-Europe Higher Education Partnerships

There are several higher education partnerships between European and African institutions of higher learning. The Swedish International Development Cooperation Agency and Sustainable Agriculture Research and Extension Center (SIDA-SAREC); *Vlaamse Interuniversitaire Raad* (VLIR), which supports collaboration between Belgian and African institutions; the Norwegian Agency for Development Cooperation (NORAD); and the Netherlands Universities Foundation for International Cooperation (NUFFIC) are some examples with specific success stories. It is not the intention of this chapter to discuss these partnerships in detail but to provide some examples. Many of these European development assistance programmes have recognised the importance of African universities' ownership of programmes as a core principle of engagement or partnership.

With financial support from the Belgian government, the VLIR-UOS (*Vlaamse Interuniversitaire Raad* and University of Stavanger) Programme for Institutional University Cooperation supports a partnership between Jimma University (JU) in Ethiopia and the University of Ghent in Belgium, drawing on expertise from five Flemish universities as well as non-university Belgian research institutes. The partnership has focused on strengthening the research and extension functions of JU, which has a unique community-based education and training programme to provide public service and community outreach. The programme has contributed significantly to strengthened research capacity and increased research output while setting the foundation of a dynamic and collaborative research culture directed toward critical social, scientific, and public health issues of direct relevance to development.

In 2006, SIDA had programmes of bilateral research cooperation with institutions in Burkina Faso, Ethiopia, Mozambique, Rwanda, Tanzania, and Uganda. Makerere University implemented the Makerere-SIDA/SAREC bilateral collaborative Research Support programme between 2000 and 2009 to assist Uganda in its efforts to promote research for attainment of new knowledge. Results of research were incorporated into teaching, ensuring that graduates will have the knowledge and qualifications necessary to effect a positive impact on the socio-economic fabric of Uganda.

A long-lasting collaboration between Sokoine University of Agriculture (SUA) and the Agricultural University of Norway, funded by NORAD, has proven rather successful in capacity-building (Royal Norwegian Ministry of Foreign Affairs, 1998). The collaboration has provided significant inputs, allowing for a comparatively high intake of students at various levels, and offered significant opportunities in teaching and research. The faculties and departments have become capable of carrying out education at various levels, managing programmes and setting priorities. An estimated 131 students have received Masters' degrees, and thirty have received PhDs, which has improved SUA's capacity to undertake problem-solving research and disseminate results to the surrounding community, as well as informing agricultural policies in Tanzania.

## Africa-U.S. Higher Education Partnerships

Various U.S. State Department programmes have made more than seven hundred partnership project awards, linking 250 U.S. colleges and universities in forty-seven states with higher education institutions in approximately 135 countries (Wiley and Root, 2003). For example, the Fulbright Educational Partnership Program provides awards that are usually used for faculty and staff development, curriculum development, collaborative research, and outreach activities. USAID has funded several broad partnership programmes and an increasing number of specialised programmes, some administered directly by USAID and others managed by non-governmental organisations (NGOs). USAID funding for programmes devoted to higher education partnerships is broadly directed to countries that have USAID Missions and that match country-specific strategic objectives.

One of the largest institution-building projects, and in quantitative terms perhaps one of the most successful, linked six U.S. universities to nine state agricultural universities in India beginning in 1952. It provided at least one thousand U.S.-trained Indian students with advanced degrees and sent 337 U.S. faculty members to serve at Indian institutions (U.S. Congress, 1991). A 1974 evaluation found that the number of Indian PhD staff increased from 251 to 1,234, and the number of professors granted advanced degrees from U.S. universities increased from 140 to 486 (U.S. Congress, 1991).

For example, during the eighteen years' cooperation between Punjab Agricultural University (PAU) and the Ohio State University (OSU), 160 Indian scientists received advanced degrees at OSU, and, in turn, PAU made very significant contributions to the Green Revolution in the Indo-Gangetic Plains (Erbaugh and Hansen, 2012). OSU supported PAU in developing academic departments, helping to develop curricula, and teaching courses. The partners also worked on establishing applied research and outreach programs. Erbaugh and Hansen (2012) stated that, although some have argued that the Green Revolution primarily resulted from the development of new, high-yielding varieties of major crops such as maize, wheat, and rice, the development of the agricultural universities with academia, research, and extension functions was equally as important.

Since the initiation of collaborative association between the U.S. Agency for International Development (USAID) and U.S. universities, land-grant institutions have been involved in development assistance (U.S. Congress, 1991). The U.S. land-grant college/university system is unique, with three integral components: teaching, research, and extension (APLU, 2012). A key component of the land-grant system is the agricultural experiment station. Through higher education partnerships since the 1960s, U.S. land-grant universities have been involved in research and technology generation, extension and technology transfer, education and training, and institution-building in developing countries.

During the 1960s and 1970s, a number of U.S. universities, with support from USAID, introduced and helped set up new universities in sub-Saharan Africa that embodied the land grant ideals in which agricultural extension and research were introduced. These are listed in Table 5.1.

The following sections highlight the accomplishments of some of these partners. They also look at the challenges involved in introducing the land-grant model of AET institutions into sub-Saharan Africa.

*Ethiopia: Alemaya University and Oklahoma State University.* Following the U.S. land-grant model and funded by USAID, Oklahoma State University's partnership with Alemaya University in Ethiopia lasted for sixteen years, ending in 1968. The partnership developed undergraduate academic programmes in agriculture as well as major agricultural research and extension programmes (Price and Evans, 1989). Alemaya's impact on agricultural technology in Ethiopia, including forestry technology, derived from its development of new techniques and adaptation of improved technologies. Alemaya's strong research programmes recommended technologies such as improved varieties of maize, wheat, sorghum, tef, potato, vegetables, dairy, and poultry (Price and Evans, 1989). Local farmers and associations continue to benefit from the university's work on improved varieties and farming practices, and also on solutions to soil, insect, and disease problems.

Alemaya has adhered to the U.S. land-grant model of interlinked programmes in teaching, research, and extension despite strong attempts to remove the extension and research functions from campus (Price and Evans, 1989). Although extension work is done by the Ministry of Agriculture, the university still maintains a small extension service. Faculty and students frequently visit farmers and cooperatives to offer technical assistance and have formalised a number of these relationships as interdisciplinary research and extension projects. The National Institute of Agricultural Research coordinates all research activity, but Alemaya also has strong research of its own. In fact, the Institute has named Alemaya University one of the centres for specific research activities.

| African Institution | U.S. Land-grant Institution |
|---|---|
| Alemaya University (Ethiopia) | Oklahoma State University (1952–1968) |
| Egerton University (Kenya) | West Virginia University (1962–1972) |
| Makerere University (Uganda) | West Virginia University (1964–1973)<br>Ohio State University (1984–1993) |
| Ahmadou Bello University (Nigeria) | Kansas State University (1962–1978) |
| University of Ife (Nigeria) | University of Wisconsin (1964–1975) |
| Bunda College of Agriculture (Malawi) | University of Massachusetts (1963–1970) |
| Njala Agricultural University (Sierra Leone) | University of Illinois (1963–1971) |
| Agricultural University of Dschang (Cameroon) | University of Florida (1982–1990) |
| University of Ouagadougou (Burkina Faso) | University of Georgia (1983–1990) |
| University of Zimbabwe | Michigan State University (1984–1989) |
| Egerton University (Kenya) | University of Illinois (1986–1991) |

**Table 5.1. Partnership between U.S. land-grant universities and institutions in sub-Saharan Africa. Source: U.S. Congress, 1991**

*Nigeria: Ahamdou Bello University and Kansas State University; University of Ife and University of Wisconsin; University of Nigeria at Nsukka and Michigan State University.* USAID provided support to establish three Nigerian universities with cooperating U.S. land-grant universities: Ahamdou Bello University with Kansas State University, University of Ife with University of Wisconsin, and University of Nigeria at Nsukka with Michigan State University. Efforts to transfer the tripartite land-grant model of teaching, research, and extension to the three Nigerian universities have had mixed success (Gamble et al., 1988). As a result of its incorporation of established research institutes at the time of its founding, Ahmadou Bello University comes closest to the model in practice. Due to a challenging national structural system that disengages research and extension from the universities, most of these institutions were not successful in implementing the land-grant model (Gamble et al., 1988).

*Kenya: Egerton University and University of Illinois:* The Institutional Development for Agricultural Training (IDAT) Project authorised in July 1986 was a USAID/Kenya project to establish permanent institutional relationships between Egerton University and the University of Illinois at Urbana-Champaign (USAID, 1999). Egerton is an agricultural education institution of excellence serving the needs of Kenya and other African countries; IDAT was designed to help Egerton adjust its curriculum from that of a diploma-granting college to that of a practical, degree-granting College of Agriculture and to expand its research and extension functions after the manner of a U.S. land-grant college. As a result of the project, Egerton University established, reviewed, and revised its academic curriculum and met the planned output regarding graduate degree training for Egerton faculty. Egerton has made an important start in realising its long-term objective of contributing significantly to outreach. Its Agricultural Resource Centre has hosted continuing education programmes and regional in-service training in crop management and research (USAID, 1999). While the introduction of the land-grant model at Egerton has been quite successful, the University still needs to further expand its efforts in research and extension.

*Malawi: Bunda Agricultural College and University of Massachusetts.* In the more than forty years since 1969, Bunda College in Malawi has granted around 860 diploma degrees and three hundred BSc degrees. Nearly all of the middle- and senior-level managers, researchers, and extension specialists in Malawi's Ministry of Agriculture and its parastatal Agricultural Development and Marketing Corporation are Bunda graduates. This is all due to a Bunda Agricultural College partnership with the University of Massachusetts, funded through USAID. The partnership has financed construction of campus buildings, provided long-term expatriate faculty, and sponsored training of Malawian faculty in U.S. universities (Welsch et al., 1987). However, the College's ability to conduct effective research and provide public service through extension has been limited. Bunda has few institutional or budgetary linkages with the Ministry of Agriculture or other potential sources of government funding that would enable it to expand its research and extension mission. Consequently, the College has little institutional support to pursue more productive research and extension.

*Cameroon: University Center Dschang and University of Florida.* A USAID-funded partnership established an agricultural university capable of providing an integrated teaching, research, and outreach programme to promote increased agricultural production in Cameroon (Corty et al., 1986). The partnership, between the University Center Dschang in Cameroon and the University of Florida, introduced the U.S. land-grant model in Cameroon, facilitated revision of academic programmes and administrative structures, and established a four-year degree programme. The land-grant model was accepted by the government and the centre with the aim of providing hands-on training in problem-solving applications of research to users. However, the university's research and extension activities were limited, mainly due to misalignment and relationship challenges between the University Center Dschang and the Ministry of Agriculture.

## The Higher Education for Development (HED) Model

USAID initiated funding and management of higher education partnerships with enactment of the 1961 Foreign Assistance Act. In 1992, the Association Liaison Office for University Cooperation and Development (ALO) was founded through a consortium of U.S. higher education associations. ALO began awarding grants funded by USAID in 1998 and changed its name to Higher Education for Development (HED) in 2006.

HED promotes and manages grants that support efforts of higher education institutions to address development challenges while also benefiting the U.S. higher education community (see Figure 5.1). Through institution-to-institution collaboration, HED partnerships build human and institutional capacities to create positive change and meet development needs. Funding for such partnerships is provided by USAID Bureaus, USAID Missions, the U.S. Department of State, and other federal government agencies.

HED is uniquely positioned with access to a large number and variety of institutions in the U.S. higher education community. It provides expertise in developing programme design and facilitates merit-based, transparent competitions and rigorous peer review processes to select partners. This ensures that only the best and most cost-effective programmes are recommended for funding.

HED-managed partnerships link higher education institutions in the U.S. with institutions in developing countries, enabling them to work together to address a wide range of challenges. These partnerships largely focus on faculty and staff development, curricula review and academic programme development, research and innovation capacity-building, community outreach improvement, and dissemination of technology and better practices. Partnerships may also focus on revitalising and enhancing leadership and administration capabilities, renovating and upgrading teaching-learning and research facilities and infrastructure, and overall institutional capacity development.

Since 1998, HED has managed over three hundred and fifty international partnerships in more than sixty countries. In 2012, HED supported seventy active partnerships involving eighty-three host-country institutions in thirty-seven countries, along with fifty-seven U.S. universities and colleges (HED, 2013b).

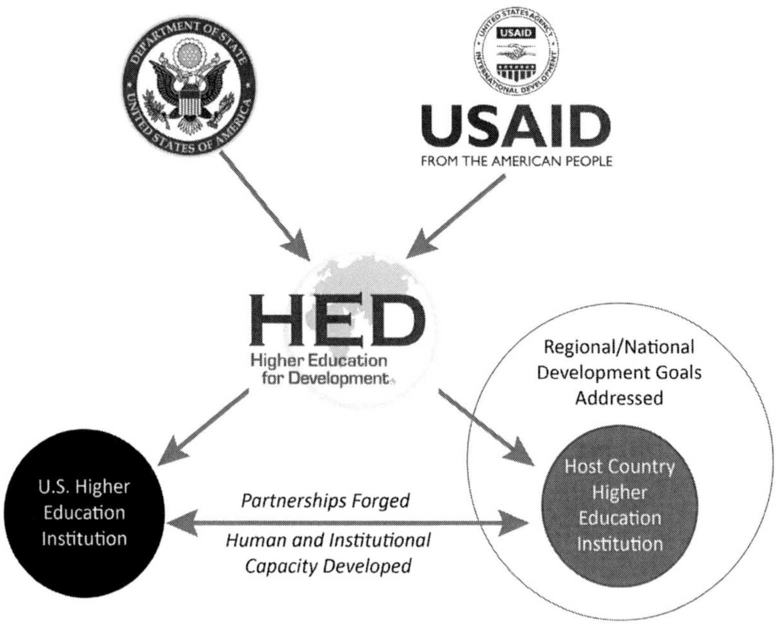


Figure 5.1. The HED higher education partnership model. Source: HED, 2013a

## HED-managed Partnerships in Agriculture and Natural Resources in sub-Saharan Africa

HED-managed partnerships have contributed to developing human capital, strengthened capabilities for sustained economic growth, increased agricultural productivity, and reduced food insecurity. To demonstrate the value of partnerships in addressing development challenges in Africa, the following sections present a select number of HED-managed higher education partnerships that highlight the importance of long-term collaboration geared toward problem-solving and African ownership of the process.

Many partnerships between U.S. institutions and AET institutions in sub-Saharan Africa are implementing activities that expand the African institutions' education, research, and extension functions to facilitate societal problem-solving and improved agricultural production and productivity.

In 2009, USAID funded eleven university partnerships in sub-Saharan Africa under the Africa-U.S. Higher Education Initiative. These partnerships established a five-year strategic plan and ten-year vision to improve agriculture and natural resource education as well as the research and extension functions of universities in the region (Table 5.2). If long-term funding is made available and the partnerships continue implementing planned activities, there will be some significant impact in the agricultural sector in these countries in terms of human resource development, knowledge generation and adaptation, innovation, and service to local communities. This section, largely based on success stories written by HED, provides recent and current examples of HED-managed partnerships in action (HED, 2013c).

| Country | Partnering Institutions | Partnership Title |
|---|---|---|
| Burkina Faso | International Institute for Water and Environmental Engineering (2iE) & Tuskegee University | *Africa-U.S. Network of Centers of Excellence in Water and Environmental Science & Technology* |
| Ethiopia | Addis Ababa University & University of Connecticut | *Sustainable Water Resources: Capacity Building in Education, Research and Outreach* |
| Kenya | University of Nairobi & Colorado State University | *The Centre for Sustainable Drylands: A University Collaboration for Transforming Higher Education in Africa at the University of Nairobi* |
| Malawi | University of Malawi & Michigan State University | *Ecosystem Services: Linking Science to Action in Malawi and the Region* |
| Senegal | Université Gaston-Berger & The Ohio State University | *Development of Agronomy and Crop Production Academic Programs, Research, and Needs-based Extension Programmes for Sustainable Food Production in Sahel* |
| Sudan | Catholic University of Sudan, University of Juba & Virginia Polytechnic Institute and State University | *Rebuilding of Higher Education in Agriculture to Support Food Security, Economic Growth, and Peace Efforts in Post-Conflict Southern Sudan* |
| Uganda | Makerere University & North Dakota State University | *Capacity Building in Integrated Management of Trans-boundary Animal Diseases and Zoonoses* |

**Table 5.2. Agriculture and natural resource-related partnerships under the Africa-U.S. Higher Education Initiative Source: HED, 2013d**

*Ethiopia: Addis Ababa University and University of Connecticut.* Jointly identifying and designing a programme to address water management challenges in Ethiopia, Addis Ababa University and the University of Connecticut established the Ethiopian Institute of Water Resources (EIWR). The institute, in collaboration with five other local universities, offers a degree-level curriculum in Water Resources and Engineering Management (WREM) and Water Sanitation and Health (WASH). EIWR has established a system that identifies local water management problems and organises field work in teams to enable fruitful discourse between Masters' and PhD-level students and the local community. A team of students with a keen interest in water issues developed research skills and an understanding of the local settings through this programme and has been striving to generate practical solutions. The programme also provides short-term training for practitioners in the areas of research and management of water-related development challenges, problem-solving knowledge, innovation, and technologies. The partnership has a summer programme through which students live in local rural areas to gain a better understanding of rural communities' major challenges.

*Kenya: University of Nairobi and Colorado State University.* In Kenya, a partnership between the University of Nairobi and Colorado State University created the Center for Sustainable Dryland Ecosystems and Societies, which has strong potential as a centre of excellence promoting education, research, and public service in Eastern and Central Africa. In South Sudan, the University of Juba and the Catholic University of South Sudan, in partnership with Virginia Polytechnic and State University, are implementing a comprehensive long-term programme to restore agricultural higher education training, extension, and research capacity in South Sudan. It is expected that the resulting improved institutional capacity will address food security and development challenges with skilled graduates and effective outreach programmes in the world's youngest country. In Uganda, a partnership between Makerere University and North Dakota State University has established a joint regional MSc degree and Graduate Certificate in International Infectious Disease Management and Bio-security. Its goal includes risk assessment to reduce transmission of zoonotic and vector-borne diseases. Partners participate in investigations of anthrax and brucellosis outbreaks in Uganda and collaborate in the emerging pandemic threats programme to synergise activities in the region. All of the partnerships discussed above focus on building capacities, as well as research and extension, of the respective African universities to function as land-grant institutions with meaningful roles in the agricultural and innovation systems.

*Malawi: Bunda College of Agriculture and Texas A&M University.* Between 2001 and 2003, Texas A&M University partnered with the University of Malawi, Bunda College of Agriculture, to strengthen the capacity of animal health training and research. The partners established basic bacteriology and haematology laboratories; developed a Master of Science programme in Animal Health with an emphasis on bacteriology and epidemiology; conducted research on the prevalence and etiologies of mastitis and East Coast fever; and developed and disseminated animal health information. The partnership also trained more than forty field veterinary technicians and more than 180 farmers on topics ranging from milk quality and proper milking hygiene to food safety, nutrition, and general animal husbandry. This demonstrates the value of higher education partnerships in building local capacities, specifically for improving their ability to conduct research and disseminate research findings through extension, training of local community, and public service.

*Namibia: University of Namibia and Pennsylvania State University.* A university partnership in Namibia established farmers' cooperative groups and conducted training programmes for rural women, providing them with educational and entrepreneurial opportunities to become environmentally aware farmers, producers of value-added food made from locally grown raw crops, and marketers and distributors of those products to individual and institutional customers. This USAID-funded, HED-managed partnership involved Pennsylvania State University and the University of Namibia (UNAM). In the Northern Region of Namibia, the partners established a food cooperative that processes nutritious foods using local crops grown by cooperative members. UNAM conducted five training sessions for the community at Omahenene on issues related to cooperative organisation, product development, costing, and recordkeeping. UNAM's focus on community outreach and capacity development will have a significant impact on improving livelihoods and agricultural productivity in Namibia.

*Burundi: Ngozi University and South Carolina State University.* A U.S. land-grant institution, South Carolina State University (SCSU) partnered with Ngozi University in Burundi to establish an innovative and unique agribusiness programme. In Kayanza and Ngozi Provinces, partners opened five research stations focused on the cultivation and testing of several rice varieties as well as testing of fruits and vegetables ranging from eggplant and cucumbers to carrots and fennel. Some of these crops include hybrids and high-yielding seeds from the United States. In line with the land-grant core principles, this partnership enabled Ngozi to strengthen its research capacity and, most importantly, to reach out to the community and fulfill its public service mission. Researchers tested the crops in different climates and soils and also tested various methods of pest management. Once the best practices and crops were determined, researchers shared their results with local farmers, who used the new-found information to improve their yields. Such a streamlined effort to link the educational missions and research activities of a university with public service and community outreach promotes fulfillment of the three-pronged functional model learned from the U.S. land-grant model.

*Malawi: Bunda College of Agriculture and Lincoln University of Missouri and Michigan State University.* A successful partnership between the University of Malawi at Bunda College of Agriculture and Lincoln University of Missouri has demonstrated the benefits of long-term engagement and institutional commitment to solving societal problems. This partnership promoted high-quality nutritional supplements containing goat's milk mixed with other high-protein supplements for malnourished and vulnerable children and orphans. Historically, goat's milk in the region was fed to goat kids, while goat's meat was the family economic enterprise. The partnership introduced cross-breeding of local goats with purebred strains from France and South Africa to improve animal health and increase milk production. As these new varieties of goats were being developed, university nutritionists were demonstrating the value of goat's milk mixed with soy supplements, which they provided to a local hospital rehabilitation centre for malnourished and vulnerable children. When the children received the enhanced milk, the mortality rate dropped considerably and the children's health improved in record time, allowing them to return to their homes.

As reported by HED, ten years later the production of these healthy, high-milk-producing goats is still supervised by Bunda College to assist families with income generation from the sale of improved goat varieties, which often fetch more than twice the normal price (HED, 2013e). Local and international NGOs frequently purchase the healthy milking goats for their orphanages and child nutrition rehabilitation centres as an on-site source of good protein. Human consumption of goat's milk has made the goats so valuable that local citizens must now worry about thievery of their highly valuable animals. The accomplishments of this partnership, in terms of human capacity development, research and extension, helped partners forge a new engagement of Bunda College, Lincoln University of Missouri, and Michigan State University (MSU) to develop the University of Malawi's capacity to contribute more effectively to the drive for agriculture-led economic growth and management of human activities to protect vital natural resources.

Through institutional collaborations, U.S. land-grant universities have helped Rwandan institutions strengthen and effectively perform their education, research, and extension functions, with a significant impact on agricultural productivity and the national economy. The National University of Rwanda and two U.S. land-grant universities, Michigan State University and Texas A&M University, improved teaching, research, and outreach capacity by developing faculty, re-designing the agriculture curriculum, and establishing a community outreach centre. Supported by USAID/Rwanda, the Partnership for Enhancing Agriculture in Rwanda through Linkages (PEARL) developed a model for the fully operational Outreach Centre in Butare, a focal point for extension activities with the surrounding farming communities. Such outreach has contributed, for example, to enabling the Maraba Coffee Producer's Association to market better quality shade-grown specialty coffees and receive fair trade certification. The Maraba, Karaba, and Gashonga districts directly benefited from the PEARL partnership programme, contributing to changing lives and improving economic opportunities for the farming community.

Between 2001 and 2003, Oregon State University partnered with University of Botswana (UB) to improve regional efforts to manage and sustain natural resources by developing the institutional and resource capacity of the Harry Oppenheimer Okavango Research Centre (HOORC) at UB. The HOORC is the only research institution serving the Okavango Delta, an environmentally sensitive area containing habitat and wildlife populations that cross national borders in migration. Through the partnership, HOORC staff expanded their technical knowledge and skills and increased awareness of how to develop community support for conservation research through alliance-building with NGOs and other agencies. The HOORC also increased its abilities to monitor ecological systems, identify regional interventions more effectively, test new technologies and approaches, and collect and analyse data.

## Lessons Learned

The above examples demonstrate how well-designed, monitored, and supported collaborations of higher education institutions have a far-reaching impact on problem solving and agricultural development in sub-Saharan Africa. HED-managed higher education partnerships, as described above, have proved to be effective modalities of development intervention through their ability to ignite and rekindle the innovation and problem-solving potential of the higher education community. By upgrading curricula, teaching methods, and resources, partnerships produce graduates that can contribute to societal transformation. Through faculty development and exchange, partnerships enhance personal and professional capacities for enhanced teaching, research, and community outreach as well as providing policy advice and leading dialogues.

If adapted and effectively implemented, the U.S. land-grant institution can serve as a great model for agricultural education and training in sub-Saharan Africa, affording AET institutions a better role in development and in improving the lives of those in the rural community. In assessing the success and impact of introducing the U.S. land-grant model into sub-Saharan Africa, one should note that the U.S. institutions are historically established with full structural and functional clarity and purpose. Although the scope and level of engagement has expanded from the initial articulation of their missions, U.S. land-grant universities still embody the education, research, and extension or public service functions that were entrusted to them at their founding. They are fully supported by state governments, the private sector, farmer associations, and industry.

On the other hand, AET institutions in sub-Saharan Africa—where the U.S. land-grant model has been introduced over the past fifty years—have largely been unable to discharge these functions effectively. Most of these institutions were established primarily to meet the demand for highly skilled human resources, with a major focus being teaching and/or training.

The most critical challenge is that the three core functions of a land-grant university were dispersed among three public entities, namely universities for education, national institutes for research, and ministries of agriculture for extension. Furthermore, governments offer no recognition of the challenge or vision to streamline formal linkages and collaborative working relationships among these three entities. As demonstrated in Nigeria, Ethiopia and Kenya, AET institutions have faced challenges from ministries of agriculture and national research bodies that have statutory mandates for extension and research. Many sub-Saharan Africa countries have not put in place enabling environments and funding for AET institutions to conduct research effectively and disseminate findings through extension.

The ever increasing enrolment of undergraduate students in most AET institutions in sub-Saharan Africa has also limited the ability of these institutions to embark on research and extension activities. Faculty members are so overloaded with courses that they are commonly unable to do research and extension. Furthermore, the lack of any institutional reward for problem-solving or applied research and extension discourages faculty from engaging in these functions. Findings from problem-solving or applied research are also unlikely to be accepted and published in most reputable journals in Europe or America, which affects the academic progress of faculty but also affects their interest in engaging in applied research and extension or public service. Most AET institutions lack formal institutional procedures and policies to reward or recognise faculty members for their involvement in extension or public service, which constitutes one of the core pillars of the land-grant model.

The weak relationships between—and lack of pressure from—the public and private sectors, NGOs, legislators, farmers' organisations, agro-industry, and business represent another challenge sub-Saharan African AET institutions face in adopting and implementing the land-grant model. Such stakeholder pressure is needed to ensure that AET institutions generate and adapt knowledge, make innovations, and disseminate findings to the community to change practices, transform the agricultural sector, and improve livelihoods. In many cases, stakeholders seem disengaged, with little or no pressure, feedback, or support for AET's quest to engage in applied problem-solving research and extension.

In terms of producing large numbers of quality graduates, the contribution of AET universities in sub-Saharan Africa is laudable. Of the three components of the U.S. land-grant model of an agriculturally oriented university, teaching is the most widely and successfully applied. However, to be more effective in public service and become more of an instrument of development through knowledge generation, adaptation and innovation, greater linkage should be formally encouraged, established, and strengthened between AET institutions and the national bodies coordinating and funding research and extension, namely research organisations and ministries of agriculture. Only then will AET universities meaningfully contribute to and become active players in agricultural development and transformation that changes lives and economies in sub-Saharan Africa.

As discussed in this chapter, even if the U.S. land-grant model is not strictly followed, there are encouraging indications that AET institutions in sub-Saharan Africa are striving to fulfill their research and extension functions effectively and meaningfully. There is evidence of institutional commitment and capacity for the AET institutions in Africa to become more involved in problem-solving, applied research, and meaningful public service through extension and outreach. Visionary leadership at the institutional level and structural changes at the policy level are urgently needed to fully utilise the immense potential of AET universities. Agricultural education and training institutions have a large pool of high calibre agricultural professionals and scientists and have a great opportunity to bring in experiences and lessons from international collaborators through partnerships.

## Concluding Comments

A growing number of policy makers, scholars, and practitioners are reaffirming the importance of higher education and agricultural development and transformation to overall socio-economic development in sub-Saharan Africa. Universities are increasingly recognised as crucial national, as well as regional, assets that have the ability to nourish transformation in Africa.

Higher education partnerships are mechanisms of collaboration between institutions toward a common goal. Such partnerships enhance shared learning and mutual benefit, with contributions from all partnering institutions. A new kind of partnership based on listening to Africa's voice and African ownership of initiatives, as well as building on the values of trust, mutual benefit, and problem-solving, will give agricultural education and training institutions in sub-Saharan Africa an opportunity to play a greater and more effective role in agricultural systems. Higher education partnerships build cross-cultural understanding, improve awareness of critical development challenges, create opportunities for joint research, and re-orient institutional missions toward problem-solving and community and national development.

Partnerships have developed capacities such as enabling rapid response to zoonotic disease outbreaks in Karamoja, Uganda (through the Makerere University–North Dakota State University partnership) in several AET universities across the region. They have also contributed to establishing research centres that can generate and adapt knowledge and expand community outreach for conservation of natural resources, cultivation and testing of crops, and developing pest management interventions, as in the University of Botswana–Oregon State University and the Ngozi University–South Carolina State University partnerships. As a result of its partnership with Lincoln University of Missouri, Bunda College in Malawi was able to reach out to communities with its research findings on the improved nutrition of soya-supplemented goat's milk. Universities in Ethiopia and Nigeria have played meaningful roles in research and extension activities, partly as a result of the introduction of the U.S. land-grant model through higher education partnerships. Long-term European support to university collaborations has shown similar results in Uganda, Tanzania, Ethiopia, and Kenya. It is important that development partners continue supporting such long-term partnerships, based on African ownership, mutual benefit, and trust.

Success and sustainability of partnerships heavily depend on the ability of the collaborations to ensure that AET institutions in Africa provide leadership, own the process, and are involved in identifying and prioritising problems, determining performance objectives, and designing problem-solving plans.

Agricultural education and training capacity development efforts in sub-Saharan Africa need to focus on improving quality and relevance of academic programmes, on reorienting missions toward more community service, and on renewed emphasis on research, innovation, and problem-solving. It is important to note that most of the higher education institutions that have introduced the U.S. land-grant model have maintained a mission to perform the three pillars—teaching, research, and extension or public service.

In order for AET institutions to be more effective in fulfilling all three missions, national governments need to recognise the critical role of these institutions and reassess current policies regarding organisational arrangements and funding for research and extension. There should be a formal relationship between the ministries of agriculture and national research bodies to provide the space and resources for AET institutions to promote research and extension or public service. In terms of policy, national structures and organisational set-ups in sub-Saharan Africa often limit the ability of AET institutions to effectively perform research and meet extension goals.

In most countries, agricultural extension is the sole responsibility of ministries of agriculture, with weak or non-existent links to agricultural education and training. Agricultural research, in many cases, is also the responsibility of national organisations outside of AET, although, to varying degrees and funding levels, these research bodies—unlike the ministries of agriculture—usually engage AET institutions in their research. In some countries these institutions are given full responsibility to coordinate research on certain crops, livestock, and forestry areas. Universities need to engage in setting goals and research priorities. Neither the national research bodies nor AET institutions effectively engage in the dissemination of research findings to local communities.

Strengthening agricultural research and extension systems in sub-Saharan Africa is closely linked to strengthening the human resource capacities and research and extension capabilities of agricultural education and training institutions. The education, research, and extension functions of AET universities can be developed and strengthened through well designed partnerships, such as those that introduced the land-grant model in the 1960s and 1970s, and those managed by HED for developing human and institutional capacities. However, in order for AET universities to become active players in agricultural development and economic growth in sub-Saharan Africa, their research and extensions functions need to be strengthened.

*The views and interpretations expressed in this chapter are those of the author and should not be attributed to USAID or the American Council on Education (ACE).*

## References

African Union, 2006. The Second Decade of Education for Africa (2006–2010). Addis Ababa.

Alemneh, T. and T. Cornick. 2012. Higher Education Partnerships: Experience and Impact in sub-Saharan Africa. Paper presented at the Third Annual Conference of the Regional Universities Forum for Capacity Building in Agriculture (RUFORUM), September 24–28, Entebbe, Uganda.

Association of Public and Land-grant Universities (APLU). 2012. The Land-Grant Tradition. http://www.aplu.org/document.doc?id=780

Association of Commonwealth Universities. 2011. Building University Partnerships for Sustainable Development. ACU Spotlight, Policy and Research Unit, No. 2.

Bailey, F. and A. Dolan. 2011. The Meaning of Partnership in Development: Lessons for Development Education. Policy and Practice: A Development Education Review 13: 30–48.
http://www.DevelopmentEducationReview.com

Corty, L., D. Ignasias, D. Snyder, and J. Washington. 1986. Threshold Evaluation of the Agricultural Education Project. University Center Dschang, Cameroon. http://pdf.usaid.gov/pdf_docs/XDAAV270A.pdf

Davis, K., J. Ekobir, W. Mekasha, C. Ocheieng, D. Spielman, and E. Zerfu. 2007. Strengthening Agricultural Education and Training in sub-Saharan Africa from an Innovative System Perspective. IFPRI Discussion Paper 00736.

Erbaugh, M. and D. Hansen. 2012. The Punjab Agricultural University–Ohio State University Long Term Partnership: Past, Present and Future Potential Collaboration. Unpublished paper presented at the 50th Anniversary of Punjab Agricultural University, India, November, 2012.

Gamble, W., R. Blumberg, V. Johnson, and N. Raun. 1988. Three Nigerian Universities and their Role in Agricultural Development. AID Project Impact Evaluation #6. USAID.
http://pdf.usaid.gov/pdf_docs/PNAAX200.pdf

Gutierrez, D. 2008. Beyond Disappointment: Transforming Ideology and Practice in North-South Research Partnerships. In NORRAG News Special Issue, The Politics of Partnership: Peril or Promise, NORRAG News 41. http://www.norrag.org/issues/41

Higher Education for Development (HED). 2013a. The HED Model.
http://www.hedprogram.org/about/model/index.cfm

—. 2013b. http://hedprogram.org/resources/loader.cfm?csModule=security/getfileandpageid=13554

—. 2013c. Success Stories. http://hedprogram.org/ourwork/partnerships/RWA-2000-09-21.cfm
http://hedprogram.org/impact/success/2011-SS-SCSU-Ngozi.cfm

—. 2013d. Partnerships Offer Sustainable Solutions to Development Issues at U.S.-Africa Higher Education Initiative Partner's Meeting. http://www.hedprogram.—org/media/news_releases/Partnerships-Offer-Sustainable-Solutions.cfm

—. 2013e. Improved Variety of Goats Enhances Nutritional Supplements, Family Income. http://www.hedprogram.org/impact/success/2009-SS-Lincoln-Bunda-Malawi.cfm

Ishengoma, J. 2011. Strengthening Higher Education Space in Africa through North-South Partnerships and Links: Myths and Realities from Tanzania Public Universities. Paper presented at Conference of Rectors, vice Chancellors and Presidents (COREVIP), May 30–June 3, Stellenbosch, South Africa.

Johanson, R. and W. Saint. 2007. Cultivating Knowledge and Skills to Grow Africa's Agriculture: A Synthesis of an Institutional, Regional and International Review. World Bank, Agricultural and Rural Development Department, Washington, DC.

Kapur, D. and M. Crowley. 2008. Beyond the ABCs: Higher Education and Developing Countries. Working Paper 139. Centre for Global Development.

Kellogg, E., A. Hervy, and T. Yizengaw. 2008. Africa-U.S. Higher Education Initiative: Empowering African Higher Education for Africa's Transformation. The Institute of International Education Magazine. IIE Networker, Fall 2008.

King, K. 2008. Editorial: The Promise and Peril of Partnership. In NORRAG News Special Issue, The Politics of Partnership: Peril or Promise, NORRAG News 41. http://www.norrag.org/issues/41

Kinglsow, E. 2007. AET in Africa: A Survey of Progress. A Report prepared for the Kellogg Foundation. Kingslow Associates LLC.

Parker, S. 2012. Lessons from a Ten-Year Collaborative: A Case Study of the Partnership for Higher Education in Africa (PHEA). Clear Thinking Communications.

Price, E. and C. Evans. 1989. Ethiopia: Alemaya University of Agriculture. AID Project Impact Evaluation Report 71. USAID. http://pdf.usaid.gov/pdf_docs/PNAAX219.pdf

Royal Norwegian Ministry of Foreign Affairs. 1998. Institutional Cooperation between Sokoine and Norwegian Agricultural Universities. Development through Institutions, Sub-Study 2.

RUFORUM. 2010. RUFORUM: The Role of the African Universities in Building Capacity to Implement CAADP. September 2010.

Samoff, J. and C. Bidemi. 2002. The Promise of Partnership and Continuities of Dependence: External Support to Higher Education in Africa. 45th Annual Meeting of the African Studies Association. December 5–8, Washington, DC.

Staatz, J. and N. Dembele. 2008. Agriculture for Development in SSA. Background Paper for the WDR.

U.S. Congress. 1991. New Opportunities for U.S. Universities in Development Assistance. Office of Technology Assessment, OTA-BP-F-71, Washington, DC.

UNESCO/World Bank. 2000. Higher Education in Developing Countries: Peril and Promise. The Task Force on Higher Education and Society. World Bank, Washington, DC.

USAID. 1999. Evaluation of Institutional Development for Agricultural Training (IDAT). WPDATA\REPORTS\1699-027\027-002.W51

Wanni, N., A. Hinz, and R. Day. 2010. Good Practices in Educational Partnerships Guide: UK-Africa Higher and Further Education Partnerships. The Africa Unit/ACU, UK/Africa Partnerships in HE/FE. http://www.hea.ie/files/good_practice_Guide-1.pdf

Welsch, D., J. Flora, H. Foth, T. Westing, and G. Hansen. 1987. Malawi: Bunda Agricultural College. AID Project Impact Evaluation Report No. 64.

Wiley, D. and C. Root. 2003. Educational Partnerships with Foreign Institutions for Increasing the Quality of International Education in the United States. National Conference on Global Challenges and U.S. Higher Education National Needs and Policy Implications, January 24.

World Bank. 2008. Accelerating Catch Up: Tertiary Education for Growth in sub-Saharan Africa. World Bank, Washington D.C.

Yizengaw, T. 2008. Challenges of Higher Education in Africa and Lessons of Experience for the Africa-U.S. Higher Education Collaboration Initiative. National Association of State Universities and Land-Grant Colleges (NASULGC). Working Paper.

# CHAPTER SIX

# SHAPING TOMORROW'S LEADERS TODAY: THE EARTH UNIVERSITY MODEL

DANIEL SHERRAD

This chapter introduces an innovative new model for agricultural higher education, the *Escuela de Agricultura de la Region Tropical Humeda*, known as EARTH University. EARTH was founded in Costa Rica in 1990; its founding was based upon the conviction that traditional programmes in agriculture and natural resource management were not equipped to prepare professionals with the skills needed to promote sustainable development in the region.

There are many ways in which the principles of EARTH' founding differ from traditional programs. Where existing programmes stressed specialisation, EARTH proposed the formation of generalists with a holistic view of agriculture. While other programmes viewed agricultural production and environmental protection as competing agendas, EARTH proposed that they be seen as compatible and interdependent. Rather than the traditional focus on the acquisition of knowledge, EARTH's model stressed the construction of knowledge and the importance of experience; and where other programmes tended to focus on training professionals for careers in the public sector, EARTH's programme included entrepreneurship as a central pillar of the curriculum. While traditional programmes selected students based on the results of test scores, EARTH sought candidates for admission based on diverse criteria including leadership potential, vocation, values, and commitment to working for positive change in their societies. Rather than the traditional location within comprehensive universities in urban settings, EARTH was purposely established in a relatively remote rural setting where students and faculty would have the opportunity to participate continuously in agricultural production and to interact intimately with the rural communities surrounding the University.

African leaders have countless times articulated the importance of considering different models of agricultural higher education. A review of the literature makes it clear that many African institutions are not responding effectively to the development needs of their countries (Adisa and Oladele, 2011; World Bank, 2007; CHEA, 2010; UNDP, u.d.; Idabacha, 2003).

Many of the professionals graduating from agricultural programmes lack the skills, attitudes, and knowledge required by potential employers; and even more, they lack the leadership qualities needed to effect change in the agriculture sector.

The purpose of this chapter is to stimulate reflection and share experience, not to suggest that the EARTH model be copied in Africa. Yet in this context it would seem to have relevance. It is a model that places a high priority on building linkages with rural communities and connecting the university with its stakeholders, while at the same time building effective partnerships that align institutional programmes with national priorities. In contrast to educational models that stress knowledge transfer through memorisation and rote learning, EARTH's use of learning models that focus on student learning and the construction of knowledge through experience and reflection also seem pertinent.

African higher education faces tremendous challenges in promoting agricultural development. Institutions of higher learning are expected to produce the human resources required to lead development efforts, the research required to power development, and the outreach and extension that will make the results of research accessible to farmers and communities. Yet these same institutions are confronted with a host of constraints that severely limit their ability to meet these expectations, such as declining enrolments, professional isolation, inadequate financial support, and aging infrastructure (World Bank 2007; UNDP, u.d.; Idabacha, 2003; Juma and Ye-Cheong, 2005; Temu, u.d.; Teferra and Altbach, 2004; Lindow, 2012).

Another set of constraints—one that has less to do with resources and more to do with the conceptual framework underlying the model of higher agricultural education—is equally important to consider. These constraints are the pedagogical approaches that emphasise teaching rather than learning and agricultural production at the expense of considering the true multi-functionality of agriculture. Teaching is presented with a strong disciplinary focus and an ivory tower mentality—features of the university model inherited from Africa's colonial era (World Bank, 2007; Hansen, 1990). Clearly there are many innovations taking place in African universities, just as there are on other continents, but old habits and structures die hard.

As noted in Chapter 1, the "traditional" model of agricultural higher education, which is still widely applied in African institutions, stresses the acquisition of scientific and technical knowledge, with a focus on disciplinary training in the agricultural sciences. In an age when the generation of knowledge occurs at an accelerated pace and information is increasingly accessible, the

***Key Messages***

- The image of agriculture has to undergo a metamorphosis. As part of this the *AET system should deliver "food systems professionals"* who are able to engage with diverse issues, from production to environmental and social concerns. They should be stewards of the planet and key players in rural development processes. AET programmes should evolve to help graduates play such roles.
- *Higher education models and programmes need to be established cognisant of,* and addressing *the context* in which the institution operates. For EARTH these challenges were widespread poverty, the poor state of the sector, and the environmental crisis.
- A *clearly defined institutional mission should direct programming and help establish a new and inspiring image* of its graduates. With entrepreneurship as a central pillar, and while maintaining a focus on the needs of employers and other stakeholders, EARTH programmes reflect the key concepts in their mission: *leadership, ethical values, sustainable development, and construction of a prosperous and just society.*
- A generation of *leaders in agriculture* has to be cultivated to meet the challenges of the twenty-first century. This requires a *focus on sustainability, generalism and holism, entrepreneurship,* viewing and treating agriculture and *the environment* as interdependent and compatible, constructing knowledge and experience in the learning environment instead of passively acquiring it, accessing information based *on inter- and multidisciplinary approaches.*
- Such graduates should be seen as leaders in terms of commitment, potential and spirit, and as citizens promoting and constructing change and development, and with a commitment to participating in positive change. They *can deal with complex and multidimensional issues, integrate knowledge from different disciplines, and focus on sustainability.*
- EARTH training is expensive, but funding can be generated in a variety of ways. *Return on investment is judged to be great if the graduates can generate new economic activity, create jobs, contribute effectively to policy initiatives, or strengthen civil society.*
- *Success factors* for this type of model include careful selection of students; excellent, committed staff with an international profile; sequenced course design; strategic incentive structures; student-centred, experiential learning; strong linkages between the university and its rural stakeholders; engagement in entrepreneurial projects, internships, and scholarships; an active alumni network; partnerships that align institutional programmes with national priorities; and funders willing to take risks.
- *In traditional universities change is difficult.* Yet piecemeal application of components of EARTH's model might not work; it is the confluence of the whole approach that creates conditions for effective implementation. New models of higher education may therefore be needed to overcome ingrained resistance to change.
- Attempts to establish a similar model in Africa have failed. While proponents of change may learn from the EARTH model, *attempts to implement the model in Africa should be led, owned, and supported by Africans.*

ability to learn autonomously, to process information critically, and to integrate information and knowledge from different disciplines is increasingly important. While there is certainly a need for disciplinary specialisation, especially at the masters or doctorate level, the challenges of rural development and the need to transform African agriculture would seem to indicate the need for a cadre of professionals with a more holistic outlook and generalist formation (Maguire, 2000; Clark, 2006).

## EARTH University in Context

In the mid-1980s, a group of Costa Rican leaders from government, agro-industry, and academia began discussing the need for an institution that would prepare an agricultural professional with a profile distinct from that of graduates of existing universities. Essential elements of this profile included technical and scientific competence, environmental and social awareness and commitment, entrepreneurial skills, and strong ethical values. With resources provided by the W.K. Kellogg Foundation, a high level task force was formed. The task force conducted a needs assessment and feasibility study in the region and determined that there was indeed a need for such professionals as well as a burgeoning interest in new educational approaches, and that it would be entirely feasible to establish such a programme. In 1990, with significant financial support from the United States Agency for International Development (USAID), the W.K. Kellogg Foundation and the Government of Costa Rica, EARTH opened its doors to its first students.

EARTH is a private, international, four-year undergraduate institution, governed by an international Board of Directors. It operates under a special law of the Republic of Costa Rica. Located on a thirty-four hundred hectare property in the lowland Caribbean region, traditionally the home of large-scale banana production for export, EARTH is home to four hundred students and forty faculty, all of whom live on campus. EARTH is accredited by the National Accrediting Agency of Costa Rica (SINAES).

The University's endowment fund finances about 43 percent of the annual operating budget. Other sources of income include fundraising, tuition, sponsored projects, and commercial activities. Commercial activities include the export of approximately twenty-eight million pounds of bananas per year to the U.S. market, along with other products such as pineapples, hearts of palm, agricultural inputs, and paper produced from banana waste. An essential element of the commercial ventures has been the University's pioneering role in championing sustainable practices in banana production: implementing buffer zones to protect waterways, reduced use of agrochemicals, and the recycling of wastes, all of which have become standard practice in the industry.

EARTH's four hundred students come from approximately twenty-six countries of the Americas and Africa (African students were first enrolled in 2001 and currently constitute about 4 percent of the student population). More than half the students are from predominately rural, very resource-poor families and study thanks to scholarships provided by governments, foundations, companies, and individuals. An additional 30 percent finance their studies through partial scholarships, loans and family payments. Fewer than 20 percent of the students pay the established cost of tuition, room and board. All students are selected through a highly individualised admissions process that admits approximately 110 students per year out of twelve to thirteen hundred applicants. Selections are based on individual and group interviews, secondary school qualifications, and an admissions exam. Personal qualities, leadership potential, and vocation for a career in agriculture and natural resource management are key considerations in the admissions process. All professors are actively involved in the admissions process and personally travel to the applicants' countries to interview candidates.

The forty members of the University's faculty are selected internationally, and while the majority are from Latin America, there is representation as well from North America, Europe, Africa, and Asia. Selection is based on academic excellence, experience in industry, and interest in participating in an educational system based on student-centred learning and leadership development.

Teaching at EARTH is not for everyone. Living in a rural location in the heart of the humid tropics, working long hours, and spending significant time in the field as well as developing close relationships with students requires a very particular profile. Over 65 percent of the faculty have doctoral degrees. The rest have master's degrees or are in the process of completing their doctoral studies.

## The EARTH Model

EARTH educational model reflects the context in which the institution was created as well as the conviction that a new approach to preparing professionals for the agriculture sector was needed. The three contextual elements that were particularly important in the establishment of EARTH were widespread poverty, the state of the agriculture sector, and the environmental crisis.

*Poverty.* When EARTH was conceived, widespread poverty characterised the region, particularly rural poverty that in large measure was responsible for the multiple civil conflicts that convulsed the region during the 1980s. The University was founded on the belief that only through equitable rural development could conflicts such as the civil wars in Guatemala, El Salvador, and

Nicaragua be prevented in the future. EARTH was envisioned as an institution that would prepare leaders capable of promoting sustainable rural development while at the same time providing a space where young people from various countries of the region could learn from Costa Rica's example how to build a strong, resilient democracy.

*Agriculture sector transformation.* Agriculture plays a key role in the economies of almost all Central American nations, the exceptions being El Salvador and Panama. In the case of Costa Rica, the 1970s and 1980s saw the traditional reliance on export crops such as coffee, bananas, and sugar begin to give way to ever increasing agricultural diversification. At the same time, structural adjustment policies implemented in the face of the 1980s economic crisis resulted in a shrinking role for government and a consequent fall in employment opportunities for agricultural professionals in the public sector. Extension services and other programmes traditionally provided by the state were being reduced and, in some cases, privatised. At the same time, agricultural diversification resulted in the creation of new employment opportunities in the private sector for graduates of agricultural programmes. The late 1980s also saw an increasing demand for "sustainably" produced commodities; and niche marketing—organic, fair trade, shade grown, and more recently "whole trade"—began to gain importance in agricultural export markets.

*Environmental crisis.* Home to one of the world's highest concentrations of biodiversity, Central America's rampant deforestation has had serious ramifications for species survival, hydrologic flows, and a host of other ecological issues. From the late 1970s through the mid-1990s, Costa Rica had one of the world's highest deforestation rates and experienced an estimated loss of about 35–40 percent of its forest cover (Bishop et al., 2002). Much of this loss occurred as a result of clearing land for agriculture and pastures. The founding of EARTH reflected a deep concern for reversing these trends and a commitment to finding ways to make agricultural production and the conservation of the natural environment compatible.

In response to these contextual elements, EARTH was charged with providing the region with leaders who would possess strong social and environmental awareness and commitment, who would be instilled with the values needed by to construct a prosperous and just society, and who would have the entrepreneurial skills to work successfully either in existing enterprises or in launching their own projects.

## Student Recruitment and Admissions

EARTH University is committed to providing educational opportunities for young people who share the institution's commitment to building a better future. In many industrialised and developing countries, agriculture is viewed as a second-class pursuit not worthy of the best and most motivated students. In far too many cases, agriculture represents a student's second, third, or even fourth choice of study. Thus, many of the students enrolled in agricultural programmes have little interest in their studies and even less intention of eventually working in the field. For this to change, admission processes need to be altered. Instead of admitting students solely on the basis of test scores and secondary school grades, attention should also be paid to candidates' leadership skills, values and ethics, and vocational aptitude for a career in agriculture.

The quality of graduates depends to a great extent on the capacity of the admissions process to attract motivated, qualified candidates. Very often, highly motivated and talented young people from low-income families and remote rural areas are not encouraged to pursue higher education, and as a result the available spaces are filled by candidates from urban areas and more privileged backgrounds who may not always share the same level of motivation. Admissions processes need to identify qualified, motivated young people regardless of their origin and develop mechanisms to provide them with opportunities.

*Recruitment strategy.* EARTH's recruitment strategy is unique in important ways. First, it is believed that for faculty members to identify completely with the students and to understand their origins, it is important for them to participate actively in the recruitment and selection process. To that end, EARTH faculty travel to the countries from which students are recruited and visit communities to encourage qualified young people to apply. Every year, they visit communities in remote and often neglected regions, such as the Atlantic Coast of Nicaragua and Honduras, Madre de Dios in Peru, Beni in Bolivia, Chocó in Colombia, and Esmeraldas in Ecuador. Thanks to generous support provided by governments, foundations, companies, and concerned individuals, EARTH is able to provide scholarships to promising young people identified during these visits. Approximately 80 percent of EARTH's students have full or partial scholarship assistance.

*Selection process.* EARTH's selection process is based largely on an evaluation of the candidate's potential as a future agent of change. While EARTH uses test scores and secondary school grades as criteria in the selection process, the critical variable is a faculty evaluation of each candidate based on group and individual interviews.

*Helping her country.* Carina Théodore is from Pétionville, a suburb of Port-au-Prince, Haiti. She heard about EARTH University while in high school. After applying and during her interview, it became clear that Carina felt a strong commitment to working for the welfare of her country. The professor who interviewed her was particularly impressed with her volunteer work with the Red Cross in first aid and in a project called Ansanm Nou Kapab (ANK) (Together We Can) promoting sexual education as well as HIV and AIDS prevention among young people.

Carina felt particularly fortunate to have been accepted at EARTH because the 2010 earthquake destroyed many of the already inadequate higher education facilities in Haiti. Without the scholarship she was awarded to study at EARTH, it seemed her dream of studying at the University could never be realised.

Together with her fellow students from Haiti, she is hoping to introduce bio-gas technology to her country when she graduates. She believes that biofuels can contribute to reversing the terrible environmental consequences of deforestation in Haiti and also that what she is learning about crop production in Costa Rica can provide opportunities for rural reconstruction.

## A New Approach to Agricultural Higher Education

The image of agriculture must undergo a metamorphosis. If higher education programmes in agriculture are to become attractive to prospective candidates, the image of agricultural professionals must evolve and embrace new realities. The new approach must focus on food systems professionals who are involved in production, processing, marketing, nutrition, research, and a host of other areas. Environmental and social concerns must also be seen as their domain.

Those who manage food system are stewards of the planet as well as key players in rural development processes. These roles need to be highlighted for young people who are considering their futures, and university programmes in agriculture and natural resource management should evolve to prepare graduates to play such roles.

EARTH has only one course of study, and all students graduate with the same degree—a *licenciatura* in agricultural science. The *licenciatura* is roughly equivalent to a Bachelor of Science (BSc) degree in other systems, although the plan of study includes a larger number of credit hours. EARTH's educational model reflects both the reality of its Central American context and the results of an analysis evaluating the state of agricultural higher education both within and outside of the region. Many of the current constraints in the African context were (and continue to be) present in the Central American region and were considered in this analysis. Existing universities' relative isolation from the agricultural and rural sector, their strong focus on disciplinary learning as opposed to a more generalist and integrated approach, their emphasis on "book learning" and memorisation without opportunities for practical experience, and pedagogy that tends to focus on the teacher while relegating the learner to a relatively passive role were among the elements considered.

EARTH's programme was designed to provide the basic skills and knowledge needed for higher learning and application. An understanding of the theory that underlies practice is essential, and courses that provide this basic store of knowledge are included in the plan of study. Nevertheless, a concerted effort has been made in the curricular design to place this material within a multi-disciplinary context in relation to agricultural production and natural resource management. The focus at EARTH is to educate students as generalists who understand the interactions of the various components making up the agricultural system.

The plan of study at EARTH provides students with an overview of agriculture and its role in society during their first year. At the same time, the development of fundamental skills and positive attitudes towards work are emphasised. In the second year, coursework focuses on the development of technical skills and knowledge building. This process continues in the third year, culminating in an internship experience where students have the opportunity to apply their acquired skills and knowledge in a "real world" situation. During their fourth and final year, particularly in their professional experience course and their graduation projects, students are expected to demonstrate their capacities for analysis and synthesis, professionalism, independence, and responsibility.

## Four-pillar Educational Model

EARTH's educational model rests on four pillars: scientific and technical competency, development of social and environmental consciousness and commitment, personal development (including values and ethics), and entrepreneurship.

## Scientific and technical competency: Imparting knowledge, skills and abilities that guarantee EARTH graduates will have the technical competence to manage agricultural systems sustainably

EARTH's curriculum is broad in order to prepare its students to be "generalists" in the agricultural sciences, although it specifically focuses on important aspects related to the management of natural resources. Graduates are expected to go to work after graduation in areas such as managing agricultural production operations, engaging in commercialisation activities, or working with NGO's in rural development efforts and related areas. The wide-ranging knowledge they receive from their university education will enable them to integrate knowledge from different disciplines and to deal effectively with complex and multi-dimensional issues.

## Social and environmental consciousness: Instilling in the student a sense of social and environmental responsibility

Raising students' awareness of and commitment to the well-being of other people and the environment is a complex process that takes place through formal academic activities, organised community activities, and interactions with the natural environment. Service learning and continuous exposure to environmental issues are critical strategies EARTH utilises to engage students.

EARTH views its relationship with the communities of the tropics as a two-way bridge in which knowledge and experience flow outward to the greater community (including local populations, institutions, business, other universities, and countries) as well as inward to the university community. In contrast to the traditional view of the academy as an ivory tower, EARTH has worked to remove barriers between the university and the larger community and established a learning institution that listens and learns from all people, striving to teach by example.

EARTH's community outreach programme has made community development an integral part of its learning system. Community outreach is considered an important means for developing critical professional skills in students while at the same time working to improve the quality of life in nearby rural communities. Frequent student interaction with rural families fosters greater understanding of local dynamics and prevailing social and economic realities, thus creating professionals who are more knowledgeable and socially aware, and better able to work effectively with rural communities.

EARTH's Continuing Education Programme provides leadership to community outreach activities involving students and faculty. The faculty's presence in the field reinforces the idea that community work is everyone's responsibility. EARTH adapts its curriculum to the local region to ensure that its community work responds most appropriately to local needs. The Continuing Education Programme, which includes a small conference center and hotel, also coordinates development projects at the national and regional level. Recent projects have included training grassroots cooperatives and producer associations throughout Central America and Mexico in financial literacy and management and the development of selected value chains in Northern Costa Rica.

EARTH has contributed to the development of sustainable agriculture through student and faculty efforts to encourage more intensive use of farm resources. Some examples include installing bio-gas digesters, converting waste materials into organic fertiliser, teaching appropriate pesticide management, promoting organic agriculture practices, and providing technical support in the form of soil and market studies. In addition, a significant amount of effort has been directed toward the establishment and support of a sustainable organic farm, which is used as a model for integrated production and a place for the exchange of ideas among farmers, students, and faculty members. The organic farm is also used as an experiential learning centre for students and farmers.

## Personal development: Development of the leadership potential of every student

Personal development involves students learning about themselves as well as acquiring skills related to building interpersonal relations. The attributes that are considered essential to leadership-building include empathy, honesty, respect for the opinions of others, tolerance, intercultural understanding, teamwork, effective communication, and development as a lifelong learner.

EARTH's model is designed to be oriented towards personal development. The relatively low student-faculty ratio, the residential setting, careful selection of professors who genuinely care about students' integral development, and the selection of students who have demonstrated a concern about the future of their communities, nations, and the planet are all critical components of this personal development orientation. While classes and other formal academic activities are important spaces for promoting students' personal development, informal student interactions with other students, professors and staff, contact with communities and other off-campus activities are all essential opportunities for students to learn about themselves and their relations with others.

### Entrepreneurship: Encouraging a proactive search for opportunities and the ability to resolve problems

Entrepreneurship requires that students develop their capacities to evaluate, plan, organise, administer, and execute. The university's Entrepreneurial Project course provides an important opportunity for students to gain practical and theoretical skills in entrepreneurship during their four-year plan of study. Entrepreneurial formation is also strengthened in other courses that stress the skills, knowledge, and attitudes required by successful entrepreneurs.

## EARTH's Educational Principles and Coursework

### Educational principles

The educational principles upon which the educational model is based include student-centred learning and experiential education.

*Student-centred learning.* At EARTH, students are expected to be active participants in their learning. No longer passive recipients of knowledge, they are converted into partners in the process. While engaged in following a structured curriculum, students are encouraged to take an active role in discussing and determining the content of their courses as well as the methods used to facilitate learning. Professors, in addition to sharing their knowledge and experience, design and apply strategies that maximise students' learning, taking into account different learning styles. When the emphasis is placed on learning, then "covering" material in a set period of time becomes relatively less important than guaranteeing that learning has occurred. Equipping students with the ability to learn autonomously is considered more important than transmitting a determined quantity of information.

*Experiential learning.* Through direct experience in the field, in the community, and in the laboratory, and through active participation in the classroom, students are able to construct knowledge, develop abilities, and derive meaning and relevance from their time at EARTH. Experiential learning focuses on process rather than content, making it particularly well suited to the complex and dynamic world of agriculture and rural development. Perhaps most important, experiential learning encourages the integration of knowledge and the acquisition of practical skills within the context of the "real world."

Experiential learning encourages the integration of information and the acquisition of practical skills within the context of real world problems (Bawden and Macadam, 1988). Experiential learning requires learners to alternate between the realms of experience and concept in attempting to understand their environment as they struggle to function more effectively within it. In EARTH's approach to learning, action is integrated from the very beginning and provides the motivation that many students require.

## Sequenced Course Design for Student Development

EARTH's curriculum integrates the institutional mission and the graduate profile with the educational principles. The plan of study can be characterized as an "upside down" approach in that students begin by focusing on systems and on gaining a holistic vision of agricultural production and the realities of rural life. From their first days at the University, students are confronted with the complexity of the production system and the role people play within that system. Later, as they advance in their studies, they increasingly concentrate on the parts of the system and the components (biology, chemistry, physics, etc.) underlying it.

The emphasis during the first year is on constructing knowledge and basic skills in agriculture and on gaining a systemic view. Communication skills and language are also emphasized, as is the development of intercultural competencies.

The focus in the second year is on the sustainable management of production, the development of knowledge and skills in the area of entrepreneurship, and intensive interaction with the communities that surround the University.

During their third year, students continue to interact with the community, and courses focus on the application of knowledge to the complexities of agricultural production and natural resource management. The third year culminates in a fifteen-week internship in which students have the opportunity to apply their knowledge and skills in a "real world" situation.

In the fourth and final year, students must demonstrate a capacity for analysis and synthesis as well as professionalism with a high degree of independence and responsibility. They manage different projects and programmes on the various campus farms and undertake a year-long graduation project. Students also have the opportunity to select elective courses that correspond to their professional interests.

While every course at EARTH incorporates the principles of experiential learning, the Work Experience, Professional Experience, Entrepreneurial Projects, and Internship courses best reflect the importance of this approach in EARTH's programme. These courses are organised and sequenced so that students develop and acquire the skills, knowledge base, and attitudes required of a successful professional in a progressive and logical manner throughout the four-year programme.

*Work Experience.* In the Work Experience course students engage in a series of modules in crop, animal, and forestry production on EARTH's academic farms, working with producers and their families in surrounding communities. In the first year, the work is largely manual and routine, and the experience centres on the acquisition of basic skills, work habits, and general knowledge and familiarity with production. In the second year, the focus changes to management strategies for these same activities. Through working with farmers and their families in neighbouring communities, students are introduced to outreach and extension. During the third year, they spend seven weeks in northwestern Costa Rica near EARTH's La Flor campus, living in small rural communities with families and actively engaging in a very different cultural and ecological environment. Students work on an individual basis with small, local enterprises, as well as in community development activities in the towns in which they live.

*Professional Experience.* During the fourth year of study, students embark on their Professional Experience course. First students identify work sites or activities on campus that in some way correspond with their career goals. Thereafter, each student is responsible for contacting the supervisors of the identified campus operation to request an interview and solicit "employment." If a supervisor agrees, student and supervisor together develop a work plan, and the student is then responsible for carrying it out, dedicating a minimum of ten hours per week to the "job."

*Entrepreneurial Projects.* The Entrepreneurial Projects course provides students the opportunity to carry out an entrepreneurial project from beginning to end. Small groups of four to six students of different nationalities decide upon a production activity, develop feasibility studies, borrow money from the University, and carry out their projects (including the marketing of their products). After repaying their interest-bearing loans the groups share the profits. Enterprise projects are crucial for 1) giving students hands-on experience in planning and carrying out commercial production projects; 2) instilling an entrepreneurial spirit in students; 3) giving students invaluable experience working in teams; and 4) bringing students into contact with the demands of the market. Projects range from production of crops and animals to processing and services. The Entrepreneurial Projects course begins in the first year and concludes in the third. As students progress through the process of developing their enterprises and implementing their plans, they simultaneously advance through a series of modules that provide the required theoretical knowledge. These modules focus on themes such as business organisation, accounting, and marketing.

*Internship.* A fifteen-week internship is required for all students in the third trimester of the third year of study and is one of the courses that exemplify EARTH's emphasis on experiential learning. The internship is an opportunity for students to put into practice all they have learned, to be exposed to a "real world" enterprise, and to function independently of the university. For many students, this is also a chance to begin making connections that may lead to employment after graduation. While many students complete their internships in their home countries, the international nature of the institution allows students the opportunity to follow their individual interests, even when those interests lead to destinations far from the student's own country.

## EARTH Graduates

The graduate profile established by the University emphasizes both "soft" and "hard" domains. Honesty, teamwork, tolerance, respect, and intercultural understanding are among the "soft" skills the University hopes to develop and strengthen in students during their four years of study; technical and scientific knowledge, communication skills, and relative fluency in English are included in the area of "hard" skills. In addition, the graduate profile stresses an orientation towards working in the private sector, ideally in the creation of new enterprises. The idea of service is an important element in the profile of an EARTH graduate, and giving back to the community is considered an essential responsibility of all graduates.

*Alumni as professionals.* As of December, 2010, EARTH had graduated 1,546 professionals. Of these, 93 percent were working in the Latin American region. This is important, because even though the issue of "brain drain" may not be as serious a problem in Latin America as it is in Africa, it nevertheless is a concern. Reflecting the programme's emphasis on preparing graduates for private sector employment or for developing their own businesses, over 72 percent of graduates are active in the private sector, with approximately 15 percent in a business they themselves created. While a significant majority of graduates (over 80 percent) are involved in occupations associated with the agricultural and natural resource sectors, the range of jobs within those sectors is considerable.

*Alumni follow-up surveys.* The University maintains an alumni relations office and attempts to track all graduates. Periodic surveys are conducted to seek graduate feedback, which informs decision-making processes including curricular modification. Based in large measure on graduate feedback,

EARTH has increased its focus on the study of English as a second language as well as the breadth of courses related to macroeconomics.

*Alumni and job creation, volunteerism and consulting services.* The 2009 graduate survey indicated that 22 percent of graduates were actively creating jobs—starting their own businesses or creating new business opportunities for their employers—and that for every graduate four new jobs had been created. The same survey revealed that over 50 percent of graduates were actively involved in volunteer activities in their communities, a number that compared favourably with 26 percent in the case of the U.S. population (U.S. Bureau of Labor Statistics, 2010).

An interesting trend that has emerged in recent years among EARTH graduates is the increasing number engaged in providing consulting services. The services they offer tend to be in the environmental area, including certification services (for example organic certification, International Standardization Organisation (ISO) social responsibility certification, Europgap), and consultancies regarding sustainable production practices and waste management for agricultural producers and municipalities.

## EARTH's Impacts

In its twenty-three-year history, EARTH has had an impact on agricultural higher education out of proportion to the size of the institution. The volume of visitors and high-level meetings and the number of symposia to which EARTH representatives are invited to present the University's model are indicators of the interest among educators, policy makers, and others in the University's experience. The increasing interest in including entrepreneurship as part of agricultural higher education programmes can be attributed in part to EARTH's pioneering of this concept. The World Bank *World Development Report 2008: Agriculture for Development* devoted a section to business education for the entrepreneurs of the "new agriculture." Likewise, EARTH has consistently insisted on other innovations such as the inclusion of the theme of sustainability and environmental conservation in agricultural higher education programmes—changes that are now becoming common.

***José Antonio Pacheco Tzul,*** who graduated from EARTH in 1994, is the fifth of seven siblings and grew up in the remote indigenous community of Patzarajmac de Totonicapán in western Guatemala. After attending a military/agricultural high school in Retalhuleu, Guatemala, José Antonio studied at EARTH University on a scholarship provided by the W.K. Kellogg Foundation.

Following his graduation in 1994, he briefly worked as the general manager of a banana operation in Tabasco, Mexico and later joined a major agricultural conglomerate in Guatemala. After approximately three years he began to seriously consider the possibility of *starting his own company.*

In early 2000, with his first product (a calcium-based fertiliser), he resigned from his job and together with a partner founded *Enlace Agropecuario S.A.* (ENLASA) with personal savings and family loans. By 2003, ENLASA had developed more than twelve product lines, and in addition to serving Guatemala and Costa Rica had begun exporting to Nicaragua, Panama, Honduras, Ecuador, and Chile. By 2010 the company offered more than fifty products for the agriculture and livestock sectors, operating offices in four countries and employing fifty administrators and technicians.

ENLASA's working capital has multiplied by a factor of 39 since the company's inception, and it boasts major multinational clients that generate hundreds of thousands of dollars in revenue for the company each year.

The company recently established the ENLASA Foundation, which seeks to improve access to higher education for economically disadvantaged young people and to promote excellence and improvements in established universities. The company already contributes to EARTH University's alumni-financed scholarship programme, known as *Hijos de la Tierra.*

The following two examples of international initiatives in which EARTH played a leading role serve to illustrate the University's impact on agricultural higher education.

## Sustainability, Education, and Management of Change in the Tropics Project (SEMCIT)

In 1999, EARTH, together with the Salzburg Seminar and Noragric of the Norwegian Agricultural University, launched the SEMCIT Project. Its principle activity was a series of seminars and workshops that brought together agricultural higher education leaders, policy makers, and private sector representatives to discuss the future of agricultural higher education and to devise recommendations for the future. The SEMCIT project produced a series of documents and papers that detailed strategic directions for agricultural higher education.

In the case of Africa, the *Jinja Consensus* developed at the SEMCIT seminar held in Jinja, Uganda (SEMCIT, 2002) called for changes in institutions of higher learning that would eventually lead to institutions that :

- focus on developing students with high ethical values as well as a vocational interest and commitment to rural activities, with graduates who are capable of independently accessing and using knowledge;
- operate agricultural programmes that provide students with a market orientation, leadership, and entrepreneurial and management skills that also embrace social and environmental responsibility;
- have student selection criteria and admission policies to ensure that agricultural faculties and universities prioritize students interested in careers related to African agriculture and rural development;
- utilize a student-centred approach to learning and discovery including: flexible and practical approaches to problem solving, effective communication skills, and strong linkages to rural communities and the developmental needs of key stakeholders such as women farmers;
- employ experiential learning methods that facilitate student development rather than simply transferring knowledge.

To achieve these goals, a three-pronged strategy was proposed: 1) develop networks of those interested in promoting change processes in existing agricultural education programmes; 2) transform existing programmes along the lines suggested above; and 3) create new institutions practicing innovative models that could serve as catalysts of change. While the SEMCIT Project formally ended in 2008, EARTH has since then been involved in various initiatives to promote the three-pronged strategy.

## Global Consortium for Higher Education and Research in Agriculture (GCHERA)

Launched in 1999, GCHERA brought together presidents, vice chancellors, deans, and other leaders of institutions of agricultural higher education to discuss issues of mutual concern. The President of EARTH presided over GCHERA from 2005–2007. In 2007, GCHERA held a conference in San José, Costa Rica, with the theme "Innovation and Leadership for Relevant

Change in Agriculture." The conference gave all participants the opportunity to visit the University and experience its educational model first-hand. During the meeting, participants were given the following challenge: "Agricultural higher education is in danger of becoming irrelevant, and change is needed. Universities must respond to society's changing needs, and agriculture must become more economically competitive, socially responsible, and environmentally sustainable in an increasingly globalized world economy" (GCHERA, 2007).

The subsequent GCHERA meeting hosted by Jomo Kenyatta University in Nairobi in 2009 provided some concrete examples of how universities are responding to this challenge. Many of the presentations at the conference focused on entrepreneurial initiatives, like those at the Faculty of Agriculture at Jomo Kenyetta University, while regional initiatives such as FARA discussed new initiatives to create university-based incubators linking with the private sector under the *Unibrain* initiative.

Current collaborative efforts include sharing the University's experience working with small producers and selected value chains and educating for entrepreneurship with the Regional Universities Forum for Capacity Building in Agriculture (RUFORUM) in East Africa. EARTH is also actively participating in efforts to develop innovative new institutions in Tanzania and Nigeria.

## Relevance of the EARTH Experience

There are a number of themes to take into account when comparing the EARTH experience and model to other institutions of higher learning in agriculture. These include institutional mission and tradition, issues related to the resistance to change so common in higher education, and cost. The model that has evolved during EARTH's twenty years of existence embodies an educational philosophy and conceptualisation of the University's role that reflect a particular context and time.

*Mission.* EARTH defines its mission as the preparation of leaders with ethical values to contribute to the sustainable development of the tropics and the construction of a prosperous and just society. The importance placed on the formation of leaders is the reason EARTH has chosen to maintain a student-to-faculty ratio of ten-to-one. EARTH promotes close interaction among faculty, staff, and students, and its plan of study emphasises the development of the skills required by effective leaders, including the formation of values and ethics.

Those who are selected to study at EARTH are considered to be among the elite—not elite as defined by income or social status, but elite in terms of commitment, potential, and spirit. Constructing a just and prosperous society implies that graduates will go beyond the typical boundaries of careers in agricultural production, processing, or commercialisation. They will assume roles as citizens promoting constructive change and sustainable development. The mission is also reflected in the amount of time students and faculty dedicate to working side-by-side with small producers, grassroots associations, and community development organizations. These experiences not only provide a richer understanding of the dynamics of rural development, but also establish empathy and a commitment to participating in positive change.

Research is an integral component of the EARTH model, particularly to the extent that it involves students and addresses the problems of agriculture and the management of the natural resources of the tropics. However, the integral development of leaders and active engagement in society are considered the most important aspects of EARTH's programme.

This approach differs from other models, particularly those that put greater emphasis on the generation of research results than on teaching and involvement in outreach. In such systems, faculty rewards and tenure tend to depend largely on factors such as publication records and grant funding received. In EARTH's system, quality of teaching, engagement with students, and involvement in outreach has a much greater relative value.

*Tradition and resistance to change.* Universities are institutions steeped in tradition, often reflecting the neoclassical model, and are very resistant to change. Traditionally, professors are seen as the arbiters of knowledge and are expected to remain aloof from society's problems; the student's role is to absorb the knowledge, values, and traditions embodied in the institution. While this description certainly does not apply to all, or perhaps even most universities, certain elements of this model are clearly evident in many institutions. A model such as EARTH's—which calls for active engagement in society's problems, professors who see themselves as facilitators of learning rather than dispensers of knowledge, and students who are engaged in active exploration and experiential learning—is very difficult to impose on more traditional systems.

*Cost.* In discussions of higher agricultural education in the developing world, cost and scalability are certainly relevant. Obviously, a low student-faculty ratio, the dedication of scarce resources to maintaining well-equipped laboratories, an emphasis on off-campus activities, and extensive opportunities for experiential learning, as well as recruiting and retaining a highly qualified international faculty, all translate into a relatively high annual cost per student. In the case of EARTH, the actual annual student cost is approximately US$30,000. Approximately 43 percent of this cost is provided by the institution's endowment, the rest through an active international fund-raising programme, commercial activities, student payments, and special projects. The 15 to 20 percent of students who actually pay the full tuition charged by the University pay approximately US$20,000 per year. While in the context of the developing world this is clearly a significant cost, it is important to consider the following.

- *Student retention:* About 85 to 90 percent of the students who enrol in the University also graduate. More than 90 percent of graduates complete their studies in four years. This statistic compares very favourably with other universities, in both the developing world and in industrialised countries where retention rates vary widely but rarely approach this level. High attrition rates directly affect cost efficiency and significantly raise the overall costs of a programme, yet are rarely factored into considerations of the cost per student.
- *Transparency:* As a private institution, EARTH receives no subsidies from governmental agencies, either in Costa Rica or internationally (although certain governments do provide scholarship funds). The per student cost is calculated simply by dividing the entire University operating budget by the number of students. In many public institutions, there are often significant subsidies involved that are not included in per student costs.
- *Other costs:* The cost per student at EARTH covers virtually all student costs for tuition, materials, room, and board. Students are responsible only for their personal expenses (clothing, personal hygiene, off-campus recreation, etc.). In addition, the academic year at EARTH begins in mid-January and continues through mid-December, with classes and field activities six days a week. Since many of the students come from impoverished families and travel a considerable distance to study in Costa Rica, the University makes arrangements for those unable to return home at the end of the year to stay on campus during vacations.
- *Return on investment:* While notoriously difficult to quantify, it is worthwhile to consider the return on investment of an education such as that offered at EARTH. If graduates of such a programme can, in the course of their careers, generate new economic activity, create jobs, contribute to policy initiatives that benefit farmers and the food system in general as well as strengthen civil society, few would disagree that it is a worthy investment. In the case of EARTH, this would seem to be the case. The University's graduates are returning to their home countries (85 percent); the majority are working in the private sector (62 percent), developing their own businesses (15 percent), and generating economic opportunities in rural areas (an average of four jobs created per graduate for those graduating between 1990 and 2005) according to the most recent graduate survey (2012).

- *Stewardship of resources:* In many cases, institutions and infrastructure are built in the developing world, often financed with international loans and grants, with little or no provision being made for routine maintenance and required renovation. While keeping short-term expenses down, the failure to invest in maintenance results in high long-term expenses as well as sub-standard conditions for students and faculty once the new facilities begin to deteriorate. Maintenance and equipment replacement represent significant expenditures in EARTH's annual operational budget and are reflected in the clean and well maintained campus that visitors frequently comment upon.

Still, US$30,000, or even US$15,000 per year is far beyond the means of most university students in Latin America or Africa. Clearly EARTH's model is costly. However, many elements of the model could be integrated into traditional institutions without necessarily altering student-faculty ratios and other characteristics of programmes in public institutions of higher learning. The utilisation of experiential learning strategies and other student-centred learning methodologies, adjustments in admissions criteria, increased opportunities for internships, and attachments in industry and rural communities are only some of the changes that could be adopted and that do not necessarily involve large cost increases.

## Adoption of the EARTH model

Over the course of EARTH's relatively short existence, there have been many conversations about starting a "new" EARTH, or "EARTH-like" institution, particularly in sub-Saharan Africa (Idabacha, 2003).

While the University is involved in two nascent efforts to create new institutions of agricultural higher education, EARTH also hosts individual faculty members and administrators from African institutions who seek to attempt changes in their own institution's practice. The incorporation of entrepreneurship in the curriculum, enhanced contact among students, farmers and the larger community, and the use of innovative learning strategies are all aspects of EARTH's programme that many visitors have studied and are attempting to implement.

## Conclusion

Many aspects of EARTH's model discussed in this chapter are not unique to EARTH. Student-centred learning has been at the heart of many educational reforms around the world over the past decades. Experiential learning was championed by John Dewey and others almost a hundred years ago, and the need to educate entrepreneurs for the rural sector is certainly not a new concept. What is new, perhaps, is the way in which EARTH has put these concepts together, applied them to the field of agricultural and rural development, and done so in the setting of a developing nation for the benefit of young people from developing nations of the tropics.

EARTH was created based on the conviction that the prevailing model for preparing the next generation of agricultural leaders was not appropriate to meeting the challenges of the twenty-first century. The University set out to develop an alternative model focused on forming leaders equipped to promote agriculture in a manner that is compatible with environmental conservation and social development. Over the course of the past twenty years, EARTH has demonstrated that a programme based on sound pedagogical principles and committed to forming a new generation of leaders is capable of garnering needed support and providing opportunities for young people interested in building a more prosperous and just society.

The EARTH model, while reflecting the context in which it was developed, has nevertheless been shown to be of interest and relevance in other contexts as well. The extent to which the model as a whole or its constituent elements can serve as guides for the change process in the African context is best determined by Africans themselves.

## References

Adisa, R. and O. Oladele. 2011. Agricultural Education and the Challenges of Poverty Reduction and Sustainable Development in Developing Countries: A Nigerian Case Study. *Journal of Agricultural Research and Development* 10: 2, 57–74.

Bawden, R. and R. Macadam. 1988. Towards a University for People Centered Development: A Case History of Reform. Unpublished paper for Winrock International, 1988.

Bishop, J., N. Landell-Mills, and S. Pagiola. 2002. Selling Forest Environmental Services: Market-based Mechanisms for Conservation and Development. Earthscan Publications Ltd: London.

Clark, N. 2006. Application of the Innovation Systems Perspective in the African Higher Education Sector: Experiences and Challenges. Paper presented at the Innovation Africa Symposium, Kampala, Uganda, November 21–23, 2006.

GCHERA. 2007. GCHERA Fifth International Conference. http://www.gchera.com/past-global-conferences/2007-san-jose-costa-rica/

CHEA. 2010. Ministerial Communiqué issued at the Ministerial Conference on Higher Education in Agriculture in Africa (CHEA). November 15–19. Kampala, Uganda.

Hansen, G. 1990. Beyond the Neoclassical University: Agricultural Higher Education in the Developing World—An Interpretive Essay. USAID Programme Evaluation Report No. 20: Washington, DC.

Idabacha, F. 2003. Creating the New African University. Paper prepared for the international seminar series on Sustainability, Education, and the Management of Change in the Tropics (SEMCIT). *www.changetropics.org*

Juma, C. and L. Ye-Cheong. 2005. Innovation: Applying Knowledge in Development. UN Millennium Project Task Force on Science, Technology and Innovation: London.

Lindow, M. 2012. Weaving Success: Voices of Change in African Higher Education. Institute of International Education, New York.

Maguire, C. 2000. Agriculture Education in Africa: Managing Change. Paper prepared for workshop sponsored by Sasakawa Africa Association: Cape Coast, Ghana.

SEMCIT. 2002. Jinja Consensus on the Need for Change in Higher Level Agricultural Education in Africa. http://www.iln-africa.net/uploads/documents/resource/ SEMCIT_Jinja_Consensus.pdf

Teferra, D. and P. Altbach. 2004. African Higher Education: Challenges for the 21st Century. Higher Education 47: 21-50.

Temu, A. u.d. The Role of Agricultural Education in Food Security for Africa. World Agroforestry Centre (ICRAF): Nairobi. http://academic.sun.ac.za/biofuels/Media%20info/August%20Temu%20-%20Food%20Security%20and%20education.pdf

United Nations Development Programme (UNDP). u.d. Skills and Institutions for Sustainable Development. Capacity Development Group: New York.

United States Bureau of Labor Statistics. 2010. http://www.bls.gov/news.release/volun.nr0.htm

World Bank. 2007. Cultivating Knowledge and Skills to Grow African Agriculture: A Synthesis of an Institutional, Regional and International Review. World Bank, Washington, DC.

—. 2008. *World Development Report 2008: Agriculture for Development.* World Bank, Washington, DC.

# CHAPTER SEVEN

# AGRICULTURAL EDUCATION AND TRAINING EXPERIENCES IN EUROPE

LUC D'HAESE

## Overview of Education in Applied Life Sciences in Europe

Life sciences education is an overarching term for five different disciplinary areas: agriculture, food, rural development, natural resources, and environment. Each discipline is composed of sub-disciplinary areas (Table 7.1).

Each university creates its own study programmes and specialisations. Since the Bologna Declaration (referred to below), higher education in Europe has been divided into two cycles: a bachelor's programme—comprising academic and professional—and a master's programme. The academic bachelor's programme focuses on study of the basics in science. Upon earning such a degree, the student generally enters a master's programme to specialise in one of the disciplines mentioned in Table 7.1. In many institutions, students choose and compile parts of their own study programmes. The master's degree provides sound analytical and creative skills, guaranteeing the graduate flexibility in the labour market. By contrast, the professional bachelor's programme is designed to educate and train specialists in a specific disciplinary field of the industry.

There are two main types of universities: multidisciplinary and unidisciplinary. Multidisciplinary universities cover life sciences research and education. Often, different faculties in the same university will be involved in life sciences research and education. At unidisciplinary universities, life sciences research and education is the university's main purpose. Table 7.2 provides examples of multi- and unidisciplinary universities.

The Bologna Process, a European initiative promoting the transparency and compatibility of higher education across Europe, defines three cycles of higher education: the bachelor's and master's cycles mentioned above, and a doctoral cycle. The European Credit Transfer and Accumulation System

(ECTS) for recognition and accumulation of credits allows students the mobility to continue their studies and obtain a more advanced degree in another country.

***Key Messages***

- The European Union strategy calls for programmes that stimulate knowledge, creativity and innovation. These are believed to be the *EU's most valuable assets,* particularly as global economic competition becomes more intense in all sectors.
- The European Union believes that international cooperation and the exchange of knowledge are of critical importance to development and prosperity and consequently promotes countries working together and learning from one another. A *common working platform* for higher education institutions was developed in support of this vision.
- Programmes such as Erasmus and Erasmus Mundus have proven to be particularly successful in enabling the *exchange of European and foreign students* among European higher education institutions. Substantial support for the mobility of teaching staff and joint scientific research facilitates *collaboration across borders,* which promotes better knowledge generation and the delivery of better-trained professionals.
- The *Bologna Process*—a series of ministerial meetings and agreements between European countries designed to ensure comparability in the standards and quality of higher education qualifications—has harmonised the teaching programmes yet also enabled comparability and benchmarking.
- *Good quality assurance* procedures and agencies are seen as keys to the success of the Bologna Process. The European Credit Transfer and Accumulation System (ECTS) sets a standard for comparing the performance of higher education students across the European Union and other collaborating European countries.
- *New and emerging issues* may affect the quality assurance and performance of such collaborative efforts. Such issues include:
  - achieving *balance between accountability and ongoing improvement* within higher education institutions,
  - achieving balance in the responsibilities shared among the different actors—higher education institutions, quality assurance agencies, and policy-makers;
  - appreciation for the *important roles of diverse stakeholder groups,* such as students and the business world;
  - the need to *provide adequate and useful levels and types of information* to diverse stakeholder groups;
  - finding ways of dealing with *increasing diversity*—pedagogies, institutions, students, expectations, missions—across the higher education system;
  - the *internationalisation of higher education* that is frequently coupled to increasing commercialisation and competition; and
  - the burden caused by *increasing bureaucracy and the cost of quality assurance.*
- The field of life sciences—and hence the agricultural sector—engages professionals in a variety of endeavours. It is constantly changing and at times, changing rapidly. Programmes and their related collaboration and quality assurance initiatives therefore have to *keep evolving* as need and demand change.

| Agricultural sciences | Food sciences | Rural development |
|---|---|---|
| • Plant science<br>• Animal science<br>• Crop science<br>• Horticulture<br>• Viticulture<br>• Fishery and aquaculture<br>• Tropical agriculture | • Food technology<br>• Risk analysis<br>• Human nutrition | • Sociology<br>• Policy<br>• Law<br>• Economics |
| **Environmental sciences** | **Natural resources** | **Biotechnology** |
| • Environmental science and technology<br>• Soil science<br>• Water science | • Forestry<br>• Wood technology<br>• Landscape archi-tecture | • Biochemistry<br>• Process technology<br>• Genetics |

**Table 7.1. Overview of the common study areas per discipline**

| University | Multi-disciplinary/ Uni-disciplinary | Faculties, life scienc-es R & E |
|---|---|---|
| Spain: Universitat Politechnica de Valencia | Multidisciplinary | 2 |
| Italy : Universita degli Studi di Pisa | Multidisciplinary | 1 |
| Hungary: Corvinus University of Budapest | Multidisciplinary | 3 |
| Austria: Universitat fur Bodenkultur, Vienna | Unidisciplinary | 15 |
| Czech Republic: Czech University of Life Sciences, Prague | Unidisciplinary | 5 |
| Denmark: University of Copenhagen, Faculty of Life Sciences | Unidisciplinary | 7 |
| Sweden: Swedish University of Agricul-tural Sciences, Uppsala | Unidisciplinary | 4 |
| The Netherlands : Wageningen Universiteit | Unidisciplinary | 4 |
| France: Rennes Agro Campus Ouest | Unidisciplinary | 11 |
| Lithuania: Lithuanian University of Agriculture | Unidisciplinary | 5 |

**Table 7.2. Overview of the fragmentation of different institutions**

# Teaching Agricultural Sciences in Higher Education in Europe

## Content of the study programmes

Life sciences are multidisciplinary by nature, combining biological, economical, ecological, and technical studies with an integrated approach that requires both technical and analytical skills. Each university listed in Table 7.2 offers a variety of subjects and programmes related to life sciences.

Within life sciences, European study programmes in agricultural science are based on a sound knowledge of base sciences and of primary production areas. For example, plant and animal production studies also include specific study areas such as agricultural economics, soil science, crop protection, animal nutrition, and genetics. Exact sciences are combined with problem-solving and creativity training. Many European universities include business and communication training skills, which are indispensable for participation in modern agro-industry, in their study programmes.

## Enhancing international experience

International experiences such as studying, interning, or conducting research in a foreign country are strongly promoted in higher agricultural education in Europe. Opportunities will vary depending upon the university itself and on how it involves itself in international networking.

There are two different approaches to studying abroad. Students can spend one semester or one year of their study programme at a foreign university, or they can participate in an international programme. In the first option, the students take part in an exchange programme—the European Community Action Scheme for the Mobility of University Students (Erasmus) or Erasmus Mundus.[1] The programme was created to enable European students to do some of their study in a different European or non-European country. The Erasmus network is based on bilateral agreements between universities. The approval of these agreements is mainly based on internal procedures and experiences, and the number of agreements varies among universities. For example, Ghent University has bilateral agreements with 116 universities or institutions.

[1] Erasmus Mundus is an extension of the Erasmus programme. The difference between Erasmus and Erasmus Mundus is that Erasmus Mundus is the international, worldwide version of the Erasmus programme, which came into effect approximately twenty years ago.

Students also can obtain international learning experience in life sciences by joining one of the joint degree programmes developed since 2007. These programmes are offered jointly by several European universities, with each participating university responsible for part of the study programme. The students generally stay at different universities and are able to combine the experience and excellence of all participating institutions. There are several international two-year MSc programmes in the EU, and some are organised within the Erasmus Mundus framework. We will highlight two of these programmes, the *Agris Mundis* and the International Master's Programme in Rural Development (IMRD).

## Agris Mundis

Agris Mundis is a two-year MSc programme mainly focused on sustainable development in agriculture. It is offered by a consortium of five leading European universities offering studies in international agriculture and rural development, namely *Centre National d'Etudes Agronomiques des Régions Chaudes*, Montpellier (France), *Universitàdegli Studi di Catania* (Italy), University of Cork (UK), *Universidad Polytécnica de Madrid* (Spain), Wageningen University (the Netherlands), and the University of Copenhagen, Faculty of Life Sciences (Denmark). Agris Mundis trains students to cope with current global agriculture and rural development concerns. Graduates are able to identify and critically analyse key factors shaping the development of crops, management of forests and animal production, management of research and rural development projects, and finally, the sustainable use of natural resources and the environment. They are also trained to formulate and provide effective and appropriate responses to complex agricultural and natural resource issues.

## The International Master's Programme in Rural Development (IMRD)

The objective of IMRD is to train specialists in integrated rural development, with a focus on development's socio-economic and institutional aspects. This programme has a worldwide approach and courses that deal with developed, developing, and transitional countries. The programme is currently offered as a collaboration of seven different European universities: Ghent University (Belgium), Agro Campus Ouest (France), Humboldt University of Berlin (Germany), the University of Cordoba (Spain), Wageningen University (the Netherlands), Slovak University of Agriculture in Nitra (Slovakia), and the University of Pisa (Italy). Under the new Erasmus Mundus II programme, IMRD universities (except for the University of Cordoba) are now joined with universities in China, Ecuador, India, and South Africa.

## Partnerships at the international level

The Association for European Life Science Universities, formerly known as the Interfaculty Committee Agraria (ICA), is the largest European educational network for life sciences education. Founded in 1988, ICA includes a network of more than sixty universities of EU member states and other European countries. Its aim is to create a platform for European universities to share experiences, cooperate in new ventures, and benefit from the resulting synergies. As an overarching network, it combines different networks based on different disciplines such as the agro-business MBA, agricultural knowledge and development, life sciences in general, agricultural engineering, forestry, and international relations.

Almost every year, a member of the ICA network organises a European Conference on Higher Agricultural Education to discuss topics related to study programmes such as competences, quality assessment, employment, and the linkages between education institutions and the industry. The general objectives of these conferences are to

- contribute to the European and global agenda for higher agricultural education;
- enhance cooperation through networking;
- share experiences and outcomes on projects; and
- improve the overall quality and relevance of the teaching and learning process.

The 2010 conference was held in June in Zagreb, Croatia, with the theme: "Bologna Declaration Ten years on—Promising and delivering quality in life sciences education across Europe."

The ICA also optimises the relationships among universities, training facilities, enterprises and NGOs. In the European vision, a university is both an educational and a research facility that should train qualified workers and perform objective scientific research for different target groups. A university provides a platform, consulting and training with industrial and the business worlds. Industry can clearly understand the needs of the market and may suggest specific research items to the university.

The Euroleague for Life Sciences (ELLS), another example of a cooperation network, links leading universities cooperating in the fields of natural resource management, agricultural and forestry sciences, life sciences, veterinary sciences, food sciences, and environmental sciences. The network offers international master's programmes, summer schools, intensive programmes, student exchanges, international networking and joint teaching, and lecturer mobility. It is a collaboration among seven universities: University of Natural Resources and Applied Life Sciences in Vienna, University of Copenhagen, Faculty of Life Sciences, Swedish University of Agricultural Sciences, University of Hohenheim, Wageningen University and Research Centre, Czech University of Life Sciences in Prague, and the Warsaw University of Life Sciences.

The case of Ghent University, Belgium, introduced below, illustrates the extent of international collaboration in life sciences education in Europe.

## Case study: Ghent University Faculty of Bioscience Engineering

Ghent University, located in Flanders, the Dutch-speaking region of Belgium, is a multidisciplinary university with 11 different faculties and 34 thousand students, of which eighteen hundred are enrolled in the Faculty of Bioscience Engineering (data as of 2010). The faculty has evolved from a purely agricultural engineering faculty to an applied life sciences faculty. Since 2004, the educational system in the faculty has changed to comply with the new standards of the Bologna Declaration. Every course has been evaluated and ECTS credits have been attributed to each. Five academic BSc degrees can be obtained at this faculty, and there are sixteen master's programmes. Of these, five are international course programmes (ICP's) and one is an Erasmus Mundus programme (Table 7.3). In addition, there is a one-year Master's in Environmental Sanitation and Management. Subjects offered in the PhD programme—agricultural science, environmental science, chemistry, forestry and land management, and biotechnology—are combined in the doctoral school's bioscience engineering programme. The doctoral school organises specialised courses, top-level training sessions, guest lectures, promotional events, and interaction with industry and other external stakeholders. A large variety of research and education takes place in this faculty, which is subdivided into 14 subunits or departments. Some departments are restricted to a specific study area such as plant production, while others, such as applied mathematics, overarch the different study areas. The faculty is very involved in international projects, both in education and research, and 32 percent of the faculty's student population is foreign. Foreign students are enrolled in one of the ICPs, the Erasmus Mundus programme IMRD, a PhD programme, or are regular Erasmus/Socrates exchange students (Socrates is an earlier version of the Erasmus programme).

Interaction between the industry and the faculty is promoted by research cooperation and internships for students. Frequent seminars and symposia bring together researchers and professionals, and the faculty organises an annual jobs fair to give students an introduction to the different possibilities on the labour market. They can meet industry representatives and listen to guest speakers who discuss their work experiences.

| BSc (offered in Dutch) | MSc (offered in Dutch) | International Master's (offered in English) |
|---|---|---|
| • Agricultural science<br>• Land and forestry management<br>• Environmental technology<br>• Biotechnology<br>• Chemistry and food technology | • Agricultural science<br>• Land and water science<br>• Forestry and nature management<br>• Environmental technology<br>• Microbiology and biotechnology<br>• Chemistry and food technology<br>• Chemistry and bioprocess technology<br>• Food and nutrition<br>• Environmental technology<br>• Cell and gene biotechnology | • Aquaculture<br>• Environmental sanitation<br>• Food technology<br>• Human nutrition and rural development with 3 main subjects:<br>• human nutrition, rural development and economics<br>Tropical agriculture<br>• Land resources engineering soil sciences<br>• Rural development<br>• Environmental sanitation and management |

**Table 7.3. Different programmes at the Faculty of Bioscience Engineering of Ghent University**

## Higher Education in Europe and the Bologna Process

### Bologna Process: a brief history

The Bologna Declaration, adopted by twenty-nine countries in June, 1999, aims to improve higher education in Europe. Today, at least forty-six countries are implementing the principles and objectives of the Bologna Declaration.

The Bologna Declaration's main objectives include (European Commission, 2003):

- adoption of a system of easily readable and comparable degrees, including the introduction of a diploma supplement to improve transparency;

- adoption of a system essentially based on two main cycles, undergraduate and graduate;
- establishment of a European Credit Transfer System (ECTS);
- promotion of mobility by overcoming obstacles of a free movement of students, teachers, researchers, and administrative staff;
- promotion of European cooperation in quality assurance; and
- promotion of the necessary European dimensions in higher education.

The predecessor of the Bologna Process, the Sorbonne Declaration, adopted by France, Germany, Italy, and the United Kingdom, drafted the main goals of the European Higher Education Area (EHEA) in 1998 and provided the necessary push toward the Bologna Declaration.

A Bologna Follow-up Group was formed with the signing of the Bologna Declaration. It was decided that ministerial meetings should take place every two years, with the first meeting held in Prague in 2001. Different countries organised the so-called Bologna Seminars to cover agreed-upon topics (listed in Figure 7.1).

## The Prague Seminar, 2001

The Prague Summit introduced several new elements:

- recognised students as full and equal partners in the decision-making process; National Unions of Students in Europe (ESIB) became consultative members of the Bologna Follow-up Group together with the Council of Europe, the European University Association (EUA), and the European Association of Institutions in Higher Education (EURASHE);
- stressed the social dimension of the Bologna Process; and
- highlighted the idea that higher education is a public good and a public responsibility.

Between 2001 and 2003, more Bologna seminars were organised. The EUA developed its Trends III report in 2003 (following two previous reports in 1999 and 2001) and started a Quality Culture Project for higher education institutions.

In addition, the European Commission supported several European projects with the aim of improving quality assurance. The joint master's programmes were launched and the ESIB completed several student surveys on the implementation of the Bologna Declaration.

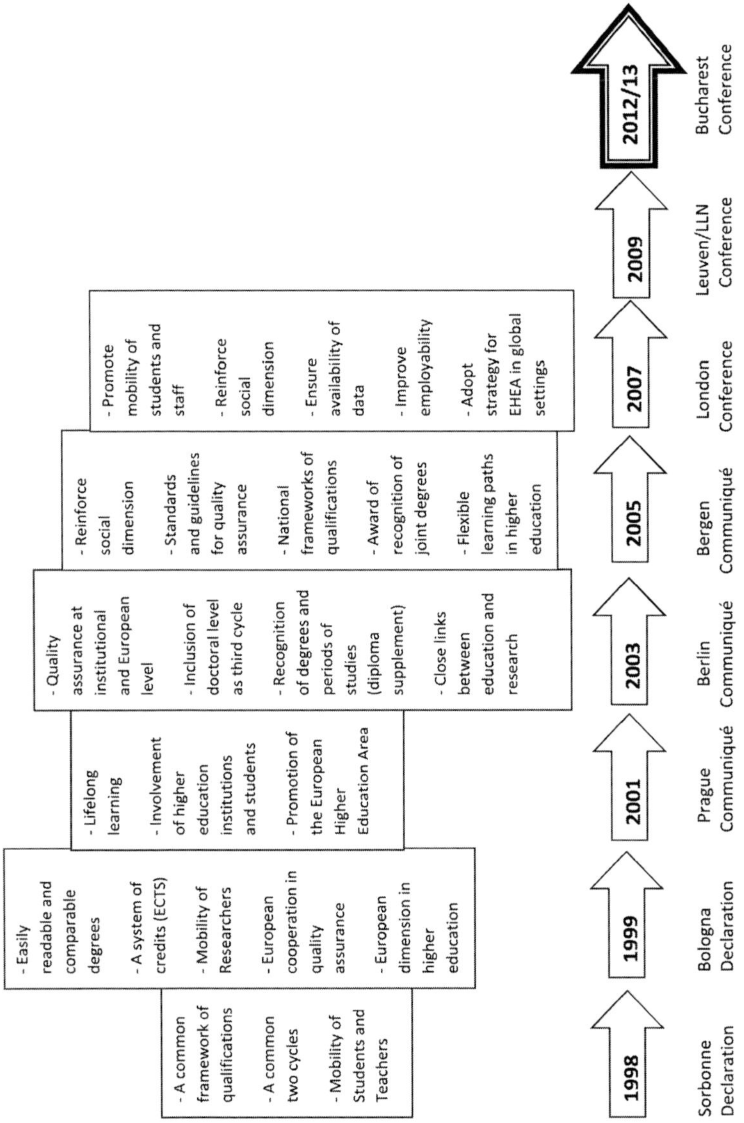


Figure 7.1. Timeline of Bologna Process. Source: Adapted from EACEA P9 Eurydice (2009)

## Berlin Seminar, 2003

At the Berlin Ministerial Conference in September, 2003, it was decided that all member states of the European Culture Convention[2] would be eligible to take part in the Bologna Process. In order to participate, the members would be required to apply for accession and submit a satisfactory plan for implementation of the Bologna goals in their higher education systems. Apart from summarizing the developments from 2001 to 2003 and setting guidelines for further work, the Berlin Communiqué also concluded that:

- research is an important part of higher education in Europe, where higher education and research are the two pillars of the knowledge-based society;
- it is necessary to go beyond the focus on two main cycles and add the third cycle—doctoral studies—to the Bologna Process; and
- ministers should take stock of progress in quality assurance, the two-cycle system, and recognition of degrees and periods of study before their 2005 meeting.

The Bologna Follow-Up Group was asked to develop instruments for quality assurance and a qualifications framework.

## Bergen Seminar, 2005

At the Bergen Ministerial Conference in May, 2005, delegates made a decision to enlarge the circle of consultative members to include the Pan-European Structure, the European Association for Quality Assurance in Higher Education (EAQA), Education International, and the Union of Industrial and Employers' Confederations of Europe.

The Bergen meeting, which confirmed the shift from future plans to practical implementation:

- added an overarching framework of qualifications for the EHEA with a commitment to elaborate national qualifications frameworks;
- adopted guidelines and standards for quality assurance and requested further elaboration of proposals for the register of quality assurance agencies;
- recognised importance of the social dimension of higher education, which includes—but is not limited to—academic mobility; and
- confirmed the necessity of improving interaction between the EHEA and other parts of the world.

---

[2] A treaty of member states for accession by European states which are not members of the European Union.

## London Seminar, 2007

During the London Seminar, May 2007, the Ministers adopted a strategy for the Bologna Process in a Global Context, which included consideration of the social dimension of the Bologna Process, the mobility of grants and qualifications frameworks, and the European Quality Assurance Register (EQAR).

## Leuven/Louvain-la-Neuve Seminar, 2009

During the Leuven/Louvain-la-Neuve Conference, April 2009, the participants analysed several topics related to higher education: social dimension (equitable access and completion), lifelong learning, employability in changing labour markets, student-centred learning and the teaching mission of higher education, education, research and innovation, international openness-mobility, and data collection quality assurance.

## Important components of the Bologna Process: a brief evaluation

An evaluation of the Bologna Process must include questions of access to higher education, study frameworks, mobility and portability and graduation (Figure 7.2).

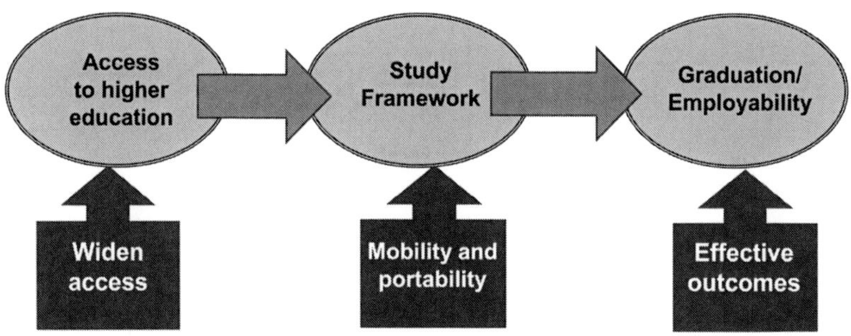


Figure 7.2. Major topics in Bologna Process. Source: Adapted from Eurostat (2009)

# Initial Conclusions

## Current status

According to Eurostat (2009), one-third of the population between the ages of eighteen and twenty was expected to enter higher education in Bologna countries in 2006. Also, half of the Bologna countries have an even higher net entry rate, and women represent 56 percent of student intake. However, in science, the intake of new students is predominately male, and female students represent only one-third of the new entrants.

The mobility of staff, students, and graduates is one of the core elements of the Bologna Process. Mobility is believed to create opportunities for personal growth, develop international cooperation between individuals and institutions, enhance the quality of higher education and research, and give substance to the European dimension. Various forms of international mobility exist, such as degree, credit, and programme mobility as well as free-movers. Yet little is known or understood about the actual situation of student mobility. According to the restricted definition used by Eurostat (2009), only a limited number of the students are foreigners, but their number is increasing by 5 percent each year. Statistical data on student mobility (duration of their curriculum abroad) are not available. Many countries only collect data on the nationality of the students. With the current information deficit, it is difficult to assess the impact of the Bologna Process on student mobility.

With regards to financial issues, public expenditures in higher education represent the efforts made by countries to ensure the basic operation of their higher education systems. These expenditures includes teaching and research as well as ancillary services to support these activities. In the Bologna countries, annual public expenditure on tertiary education increased at the same pace as the gross domestic product (GDP) between 2001 and 2005. In 2005, half of the Bologna countries spent more than 1.1 percent of their GDP on higher education, accounting for more than 2.8 percent of total public expenditure. In 2005, an average Bologna country spent €8.300 per full-time equivalent student per year, of which nearly 30 percent was devoted to research and development (R&D) and ancillary services. Expenditure per student varies on a scale of one to seven in the Bologna countries (Eurostat, 2009).

Generally, students rely on their families or jobs as a main source of income. In many countries, state support is provided to students to alleviate financial barriers to higher education. Various public support schemes with direct monetary support (including grants and loans) to students are put in place in the Bologna countries. Grants and loans accounted for a range of less than 5 percent to more than 20 percent of expenditures on tertiary education dedicated in 2005. Nearly half of the Bologna countries offer public-subsidised loan schemes, which account for around 7 percent of the public expenditure on tertiary education at the EU-27 level. In the Bologna countries, the share of public expenditures allocated as loans is particularly high in

Sweden, the United Kingdom, Iceland, Norway, Australia, New Zealand, and Japan. (Eurostat, 2009).

High qualifications play a major role in securing a job, but a high level of education is not always an immediate asset in all labour markets. Unemployment is also a matter of age. Recent graduates often experience problems entering the labour market. Yet, the rate of unemployment decreases with the level of study for men and women. Based on data from 2008, 16.5 percent of persons with low education, 8.3 percent with medium-level and 5.8 percent with a high level of education are unemployed in Bologna countries. Across levels of study, the unemployment rate for men remains lower than that for women. In most countries, gender plays a role, but the field of study also affects the graduate's chances of entering the labour market. Ten percent of young EU graduates in the fields of the humanities, languages, and the arts are unemployed, while in half of the Bologna countries, 20 percent or more of young workers who completed tertiary education are employed below their theoretical skills level. In this case, men are slightly more mismatched than women (Eurostat, 2009).

## Future Challenges in the Bologna Process

*Harmonisation of BA/MA cycles.* Higher education institutions in all Bologna countries have introduced Bologna cycles, but the models vary among countries. With regard to the bachelor programmes, two main structural models have been adopted in nineteen countries. Bachelor programmes have been commonly designed on the basis of 180 ECTS credits (three years), but eleven countries require 240 ECTS credits (four years). In some countries, no single model emerges as a reference, and institutional practice tends to draw upon both the two preceding models.

For master's programmes, the 120 ECTS credit model (two years) is used in the majority of Bologna countries (see Figure 7.3). This model is the most common, even though some master's programmes have fewer credits—as low as ninety in some countries.

## Implementation of ECTS

Although the ECTS system is in place in most countries, there are different understandings of the content and implementation of the ECTS system. The important systems used are:

- ECTS based on learning outcomes and student workload, or based only on student workload;

- ECTS based on contact hours, or contact hours and student workload;
- ECTS with a variety of credit definitions; and
- ECTS mainly used for transfer with national credit systems.

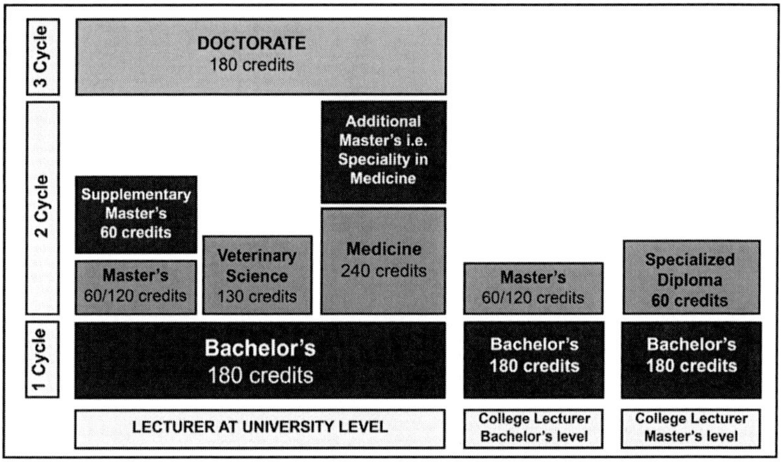


Figure 7.3. Structure of higher education. Source: Kaufmann (2009)

## Diploma supplement (DS)

The education ministers in the Berlin Communiqué determined that the Diploma Supplement should be issued automatically and free of charge to graduates at the end of their study programmes. However, this is not yet the case in all countries, and considerable variations are reported.

Many countries are lagging behind in implementing what is essentially a simple and practical measure, albeit one with organizational and financial implications. Only a limited number of countries have undertaken monitoring exercises to find out if the diploma supplement is actually used by employers.

*Mobility and portability (students)*

It is difficult to determine the relationship between mobility and portability of student support. National grants and loans are fully portable in some countries, while specific grants and loans for mobile students have been introduced in other countries, and still others combine both measures. It is difficult to assess the importance of the financial support given because the decision to study abroad is complex.

Lack of data is still a problem in assessing student and staff mobility. The realities of student mobility are often not well understood, and little is known even after ten years of experience in the EHEA. Many countries still only gather data on student nationality rather than tracking movements between countries for the purpose of study. It is important to stress the significance of enabling equitable opportunities for mobility in the EHEAs. As many countries are currently experiencing financial difficulties and shifting a greater share of costs to individual learners, developing equal opportunities should remain high on the agenda.

## Future Challenges in the Development of the EHEA

The European University Association identifies three key challenges for the future in its Trends V Report.

- Governments, higher education institutions, and other societal stakeholders need to strengthen their relationships. The debate among employers, students, parents, and other stakeholders should be encouraged in order to enhance trust and confidence in the quality and relevance of institutional engagement. It is also important that institutions and governments increase communication of the results and the implication of the structural and curricular reforms that are taking place.
- Institutions need to develop their capacities to respond strategically to the lifelong learning agenda. Opportunities for structural changes and tools related to the Bologna Process should be used correctly in order to enhance student-centred and flexible learning as well as greater mobility. The ever-changing needs of a society and economy should be detected in a timely manner by increasing dialogue with employers. Institutions should be motivated to offer constant training and re-training.
- Institutions must begin to think through the implications of the existence of the EHEA. Governments and institutions should recognise the need for adaptation to future changes in the EHEA. The Bologna Process is not yet completed, and some aspects still need to be implemented or changed. Moreover, the EHEA is part of an increasingly inter-connected global context, and changes in it should be accounted for.

### Quality assurance in higher education

Quality assurance has received much attention in the EHEA since 2003. A significant impulse for new developments took place at the Berlin Ministerial meeting in 2003 when ministers declared that "the primary responsibility for quality assurance lies with each institution itself and this provides the basis for real accountability of the academic system within the national quality framework." A number of national systems had already been in operation, but the explicit requirements led to a significant change of attitude and perception in many countries across Europe, as well as in many academic and institutional European networks.

Standards and guidelines were formally adopted by the European Ministers in Bergen in 2005 and have since been widely disseminated, discussed, and promoted. The agreement on European standards and guidelines for internal and external quality assurance has been the result of intensive efforts of quality assurance agencies, higher education institutions, and student representatives.

New quality assurance agencies have been created in almost all of the forty-six countries of the EHEA. They differ in size, scope, statute, focus, and international capacity.

## European standards and guidelines on quality assurance: basic principles

The European Standards and Guidelines on quality assurance adopted in 2005 consist of three parts: i) principles for the institution's own internal quality assurance system, ii) standards for the external evaluation of the institutions and their programmes by agencies, and iii) standards to be applied to quality assurance agencies themselves.

The basic principles on quality assurance, both within and external to higher education, include (EAQAHE, 2005):

- providers of higher education have the primary responsibility for the quality of their provision and its assurance;
- the interests of society in the quality and standards of higher education should to be safeguarded;
- the quality of academic programmes should be developed and improved for students and other beneficiaries of higher education across the EHEA);
- efficient and effective organisational structures are needed, within which those academic programmes can be provided and supported;
- transparency and the use of external expertise in quality assurance processes are important;
- a culture of quality should be encouraged within higher education institutions;
- processes should be developed through which higher education institutions can demonstrate their accountability, including accountability for the investment of public and private money;

- quality assurance for accountability purposes should be fully compatible with quality assurance for enhancement purposes;
- institutions should be able to demonstrate their quality at home and internationally; and
- processes used should not stifle diversity and innovation.

Thousands of people, including university staff, experts, students, and representatives of business and society, have undertaken self-evaluation procedures, participated in evaluation panels, or have been interviewed during evaluation visits. Students have a key role in quality assurance, although this involvement often has been limited to certain aspects or procedures. The involvement of business representatives and of non-nationals varies significantly in internal and external quality assurance procedures—it is generally stronger in agencies' evaluation panels than in their decision-making bodies.

The standards may include a checklist of formal requirements, a code of good professional practice, or soft guidelines. While having overall compliance rather than the obligation to meet each of the standards seems reasonable, it allows for a wide range of interpretations.

## Quality assurance of teaching staff

The teaching staff in higher education is expected to have full knowledge and understanding of the subjects taught, to have skills and experience in transmitting knowledge, and to be able to understand the students in a range of teaching contexts. However, teachers also need feedback on their teaching. It is the role of the institutions to develop procedures for recruiting and appointing the best qualified academic staff to ensure quality teaching and research. Institutions should provide staff with the opportunities to develop and extend teaching capacity. However, institutions should also have the right to remove poor performing teachers from their teaching duties.

## Quality of learning resources and student support

Students need to have sufficient access to physical resources, such as libraries and computers, as well as human support in the form of tutors, counsellors, and other advisers. These resources need to address specific student needs and be responsive to feedback from those who use the services provided. It is the task of the institutions to routinely monitor, review, and improve the effectiveness of the support services available to their students.

## External quality assurance

External quality assurance is an essential part of the process of implementing standards and guidelines. This may vary from system to system and may include institutional evaluations; subject or programme evaluations for accreditation at subject, programme, and institutional levels; or a combination of these. To be fully effective, the external evaluations largely depend on the level of implementation of an internal quality assurance strategy and on the use of mechanisms and methods aimed at achieving quality assurance objectives within the institutions. It is worth mentioning that external quality assurance agencies differ across Europe because of the environments in which they operate. However, there are general principles of good practice for external quality assurance, which include:

- respect for institutional autonomy;
- holding the interests of students and other stakeholders such as labour market representatives at the forefront of the external quality assurance process; and
- using results from internal institutional quality assurance assessments wherever possible.

## New challenges

One of the challenges in the European Higher Education Area is the implementation of the quality assurance agencies mentioned above. Quality assurance agencies need to demonstrate their independence and professionalism in order to build trust among stakeholders. The European Commission is also aware of the risk of fragmentation of these new agencies. This may be solved through agency mergers, which could be enabled through the existing European quality assurance network. Merged agencies could then consider broadening the scope of their activities in order to deal more adequately with lifelong learning, distance, online, vocational, transnational, and private higher education.

New and developing areas may affect quality assurance *inter alia* in the following areas:

- the balance between accountability and improvement within higher education institutions and the shared responsibilities of higher education institutions, quality assurance agencies, and policy-makers;
- the important roles of different stakeholder groups, including students and the business world;
- the provision of an adequate level of information;

- ways of dealing with increasing diversity across higher education (pedagogies, institutions, students, expectations, missions);
- the internationalisation of higher education, often in combination with growing commercialisation and competition; and
- awareness of the burdens caused by increasing bureaucracy and the cost of quality assurance.

In addition, increased international cooperation in quality assurance should contribute to better international outreach and the credibility of the EHEA. This was recognised in the first Bologna Policy Forum in Louvain-la-Neuve, Belgium, which specifically mentioned quality assurance as an area for cooperation between the Bologna countries and countries from across the world.

## Agricultural Higher Education in a Changing World: New Challenges

Changes in the global agricultural environment are multiple. The world is changing at an increasing pace, which constantly leads to new challenges for higher education in life sciences. Sustainability is the focal point of today, linking issues from energy security to national security, from human health to health of the planet.

Higher education in life sciences needs to take into account changes in policy and technology environments. For example, with the structural adjustment programmes of the 1980s, most countries started to propose market-oriented economic policies and drafted political structures and institutions to promote and support them (FAO, 1997). The methods and contents of agricultural education and training (AET) should accommodate these changes in order to meet the needs and realities of rural societies. Moreover, new technologies have implications for agricultural education. Rapid advances in communication technologies enable collaboration among institutions within and among countries. New developments in science and technology, such as developments in molecular biology, microbiology, and food technology, open up new areas for research and teaching.

The following are the major factors that affect teaching of agriculture at all levels, as identified by the Food and Agriculture Organisation (FAO) in 1997:

- population growth issues and the need to incorporate these issues into agricultural curricula;
- rapid scientific progress and the pace of technical change;

- changing employment opportunities;
- globalisation;
- increased awareness of gender issues and the roles of women in the agricultural sector;
- increased awareness of environmental issues; and
- the need to revise the education of extension workers to make their knowledge, skills, and attitudes more relevant to current development needs.

While all of these factors remain valid, others should be added—namely those that relate to food and agriculture, such as climate change, food security, and energy security (as shown in Figure 7.4). These factors should also be addressed by higher education institutions, which provide a mix of basic, applied, and social sciences and have the fundamental—and historical—capacity to respond to complex issues. For example, they have potential to develop methods for biologically-based energy production, secure and safe food production, protection of the environment, and efficient natural resource use.

As in all branches of science, professionals in life sciences need the knowledge and skills to respond to changes in the environment. Academic institutions need to monitor and adapt to realities of twenty-first century agriculture and agribusiness, including issues such as the integration of global agricultural markets, the growing importance of environmental studies in agriculture, the scientific redefinition of agriculture, the effects of growing consumer influence, the push for local and organic foods, the need to respond to increasing rates of food insecurity as well as obesity, and the changing demographics of the agricultural workforce. These new challenges require a permanent flow of information, interaction and partnership among all stakeholders (as identified in Figure 7.4).

In higher education, life sciences need to refocus on the curriculum and student experience required to ensure that agriculture graduates will have the necessary skills and competences to meet the needs of an ever-changing workplace and world (National Research Council, 2009).

*Changes in student profiles and careers*

Fewer of today's students have farm and rural backgrounds. Due to increasingly urban and suburban populations, students often lack even basic awareness of agricultural sciences. Furthermore, in order to attract undergraduate students in agriculture and life sciences, it is important to change the public perception that agriculture equals farming. Instead, it should be clear that agricultural studies tackle a diverse array of questions and approaches.

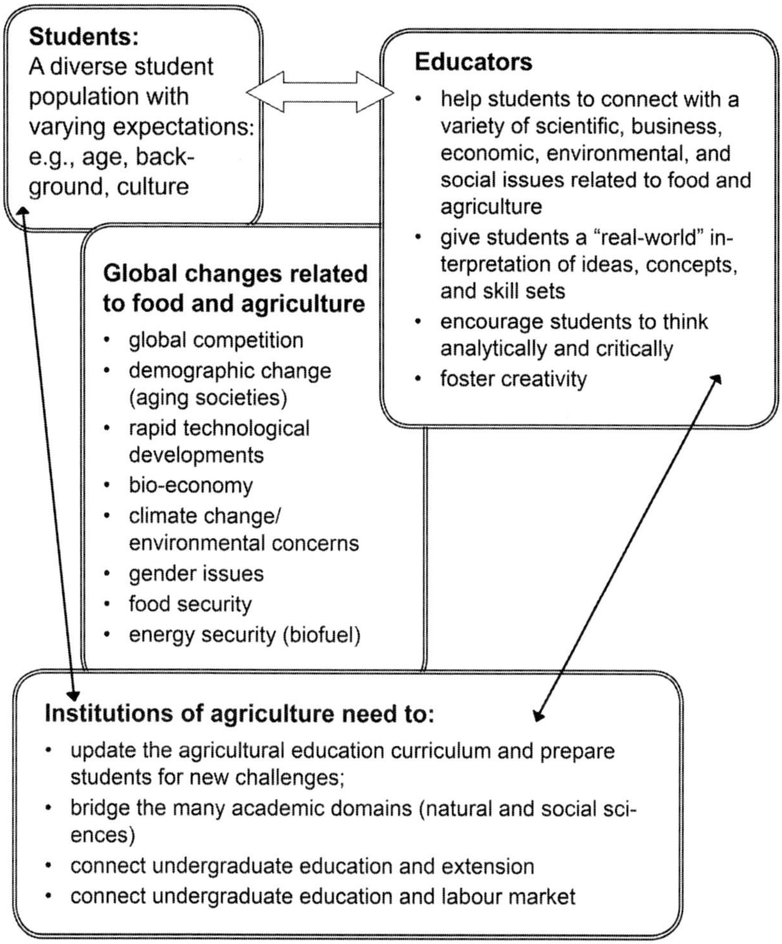


Figure 7.4. Changes in society and their impacts on higher education

Students may not be attracted to studying life sciences. They tend to choose a study that leads to a career that will provide steady income, be personally and professionally rewarding, provide an appropriate work-life balance, and be aligned with their values and interests. Careers in agriculture may appear to be old-fashioned, may not be well paid, or may not be perceived as offering sufficient opportunities for creativity. This challenges the agricultural community to find mechanisms that will attract talented students to degrees and careers in agriculture.

National governments are responsible for establishing standards to ensure the quality of education and training. Challenges that are common within the European higher education institutions need to be addressed at the subsector level. Students need to learn and understand the systems nature of agriculture and life sciences in general. They need to be exposed to the breadth of agriculture and to have opportunities to integrate what they are taught in their different courses.

Agriculture has become increasingly global, requiring agriculturalists to work with a complex regulatory regime and transportation logistics, as well as with different cultures and laws. To be properly equipped for this reality, professionals should acquire international exposure and understanding of national sensitivities. International experience early in students' training can broaden the scope of their curiosity and prepare them for future work in an international marketplace. Yet at the same time that agricultural is becoming internationalized, it is increasingly localized as well. Locally produced products and the interconnection of agriculture with the development of rural communities and environmental stewardship have become important. This means that agriculture graduates will also need to understand the ways that agriculture interacts with local environments.

Strong employment growth in many management and business occupations is expected in some developed countries, especially in such careers as technical sales, accounting and financial management, market analysis, landscape management, and international business. In contrast, weaker opportunities are expected for the agriculture sector that provides services to farmers. This situation is related to the students' study choices for higher education in Flanders (Ghent University Association) as shown in Figure 7.5.

The Bologna Declaration has increased awareness that employability is an issue across Europe. Several countries now train "professional bachelors." The change to such a two-tier structure does not necessarily come with immediate in-depth renovation of the underlying curricula. There are various ways in which first degrees can be relevant to the European labour market. A university bachelor's may, in some countries, be interpreted as a preparation and a platform for the choice of postgraduate studies, but this is less a problem where a strong college sector produces a significant number of holders of professionally oriented bachelor's.

Academic institutions need to alter the focus of their academic programmes and the experiences that they offer to students to keep up with the changing careers and opportunities available to their graduates.

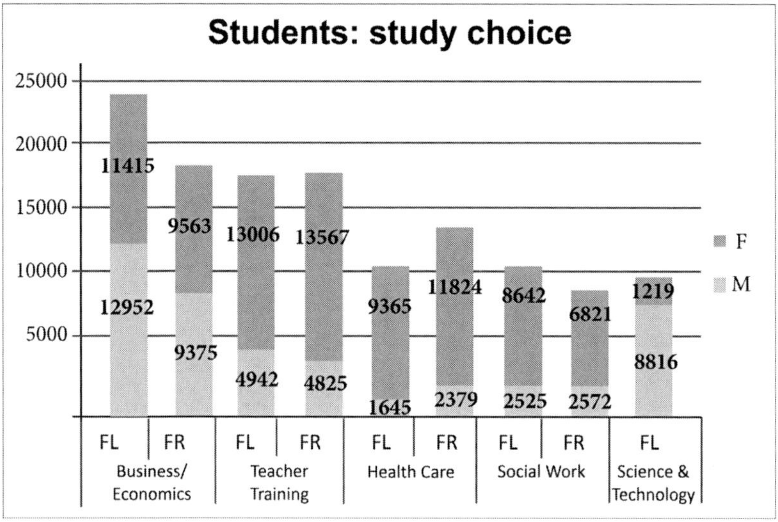


Figure 7.5. Study choice in high education in Belgium. Source: Dekeyzer (2009)

## Concluding Comments

Among European politicians, there seems to be a consensus that higher education and training are essential to the development and success of today's knowledge-based society and economy. Thus, the EU strategy calls for supporting programmes that stimulate knowledge, creativity, and innovation, which are believed to be the EU's most valuable assets, particularly as global economic competition becomes more intense in all sectors. The EU believes that international cooperation and exchange of knowledge are of key importance; consequently it promotes countries' working together and learning from each other.

This chapter has presented an overview of the way in which the European Union has developed a common working platform for higher education institutions. Different programmes, such as Erasmus and Erasmus Mundus, have proven particularly successful in enabling exchange of (European and foreign) students among European higher education institutions. Substantial support to the mobility of teaching staff and joint scientific research enables collaboration across borders, also in the field of life sciences, towards better knowledge and better trained professionals. Furthermore, the Bologna Process has harmonised teaching programmes, but also enabled comparability and benchmarking. The ECTS system and good quality assurance procedures and agencies have been key to the success of the Bologna Process.

Yet, it must be recognised that the life sciences sector (including agriculture), is the working area of trained professionals, and is changing quickly and constantly. New challenges noted in this chapter suggest the need for specific rethinking of higher education curricula and methods. It is therefore important that the European higher education institutions continuously and critically evaluate their focal points, ways of working, curricula and organisation. It is also important that this be done within a European context.

## References

Dekeyzer, T. 2009. *Study in Flanders: Higher Education in Flanders.* Paper presented in Bologna Ministerial Conference, April 28–29, Leuven/Louvain-la-Neuve.

EAQAHE. 2005. *Standards and Guidelines for Quality Assurance in the European Higher Education Area.* Helsinki, Finland. http://www.enqa.eu/pubs.lasso

Education, Audiovisual and Culture Executive Agency (EACEA) P9 Eurydice. 2009. *Higher Education in Europe 2009: Developments in the Bologna Process.*

European Commission. 2003. *The Bologna process: from Berlin to 2005.* Conference on the Implementation of the Bologna Process in South-East Europe. Council of European Seminar, December 2–3, Strasbourg. http://www.untz.ba/bolonjski_proces/bologna_proc_from_berlin_to_2005.pdf

EUROSTAT. 2009. *Higher Education in Europe and the Bologna Process. Key Indicators on the Social Dimension and Mobility.* Presentation in Bologna Ministerial Conference, April 28–29, Leuven/Louvain-la-Neuve.

FAO. 1997. *Issues and Opportunities for Agricultural Education and Training in the 1990s and Beyond.* Prepared by the Agricultural Education Group of the Extension, Education, and Communication Service (SDRE). FAO Research, Extension and Training Division. http://www.fao.org/sd/EXdirect/EXre0003.htm

Kaufmann, C. 2009. *L'enseignement supérieur du 21ème siècle en Communauté française de Belgique Journées Bruxelles 2009 Colloque.* Enseignement supérieur. Bruxelles, EPHEC, January 23, 2009.

National Research Council. 2009. *Transforming Agricultural Education for a Changing World.* National Academies Press: Washington, DC. http://www.nap.edu/catalog/12602.html

# CHAPTER EIGHT

# ROLE AND CONTRIBUTIONS OF THE CGIAR TO AET

PONNIAH ANANDAJAYASEKERAM

The context of agriculture in many developing countries and its role in broader economic development is changing very rapidly, as are the challenges facing research and education systems. The nature and dimensions of the challenges confronting the agriculture sector are becoming increasingly complex, requiring new strategies and innovative approaches to address them. The agricultural research and education system is the primary organ for managing and enlarging the knowledge base required for improving efficiency and productivity in the agriculture sector, and growth of the agriculture sector will continue to be a significant factor in improving livelihoods of the poor in sub-Saharan Africa. There is now common agreement in the development community that without adequate human, scientific, and organisational capacity, the Millennium Development Goals cannot be realised. In the past, the agricultural education and training (AET) system emphasised providing skilled manpower for techno-scientific agricultural production to assure food security. The emerging focus is now on developing and promoting an environmentally sustainable, socially equitable, and ethically defensible agricultural development process that fosters the well-being of rural communities and of the biophysical and socio-cultural environments in which they live (Bawden, 1998; Maguire, 2000).

The Consultative Group on International Agricultural Research (CGIAR), established in 1971, is a strategic partnership whose sixty-four members currently support fifteen international agricultural research centres working in collaboration with many hundreds of government and civil society organisations as well as private businesses around the world. CGIAR members include twenty-one developing and twenty-six industrialised countries, four co-sponsors, and thirteen other international organisations. CGIAR's mission is to "achieve sustainable food security and reduce poverty in developing countries through scientific research and research-related activities in the fields of agriculture, forestry, fisheries, policy, and environment." This includes research-based capacity-strengthening and outreach.

CGIAR is one of many actors that contribute to strengthening the capacity of the Agricultural Innovation System (AIS). Since CGIAR's inception, capacity-strengthening has been an integral part of the priority activities of its International Agricultural Research Centres (IARCs). Given its research mandate, most of its capacity-strengthening activities are research based and delivered through strategic partnerships with the various actors within the national and regional innovation systems.[1] The primary clients of CGIAR's capacity-strengthening are the National Agricultural Research Systems (NARS) actors, specifically the public sector institutes. This chapter begins with a definition of the changing context of the AIS and offers a broader definition for capacity-strengthening; this is followed by a section dealing with the role of CGIAR in education and training. The past experiences and the different approaches employed by CGIAR Centres in building individual and organisational capacity are then outlined. Finally, the lessons learned are highlighted to guide the future capacity-strengthening initiatives under the ongoing change process of CGIAR. Although the CGIAR Centres have a global mandate, this chapter deals primarily with their role in capacity-strengthening in sub-Saharan Africa.

## The Changing Context for CGIAR Education and Training

Capacity-strengthening interventions are affected by a number of factors, including the economic, social, and political environments, the institutional context, the task network and organisational structure of the different institutes, and the innovation systems under consideration. As capacity-strengthening is context-specific, an in-depth understanding of the context is vital to the design and implementation of effective capacity-strengthening programmes.

The policy and institutional context in which agricultural research and innovation occur has changed dramatically over the years. Rapid changes

[1] An innovation system is a group of individuals and organisations involved in the generation, diffusion, adaptation, and use of new knowledge and the context that governs the way these interactions and processes take place. Individuals and institutes working on similar issues, be it in a specific commodity sector (maize, beans, etc.), at a particular geographical location (global, regional, national, sub-regional), or in any problem area (drought tolerance, soil fertility management, climate change, etc.) tend to form a chain or network that can be described as an innovation system.

***Key Messages***

- Human, organisational, and societal capacity is widely recognised as being critical to development. A focus on capacity-strengthening is therefore a vital prerequisite for agricultural transformation.
- To have an impact on the complex and multidimensional issues of poverty and development, capacity must be built on competencies far beyond sound technical knowledge and disciplinary know-how. In a transformed AET system, agricultural professionals need to become change agents, going beyond technical expertise to becoming co-learners and reflective practitioners in finding solutions to problems that exist or will emerge in society.
- The CGIAR system can contribute greatly to this transformation process. CGIAR centres have over the years played an important role in formal and informal capacity-strengthening, and several studies have confirmed their impact through various partnership models that deliver capacity-strengthening services to developing country professionals.
- The CGIAR change process launched in 2008 has provided an opportunity to address some of the agency's weaknesses, including decreased core funds for teaching and learning, the absence of systematic training-needs analysis, weak quality assurance, a severe reduction in the capacity of centres to collect, adapt, translate, and widely disseminate training materials, and lack of collaboration among centres in this area.
- The new CGIAR will have a cross-cutting capacity-strengthening platform positioned to contribute to AET in more intensive, effective ways: participation of national scientists in global innovation systems; participation of National Agricultural Research Systems (NARS) in mega-programs; further developed ICT and knowledge management tools for innovation; and greater engagement of universities with the CGIAR system and in the agricultural innovation processes.
- For greater effectiveness in further developing the mega-programmes and capacity-strengthening platform, crucial foci could be:
  - an explicit capacity-strengthening component in each mega-programme;
  - coaching and mentoring integrated into programmes;
  - contributions to capacity-strengthening as criteria in staff performance assessment;
  - a focus on sustainability in project design, including strengthening the partner institutes' capacity;
  - an open competitive fellowship programme to strengthen the capacities of smaller, weaker NARS; and
  - integrated planning, monitoring, and evaluation of impacts;
  - meta-evaluation of past capacity-strengthening efforts. Experiences and lessons should be documented and used to inform future strategies and operations.
- As CGIAR centres are a small but a significant player in the challenging task of capacity-strengthening, a much closer linkage with the formal AET system is critical in order to increase its effectiveness.

continue to take place in the structure and authority of governments, the global economy, the structure of the farming sector, and in global and local food industries and retail businesses. The organisational landscape is also changing dramatically. Third parties such as civil society, farmers' organisations, and non-government organisations play increasingly important roles in agricultural research and development. Cross-sectoral linkages between agricultural and other sectors such as water, health, energy, and education are becoming increasingly important. The agriculture sector is expected to play a significant role in poverty alleviation and food and nutrition security while at the same time protecting the environment. All these changes have far-reaching implications for AET, as well as for the skills needed to address the emerging challenges. It is worth noting that agricultural education is a multi-sectoral concern that does not fall neatly under either the education or the agriculture sector.

Some of the key changes in agricultural research for development that have implications for AET are highlighted below.

- *Research and development (R&D) reform.* The ongoing reform agenda within the R&D arena has implications for the skills needed to support the emerging demands on the R&D community. These reforms include: redefinition of government's role in agricultural R&D; decentralisation and privatisation of agricultural R&D activities; broader and more active stakeholder participation and pluralism in service provision; increased use of networks and partnerships in service delivery; an outward-looking orientation for R&D, client-oriented and impact-driven research; and an innovation systems perspective.
- *Development methods.* Over the years, the agricultural R&D system has been testing, adapting, and adopting a number of concepts and procedures to make it more relevant, effective, and efficient. These concepts include: a farming systems approach, participatory research methods, rural livelihoods, agricultural chain or value chain analysis, action research, and integrated agricultural research for development (IAR4D), amongst others. The existing R&D systems in many African countries have incorporated elements of these different paradigms.
- *National development support.* The organisational side has moved from National Agricultural Research Institutes (NARIs) to National Systems Frameworks (NSFs), including National Agricultural Research Systems (NARS), the National Agricultural Extension System (NAES), the National Agricultural Education and Training System (NAETS), the Agricultural Knowledge and Information System (AKIS), and finally the Agricultural Innovation System (AIS).

- *Agricultural innovation.* An agricultural innovation system blends capacities of different institutes, mechanisms for coordination, communication networks, and policy incentives; in turn the system fosters innovation-led gains in agricultural productivity, leading to economic growth. Adoption of an innovation system perspective has meant simultaneously strengthening the capacity of a number of actors engaged in the system.
- *Global agricultural challenges.* The agriculture sector faces emerging and increasing challenges from a variety of sources, including:
  - crises in the global financial markets;
  - the emerging food and energy crisis;
  - technological advances in biotechnology and information communication technology;
  - climate change;
  - trade, markets, and the emerging agri-food system;
  - emerging diseases; the growing need for inter-sectoral links;
  - changing expectations in science, technology, and innovation;
  - the globalisation of private agricultural research and innovation;
  - the need to meet the various commitments of the MDGs;
  - concern for the environment; and
  - under-investment in agriculture in general and in agricultural research and education systems in particular.
- *Inter-country NARS support.* Developing country NARS, such as those in India, China, Brazil, Thailand, Vietnam, and South Africa, have significantly strengthened their research capacities and can now contribute to strengthening the capacity of smaller, less advanced NARS in their respective regions.
- *Depleted research resources.* Economic structural adjustment programmes in the 1990s led to significant reduction in resources for agricultural research in sub-Saharan Africa, and in some countries this trend led to the near collapse of NARS and some public universities. HIV and AIDS also took its toll in some countries, while civil strife and political conflicts have had greater impact in sub-Saharan Africa and Latin America than anywhere else in the world. As a result current agricultural research and training capacity in many developing countries (especially in sub-Saharan Africa) is characterised by the erosion of staff, skewed age distribution, weak linkage between higher education (the supply side) and R&D institutes (the demand side), poor organisation and management of existing capacity, and lack of systematic reinforcement.
- *Regional and global entities.* Emergence of sub-regional, continental, and global organisations within the Global Agricultural Research System has played a significant role in articulating regional priorities for resource

mobilisation as well as in the planning and implementation of priority research and capacity-strengthening activities.

- *Core funding shift.* Within the CGIAR system, core funding is declining, and support for capacity-strengthening has shifted from core resource to project-based funding. The change—in both volume and categories of funding—has had large effects on how training is organised, funded, and implemented across CGIAR (CGIAR Science Council, 2006a). Declining core support for training has resulted in decentralisation of training for research scientists and programmes at national and regional levels. Currently the CGIAR system itself is going through a change process that will have implications on the training and capacity-strengthening activities of the CGIAR Centres.

Adapting to these changes within a more complex innovation system and value chain framework, where there are a greater number of actors and linkages to consider, calls for a much broader capacity-strengthening strategy. A number of recent training needs assessment studies also identified the need for developing soft skills in addition to technical competency. Given these changing circumstances, one of the major challenges confronting the AET system is how to avoid stagnation and irrelevance (Hansen, 1990).

## Current Status of Agricultural Education Systems in sub-Saharan Africa

It has become increasingly difficult to view agricultural education solely from the perspective of a university, diploma-granting college, or farmers' training centre. In today's context, African universities represent the primary source of human capital for agricultural R&D agencies, as well as for academic staff of the learning institutes. University graduates are influential in the staffing, management, funding, and operation of colleges, training centres, research centres, and extension and other rural service-providing institutes, suggesting that education and training are part of a broader system that serves the agriculture sector. Increasingly, the pressure is on the system to serve not only the needs of agriculture but also of the rural sector. The AET system stretches from adult rural education to the university.

AET comprises four interrelated subsystems—schools, colleges, non-formal educational organisations, and universities. The AET system itself is a subsystem of rural development—or of the agricultural innovation system, which includes the clientele (the ultimate users of outputs), organisations, and institutions in the private and public sector, as well as non-governmental (NGOs) and community-based organisations (CBOs) (Bawden, 1998).

The post-secondary AET system in Africa dates back to the 1920s, but the current extensive institutional infrastructure for AET had its beginnings more recently, in the 1960s, when the majority of African countries gained independence. Africa now has about two hundred public universities (compared to twenty in 1960), of which about one hundred teach agriculture and natural resource management. In addition, private universities are beginning to complement these capacities with their own offerings (World Bank, 2007a).

Strong AET systems are necessary to increase agricultural productivity, which is a pre-condition for the much-needed growth and development in most African countries. AET raises agricultural productivity directly by developing producer capacities, and indirectly by generating human capital for support services. Higher education in agriculture also contributes directly to research and advisory services (World Bank, 2007a). A 2002 NEPAD document states that "the quality of higher agricultural education is critical because it determines the expertise and competencies of scientists, professionals, technicians, teachers and civil servants, and business leaders in all aspects of agriculture and related industries."

For many years, formal AET has been largely abandoned by governments and donors in Africa. Donor assistance to African agriculture has declined sharply, and the total support of AET in Africa has largely disappeared. Assistance to formal AET declined to just 0.7 percent of agriculture sector aid between 2000 and 2004 (World Bank, 2007b). Similarly, government funding for AET in many African countries has declined due to the Economic Structural Adjustment Programmes (ESAPs). As a result, African professionals in the AET system have become isolated from knowledge networks, global information sources, and cutting-edge technology development and dissemination processes. Many African universities are ill prepared to train the continent's next generation of agricultural scientists (World Bank, 2007a). This trend has gradually been changing, however, and there is renewed interest in agricultural research, training, and development.

The way science and education are organised often promotes individual rather than team effort at finding solutions to complex and multi-faceted problems. Increasingly, the problems confronting the agriculture sector call for different disciplines and organisations to work together. If agricultural education systems do not produce employable graduates at all levels, they risk becoming irrelevant as educational institutes. The key challenges confronting the AET system are:

- delivering quality education and training in an era of decreased funding, high student enrolment, and loss of key experienced teaching and research staff; and

- providing education and training for rural development rather than for agriculture alone.

It is clear that the older curriculum, which concentrated on production agriculture, is not up to the task of educating graduates to cope with the wider problem of rural development (Maguire, 2000). There is a real need to bridge the gap between the training offered by learning institutes and the skills and capacity needed to address real world problems.

## Constraints to Building African AET Capacity

A number of factors have contributed to the current status of the AET systems in many African countries. For example, organisational responsibility for building these systems is fragmented, and there is a lack of funding and incentives, which has led to severe brain drain. A World Bank (2007a) study of AET systems in Africa identified the following as key constraints facing the system: distorted and declining enrolment patterns; isolated and fragmented AET institutions; curricula that are often obsolete and disassociated from the economy; crises in staffing; inadequate teaching facilities and outdated teaching methods; unsustainable patterns of expenditure for higher education; decline in the quality of education; in certain countries, an unsustainably high number of universities given existing budgets; high unemployment among university graduates, often due to lack of relevance of the curriculum (discussion of these constraints is detailed in Chapter 1).

### Capacity-strengthening defined

"Capacity" refers to the ability of individuals, organisations and society at large to perform appropriate tasks effectively and efficiently in a sustainable manner. These tasks are either defined by history or by a situation in a specific context. Capacity is the engine for enhancing performance.

"Capacity-strengthening" is the process by which individuals, groups, organisations, and societies increase their abilities individually and collectively to perform core functions, identify and solve problems, and define and achieve objectives and desired results. Capacity-strengthening enhances a society's ability to understand and manage the development process in a sustainable manner. This definition highlights two important points: that capacity-strengthening is largely an internal process of growth and development, and that capacity-strengthening efforts should be results oriented (Morgan, 1997).

There are three interlinked dimensions of capacity building: individual, organisational and community-level. The CGIAR Centres, given their mandate, operate on the premise that they can make effective contributions directly at the individual and organisational levels and indirectly at the community level. It has been argued that organisations learn through individuals who learn—hence organisational learning is an outcome of individuals learning together (Kibwika, 2008).

Capacity-strengthening is a dynamic development phenomenon that should truly reflect changing conditions, needs, and ongoing transformations, and should identify and address both immediate and future needs. As people develop themselves, new expectations, problems, and issues may arise that demand a new set of competencies. Capacity-strengthening activities are usually based on improving the utilisation of existing capacity and on restoring capacity that has been eroded or destroyed. Thus capacity-strengthening deals with creating new capacity (capacity creation), effective mobilisation and utilisation of existing capacity (capacity utilisation), and sustaining the created capacity over time (capacity retention). These dimensions of capacity-strengthening are interactive and dynamic.

Within the CGIAR Centres, capacity-strengthening is viewed in terms of its contribution to the AIS: enabling actors to undertake agricultural research and realise the goals of poverty alleviation, food and nutrition security, and environmental sustainability. Strengthening the capacity of institutes has three possible dimensions: strengthening individual capacity and skills, strengthening organisational capacity including management skills, and strengthening inter-institute capacity including networking. Most CGIAR Centres focus on strengthening capacities at all three levels.

## Role of the CGIAR in AET

Each of the CGIAR Centres is autonomous. Mandates vary according to the Centre's focus; these may include specific types of commodities, farming systems, or institutional innovations including policy research and advocacy. The needs of the host countries and regions guide each Centre's research efforts and the nature of the training and capacity-strengthening programmes it conducts.

CGIAR's approach to capacity-strengthening has evolved considerably over the past few decades as agricultural research has come to focus more sharply on innovations and broader agricultural and role development.

The evolution and training philosophy; modes and types of training; links with learning institutions; training outputs and outcomes; weaknesses of the capacity-strengthening initiatives; and past experiences and approaches used by CGIAR are discussed in this section.

## Evolution and training philosophy of CGIAR Centres

Training in CGIAR Centres started in 1962,[2] when a group of young agriculturalists from developing countries gathered at Los Baños, Philippines, for the first Rice Production Training course offered by the then two-year-old International Rice Research Institute (IRRI). Over the years, as other Centres established training programmes, their goals were mainly to strengthen the capacities of developing countries to use the outputs of these unique agricultural research organisations.

The initial focus of the CGIAR Centres was food crop research, mainly because of threatening food shortages in many tropical, arid environments where populations were expanding rapidly. The research-oriented strategy for food crops was fairly simple in concept: first, identify several crops and livestock species that play key roles in meeting the food needs of fast-growing populations and contribute to poor smallholders' livelihoods; bring together a critical mass of agricultural scientists to seek out the constraints to improved production and devise ways to overcome these constraints; and finally, provide ample funds and facilities to speed these scientists in their work. In line with the global mission of CGIAR, training activities were to meet the "international public goods criterion."

The approach to training by the CGIAR Centres evolved from the philosophical base of the Rockefeller Foundation's programme of education for development through research. Its cooperative work with the Government of Mexico had illustrated that research alone was not enough to enhance productivity gains. It was essential to improve national capacity through training in order to facilitate the adaptation and adoption of the technologies generated. Thus, development of human resources in agricultural research has accompanied development of genetic materials and technologies, combining intellectual content with hands-on training. Early in the work of the Centres, most training was conducted at the Centres' headquarters and on experimental farms in the host countries. This is still true today for most major training categories, but the growing trend is toward decentralisation—that is, for appropriate courses to be taken directly to specific national and regional settings.

---

[2] Although the CGIAR was formally constituted in 1971, member centres such as IRRI and *Centro Internacional de Mejoramiento de Maiz y Trigo* (CIMMYT) have been in existence since the early 1960s.

Since its inception CGIAR has supported institutes to strengthen the evolving agricultural innovation community and enhance the professional development of agricultural scientists in developing countries. This commitment was well articulated in the research priorities of the Science Council (CGIAR, 2005):

> The CGIAR priorities maintain the focus of the systems on research. However, the conduct of international research combined with the provision of world-class opportunities for capacity-strengthening is a comparative advantage of the CGIAR. Enhancing capacity in developing countries has been a major accomplishment of the CGIAR in the past. This approach will continue through programme-related opportunities and through involving appropriate partnerships to enhance innovation and learning. Additionally, specific research on institutions is designed to identify the best means for policies and institutions to support new agricultural research and create pro-poor benefits.

The primary purpose of the CGIAR Centres' training activities is to enhance the developing country organisations—predominantly the National Agricultural Innovation System (NAIS) and especially NARS—to be more effective in independently and collaboratively conducting research for solving problems primarily related to agriculture, the environment, and the economy. Training activities of the CGIAR Centres are research-based and therefore largely dictated by the mandate and focus of their research priorities. The Centres aim at improving researchers' skills and knowledge about technologies, enabling some to become trainers themselves in the future. In the recent past, specific emphasis has also been placed on strengthening the capacity of agricultural higher education institutes and building the soft skills of research and development practitioners.

The learning and capacity-strengthening initiatives of the CGIAR Centres are holistic, needs-based, and mission-driven and are based on a number of principles for effectiveness. These include broad-based participation, building on local capabilities, ongoing learning and adaptation, integration of activities at various levels to address complex problems, subsidiarity and value addition, partnership and networking based on the comparative and competitive advantage of the Centre, and impact orientation.

## Modes and types of training at CGIAR Centres

*Training modes.* In addition to formal training courses, learning also takes place informally through learning-by-doing interaction with fellow students in courses, seminars, and workshops; in policy dialogue; and through research mentoring, interacting with supervisors, and in practitioner networks. The learning processes adopted by the different Centres are summarised in a recent report of the CGIAR Science Council (2006a) as shown in Table 8.1. The training and learning strategies progress from basic training, where those who need to learn are instructed, to more facilitated and network-based strategies where there is less inequality between teachers and learners.

The learning mode, as shown in column two, progresses from transmissive or didactic, where teachers structure and deliver what they know, to the more experiential and collaborative modes of learning that take place in work settings and collaborative networks. These different modes are associated with different settings. Transmission is common in classrooms, but advisory missions and joint seminars between centres and NARS partners are more commonly associated with collaborative learning and exchanges amongst peers. The third column focuses on who learns. Once again there is a progression from an instructional strategy where it is mainly the 'trainee' who learns, to more mutual learning that happens when centre-based researchers work with NARS partners in networks and joint research projects. Some examples of each mode of training and learning are presented in the last column. The different settings in which the CGIAR Centres operate offer a unique learning experience that is difficult to replicate. The appropriate strategy and mode is often dictated by the context of the partners.

*Training types.* The term "training" covers a wide array of activities in CGIAR Centres. It can refer to one- or two-week courses on specific techniques sponsored by a Centre, or it may refer to long and intensive (MSc and PhD) degree-oriented education. 'Training' may refer to visiting scientists who spend weeks or months becoming conversant with specific procedures or working with Centre scientists on particular problems; or it may refer to young persons who have just obtained advanced university qualifications and have one or two years in postdoctoral positions at the Centres. The formal training offered by the Centres can be grouped under individual training and group training.

*Individual training.* Individual training is focused on harnessing and strengthening the skills and abilities of individuals to contribute to the realisation of organisational developmental goals.

| Learning/ training strategies | Learning modes and settings | Who learns | Example |
|---|---|---|---|
| **Instructional** The Centre knows and the trainee needs to learn. | **Transmissive or didactic courses in specialised settings** –At HQ and regional bases with experienced teachers. | • Usually the NARI scientist. | Germ plasm management. Biotech techniques. |
| **Learning manager** The Centre manages opportunities for learning. | **Mixture of didactic and experiential learning** Learning by doing. Setting is likely to be in country where training is offered | • The NARI scientist, NARS and Centre scientists through research. • Centre scientist set agenda. | Plant breeding, w/classroom course followed by on-station opportunity to apply new knowledge. Research assignments designed or allocated by Centre to NARI. |
| **Mentor or advisor or colleagues** The Centre supports learners. | **Collaborative, peer learning** Joint research, activities and projects. Mutual exchange with individual visits and collective events such as seminars. Technical advice. | • NARS and Centre scientists. • The learning agenda is initiated by both. | Farmers' participatory selection of germ plasm. Collaborative design and customisation of 'tools', methods or models. |
| **Network manager** The Centre brings related project staff together. | **Sharing experiences.** Linking diverse research and development projects to help participants learn from each other's experience or context. Meetings, workshop and conferences setting. Responsive or dialogical. | • Network members. • Initiated by CGIAR Centres. | Networks of projects and scientists in different countries to which CGIAR scientists are attached. |

**Table 8.1. Learning process adopted by the CGIAR Centres. Source: CGIAR Science Council, 2006a.**

- *Graduate fellows.* This degree-related training mainly takes place in partnership with universities, where the fellows undertake practical training or conduct their research at or in collaboration with the sponsoring Centre. The Centres cooperate with a number of universities both in developed and developing countries at the student's thesis research stage.
- *Postdoctoral fellows.* These fellows have recently completed their PhDs and work within the Centres' research programmes, usually for a period of one or two years. They gain experience alongside senior scientists and contribute their own expertise to the Centres' programmes. Some subsequently join the research staff and some renew relationships as visiting scientists.
- *Visiting scientists, research fellows.* These are accomplished scientists coming to the Centres for a fixed period. These assignments may range from a few weeks to several months and accommodate a variety of arrangements that benefit both the visitors and the host Centre. The research fellows in some cases may be the staff scientists of universities or research institutes in developing countries, who undertake non-degree related training on research issues and methodologies.
- *Attachment associates, student associates.* These are young students from either developed or developing countries who come to the Centre to work on an established research programme as a requirement of their academic training.

*Group training.* Group training programmes of the Centres are focused on the staff of the NAIS, particularly those working with the International Agricultural Research Centres' networks and institutes with a mandate to train scientists and develop practitioners and policy makers. Group training courses are often conducted in close collaboration with regional and national institutes. The CG Centres offer three categories of group training courses.

- *Core courses.* These courses cross programme boundaries to build management skills, enabling managers to enhance the effectiveness, efficiency, and performance of R&D investments. CGIAR partners are encouraged to adopt or adapt these courses for their own use. Core courses are often jointly organised and conducted by a number of Centres.
- *Programme and project courses.* These courses are aimed at scientists and trainers who are involved with a Centre's associated research themes, programmes, or projects. The course content is specific to the purpose and objectives of a particular programme.
- *Network courses.* These are targeted and tailor made for network partners. The courses are often initiated by a specific network or networks and implemented by the Centres. Depending on the need, the network may decide

to provide training on topics related to the mandate of the Centres but often specifically dealing with the purposes and objectives of the network.

| **Group Training** | | | **Individual Training** | | |
|---|---|---|---|---|---|
| *Theme* | *per-cent td* | *per-cent p* | *Theme* | *per-cent td* | *per-cent p* |
| Crop Production | 25.6 | 13.5 | Crop Protection | 18.7 | 17.3 |
| Breeding | 13.4 | 5.8 | NRM | 17.3 | 12.5 |
| Social Science | 12.1 | 9.2 | Breeding | 15.6 | 14.8 |
| Methods | 11.3 | 16.2 | Biotechnology | 11.1 | 12.5 |
| Crop Protection | 6.4 | 8.5 | Social Science | 9.5 | 8.3 |
| Bio-technology | 6.1 | 3.1 | Crop Production | 7.9 | 6.5 |
| Livestock | 6.0 | 1.2 | Agro-forestry | 6.0 | 3.4 |
| NRM | 5.9 | 12.7 | Genetic Resource | 5.8 | 7.3 |
| Seed | 4.6 | 8.7 | Methods | 3.1 | 8.2 |
| Post-harvest | 3.2 | 3.8 | Livestock | 2.1 | 3.2 |
| Genetic Resources | 2.7 | 6.5 | Post-harvest | 1.5 | 2.0 |
| Agro-forestry | 2.0 | 9.4 | Others | 0.7 | 0.7 |
| Others | 0.7 | 1.5 | Seed | 0.7 | 2.9 |

**Table 8.2. Relative importance of different themes in individual and group training in terms of trainee days (td) and number of participants (p), measured as percentages (1990–2004). Source: CGIAR Science Council, 2006a.**

*Training themes.* The various types of training offered by the CGIAR Centres are classified under twelve general theme categories: agroforestry, breeding, biotechnology, crop production, crop protection, genetic resources, livestock, methods, natural resources management, post-harvest, seed, and social sciences. The relative importance of the various theme categories in terms of trainee days and number of participants (in percentage terms) are presented in Table 8.2. According to this study, the most important changes over time were the relative increases in the coverage of people trained in seed, social sciences, and natural resources management, with decreases in training for crop production, breeding and crop protection.

In terms of training related to methods and approaches, the CGIAR Science Council report (2006a) indicates that in the past much emphasis has been placed on research methods, management, and processes; experimental design; and statistics and data management, where the Centres may have had some particular expertise and relevant orientation.

Although data on informal training and learning have not been documented, CGIAR researchers estimate that they spend an average of 12 percent of their total time on these activities, as compared to 13 percent on formal training.

## Links with learning institutions

The CGIAR Centres operate on the premise that agricultural higher learning institutes will ultimately be responsible for replenishing the stock of human capital in research, extension, and educational systems and for providing students with a broad set of skills. Although the primary focus of the CGIAR is on research, every effort is being made to establish a closer working relationship with universities and to assist in strengthening their research and training capacities.

The CGIAR Centres are typically located near agricultural universities. Several Centres have helped directly to build and strengthen educational institutes in their host countries, particularly those institutes that have needs related to the Centres' mandates. For example, the International Rice Research Institute was developed on the campus of the University of the Philippines at Los Baños (UPLB), and many of the Institute's top scientists hold appointments on the University's faculty—and vice versa. The Postgraduate College of Mexico's National School of Agriculture at Chapingo has cooperated over the years with nearby CIMMYT, to the benefit of both, and the International Centre for Agricultural Research in the Dry Areas (ICARDA) played a supporting role in enlarging and strengthening the agricultural faculty of the University of Aleppo in Syria.

Similarly, relationships have developed between the International Institute of Tropical Agriculture (IITA) and the Nigerian universities; the International Livestock Research Institute (ILRI) and a number of Universities in Kenya and Ethiopia; the International Crops Research Institute for the Semi-arid Tropics (ICRISAT) and several Indian universities; between *Centro Internacional de Agricultura Tropical* (CIAT) and several Columbian faculties; between *Centro Internacional de la Papa* (CIP) and Peru's National University of Agriculture at La Molino; and between the World Agroforestry Centre (ICRAF), the Centre for International Forestry Research (CIFOR), and a large number of universities involved in teaching forestry and agro-forestry. Many highly reputed agricultural universities in Europe, North America, and Oceania have long records of association with participants sponsored by the CGIAR Centres.

Evidence on the ground clearly demonstrates that there is a strong link between universities and the CGIAR Centres in building the skills of the participants of the AIS in sub-Saharan Africa. The areas of collaboration include student supervision, joint training, collaborative research, thesis research (through various forms of attachments), and participation in delivering courses. Strengthening the capacity of the learning institutions has become an important strategic objective of the capacity-strengthening activities of some CGIAR Centres, as demonstrated in the recent capacity-strengthening strategy of the International Livestock Research Institute (ILRI, 2008). Thirteen CGIAR Centres are partners in the CGIAR Agricultural Open Curriculum and e-Learning Initiative (AGROCURI), led by the International Food Policy Research Institute (IFPRI) and the International Crops Research Institute for the Semi-arid Tropics. It is another example of CGIAR's strengthening the capacities of the agricultural learning institutes in Africa and South Asia where poverty is still prevalent. A number of Centres are also collaborating with the Regional Universities Forum (RUFORUM) for capacity building in agriculture in sub-Saharan Africa.

Recognising the importance of universities in producing the next generation of agricultural professionals, the ongoing CGIAR change process (discussed later in this chapter) explicitly identified greater engagement with universities as one of the key activities of the proposed capacity-strengthening platform.

*Training outputs and outcomes.* The cumulative efforts of the fifteen CGIAR Centres over the decades have been realised through their contributions to overall developmental goals of food and nutrition security, improved livelihoods, and environmental sustainability. More specifically, CGIAR's work has given the world improved germ-plasm; improved technologies and practices of growing, harvesting, and handling of output; methods, tools, research approaches, and best practices for enhancing the efficiency and impact of the AIS; and trained personnel.

These outputs have contributed significantly to poverty reduction, enhanced food and nutritional security, and environmental protection in many countries. Currently, the CGIAR Centres collectively contribute to capacity-strengthening in the fields of agriculture, forestry, fisheries, policy, and environment in a number of ways. These include offering thesis research opportunities for MSc and PhD candidates, building research partnerships among trainees and their parent institutes, building skills through short courses, contributing to curriculum development, and developing and disseminating training materials and learning approaches (see Box 2 for more details).

**Box 8.1. Contribution of CGIAR Centres to capacity-strengthening**

*Thesis research for MSc and PhD degrees*
Every year, a large number of young African scientists work alongside CGIAR scientists to complete their MSc or PhD research, often working in multi-disciplinary research teams. This scheme helps build the research capacity of African scientists and contributes significantly to research outputs.

*Research partnerships*
CGIAR research is done in partnership with African institutions. University faculties are part of research teams, often leading to long-lasting partnerships. In addition, CGIAR Centres offer hands-on training to university researchers through such arrangements as internships, visiting scientist schemes, and postdoctoral opportunities.

*Skill building through short courses*
Developing the capacity of national scientists, including university lecturers, through short training courses has traditionally been a key CGIAR activity. While the volume has dropped due to changed funding arrangements, specialised training courses are still regularly organised, both on technical matters and soft skills.

*Curriculum development and review*
CGIAR works with universities to keep curricula up to date with current knowledge. This work focuses especially on multi-disciplinary and participatory fields traditionally not included in faculties' programme offerings, such as agroforestry, agricultural biodiversity, integrated natural resources management, participatory disease surveillance, and preparedness.

*Training materials*
The CGIAR Centre is a global library of training resources on agriculture, forestry, fisheries, policy, and environment. This includes tailor-made training modules based on CGIAR research outputs, databases, tools on research methods, etc. Materials are also being adapted for e-Learning through AGROCURI, a CGIAR and university joint venture.

Over the years, there has been a significant increase in the number of people trained by CGIAR Centres, as shown in Table 8.3.

| Trainee Type | 1962–1984 | | 1990–2004 | |
|---|---|---|---|---|
| Individual Trainees | Global | sub-Saharan Africa | Global | sub-Saharan Africa |
| | 19,364 | 6,189 | 12,951 | 3,240 |
| Group Trainees | 14,000 | 5,161 | 90,000 (189)[a] | (NA) |

[a] Figure in parenthesis indicates the number of countries involved.

**Table 8.3. Number trained by CGIAR Centres. Source: TAC 1986, CGIAR Science Council, 2006a,b.**

With the limited statistics available, based on two studies (TAC, 1986; CGIAR Science Council, 2006a) conducted by independent panels for CGIAR, a number of observations can be made with respect to the capacity-strengthening activities of CGIAR since its inception.

- There has been an increase in 'informal' and short-course training (≤ ten days) linked to collaborative research.
- During the period 1962–1984 more than 19,300 agricultural workers from developing countries took training courses on various crops, livestock, or farming systems at the CGIAR Centres (TAC, 1986). According to statistics available in fourteen Centres, approximately ninety thousand people from 189 countries attended the CGIAR group training events from 1990–2004. During the same period the CGIAR Centres trained about thirteen thousand individuals. The distribution of individual trainees by regions is shown in Figure 8.1.
- The percentage of degree trainees has declined over the years from 52.1 percent in 1990–1992 to 35.8 percent in the 1999–2004 period.
- In general, the percentage of women trained by Centres has increased over the years. During the period 1990–2004, the proportion of women in group training increased from 17.1 percent to 20.7 percent. With regard to individual training, the proportion of women increased from about 30 percent in 1993–1998 to about 40 percent in 1999–2004.

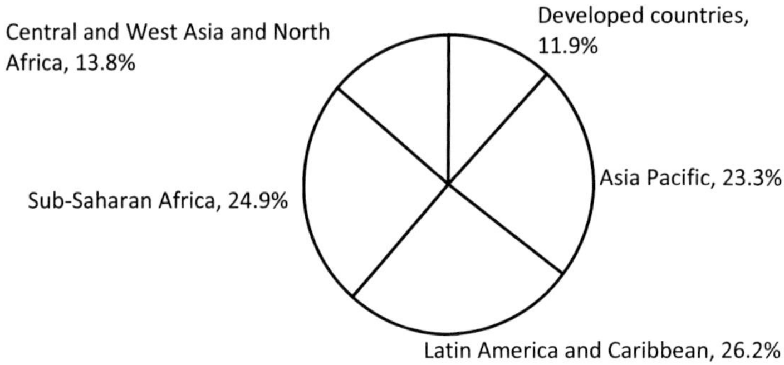


Figure 8.1. Distribution of participants by regions. Source: CGIAR Science Council 2006a.

- The thematic areas covered by the Centres have expanded over the years. There has been a significant increase in project funding and a reduction in unrestricted funding for capacity-strengthening.
- Capacity-strengthening activities have had a significant impact on the performance of the individuals and organisations as well as on the performance of the agricultural economy. In general, a number of reasons were suggested for this positive impact on the participants and their work:
  . practical experience, particularly in production and breeding courses;
  . international environments provide for exchanges with participants from different background and different disciplines;
  . multidisciplinary and interdisciplinary character of the work of the Centre as a whole—a new and broadening experience for many participants who may have narrowly defined working roles and discipline-focused training;
  . the opportunity for a period of time to concentrate efforts away from the usual pressure of an official service post;
  . satisfaction of working long hours in the field at practical, usually hands-on tasks, alongside their own peers, as well as with senior scientists from the Centre who hold advanced degrees, often have international reputations, and treat them as colleagues rather than as subordinates;
  . exposure to advanced ideas, modern facilities, latest equipment, and up-to-date scientific literature and libraries; and
  . relationships that continue when training ends because Centres make every effort to maintain a flow of current information to the former trainees—through individual correspondence, newsletters, annual and

special reports, staff visits, collaborative research, conferences, and in some cases return visits to the Centres.

It is worth noting that CGIAR Centres cannot be given full credit for all that its participants do in their later careers. However, their CGIAR experience can be cited as one powerful influence, amongst various others.
*Weaknesses of capacity-strengthening initiatives of CGIAR*

*Weaknesses.* A 2006 study of the evaluation and impact of training identified a number of weaknesses related to the current capacity-strengthening activities of the various CGIAR Centres:

- increased reliance on project funds and reduced allocation of unrestricted funds for teaching and learning;
- incomplete and incompatible records of training volume and type;
- absence of systematic needs analysis;
- weak quality assurance;
- poor records of the investments in capacity-strengthening activities;
- mismatch between training offered by the Centres and poverty on a country basis;
- severe reduction in Centre capacity to collect, adapt, translate, and disseminate existing training materials for wider use; and
- little collaboration among the CG Centres in planning and implementing capacity-strengthening activities.

*Recommendations made for strengthening.* The same impact study also made a number of recommendations to enhance the effectiveness and impact of CGIAR's capacity-strengthening activities:

- Centres should all develop and apply appropriate quality assurance protocols at all stages of both formal and informal training;
- training activities should be subjected systematically to the same type of planning, monitoring, and evaluation procedures as those applied in research;
- Centres should improve the efficiency and effectiveness of their training functions by taking advantage of opportunities for sharing their experiences, best practices, functions, and activities among themselves;
- Centres should exploit the advantage of information communication technology, such as embracing e-Learning and catering to the heterogeneity of the NAIS;
- Centres should improve coordination and strategic aspects when planning training, assembling databases, or developing course materials.

To respond to the weaknesses identified and to implement these recommendations, the capacity-strengthening groups of the CGIAR Centres have embarked on a number of collective action areas, such as: enhanced communication and coordination, partnerships for education, a common framework for online learning and distance education, and common quality assurance standards and collective action for capacity-strengthening in sub-Saharan Africa. In addition, the CGIAR Science Council is currently developing indicators for activities related to capacity-strengthening that focus on training outputs and evaluation practices.

## Past experience and approaches used by CGIAR Centres in capacity-strengthening

Early in the work of the Centres, most training was conducted either at headquarters or at the Centres' experimental farms in their host countries. While still true in some cases, the growing trend has been to take appropriate courses to the relevant regions, i.e., to decentralise away from headquarters. In addition to being cost effective, this measure has contributed to the development of national and regional capacities to continue offering training in the various regions.

In building the capacity of its partners, CGIAR Centres have been using a number of approaches. Some are for short-term skill building and some are for degree-oriented education. This section presents a sample of such approaches, identifying best practices for continued sustainable capacity-strengthening in sub-Saharan Africa.

*Short-term skill-building approaches.* These approaches largely focus on research methods and ways to address farm-level constraints, policy advocacy, and research management issues.

- Regional training in on-farm research with a farming systems perspective—a collaborative activity between a CGIAR Centre (CIMMYT), University of Zimbabwe, and the National Agricultural Research and Extension Services of Zimbabwe (Rukuni, 1987; Anandajayasekeram, 1993). This is short-term (five-week) training with a focus on research processes and approach.
- In-country 'call' system—training in on-farm research, with a farming systems perspective, for near-term technology generation and dissemination (Tripp et al., 1990). Offers a partnership approach between CGIAR Centres and national research and extension systems, focusing on research processes and ways to generate appropriate technologies and recommendations.

- Policy research and analysis—a capacity-strengthening activity between a CGIAR Centre (IFPRI) and the University of Malawi (Bunda College), with focus on building analytical and advocacy capacity related to policy issues (Babu, 1997).
- Regional crop management training—a model for decentralising headquarters-based training activities; a partnership initiative between CGIAR Centres (CIMMYT and CIAT), a national university, and NARS in Kenya, with a focus on strengthening capacity of the institute and research processes (Anandajayasekeram, 1993).
- Sub-Saharan African Challenge Programme—for developing best practices and an approach for implementing integrated agricultural research for development at the continental level; this is a partnership initiative of the Forum for Agricultural Research in Africa (FARA) and sub-regional (self-regulatory) organisations (SROs) such as the Association for Strengthening Agricultural Research in Eastern and Central Africa (ASARECA), *Conseil Ouest et Centre Africain Pour la Recherche et le Développement Agricoles* (CORAF), the West and Central African Council for Agricultural Research and Development, and the Southern African Development Community (SADC), as well as NARS and CGIAR Centres.
- Agricultural research management training—a collaborative initiative between the former Service for National Agricultural Research (ISNAR),[3] the South African Centre for Cooperation in Agricultural Research and Training (SACCAR), and the Eastern and Southern African Management Institute (ESAMI) Tanzania, designed to improve the effectiveness and efficiency of research management in NARS.

*Long-term degree-oriented approaches.* A number of approaches are being used by the CGIAR Centres to contribute to long-term degree-oriented education. These projects are at various stages of implementation, and the list, although not exhaustive, clearly demonstrates that CGIAR Centres do have significant experience in collaborative capacity-strengthening initiatives in sub-Saharan Africa.

- African Network for Agriculture, Agro-forestry and Natural Resources Education—a network approach of colleges and universities in Africa offering education programmes in agriculture, agro-forestry, natural resource management, and the environment, including related topics in the social sciences.

---

[3] Currently merged with IFPRI.

- National Model for Collective Action in Building Capacity—the case of Ethiopia. A collaborative venture between the Ministry of Agriculture and Rural Development and the CGIAR Centres based in Ethiopia.
- Agricultural Open Curriculum and Learning Initiative (AGROCURI) — for supporting postgraduate training focused on South Asia and sub-Saharan Africa.
- International Partnership for Forestry Education—a global partnership to develop forestry education and share information, experiences, and resources.
- CGIAR Gender and Diversity Program—aimed at building the talent pool of promising African women in agricultural sciences.
- Biosciences Eastern and Central Africa (BECA)—a regional programme endorsed by New Partnership for Africa's Development (NEPAD) and located at the International Livestock Research Institute, Nairobi, Kenya.
- Strengthening Postgraduate Training in Agriculture—A SADC/GTZ (*Gesellschaft Technische Zusammenarbeit* / German Technical Corporation) regional MSc programme coordinated by one of the SADC organs of the Food, Agriculture, and Natural Resources Sector.
- Collaborative Master of Science in Agricultural and Applied Economics (CMAAE)—for Eastern, Central, and Southern Africa regions.

## Some observations on capacity-strengthening approaches at CGIAR Centres

A critical review and assessment of these experiences distilling lessons and best practices is certainly beyond the scope of this chapter. However, based on the author's personal experience and knowledge of some of these initiatives, a number of observations can be made.

- All initiatives are supported by special project funds with very little co-funding support from the Centres.
- In almost all cases, the concept originated from either the Centres or the donor. The other local collaborators were brought in at the design stage.
- Assessments and evaluations, when carried out, were completed as part of the accountability commitment of the project.
- All projects were successful in achieving their immediate objectives. In cases where impact studies were conducted, there was evidence to show positive impact.
- Continuity and sustainability of the activities beyond the project period was a problem in all cases. This was largely due to lack of an exit strategy

in the project design and failure to institutionalise the various project activities within the partner organisations.

- There is limited evidence to demonstrate that the lessons learned were being utilised fully in the planning and implementation of the next generation of projects.
- While the decentralization of responsibilities for capacity-strengthening has weakened the role of training units, it has on the other hand allowed research teams to develop and foster strong research and training partnerships with a wide range of national and regional institutes.

The recent CGIAR impact study (CGIAR Science Council, 2006) concluded that the greatest future demand from AIS will be for capacity-strengthening through specialised short courses and individual non-degree and higher degree training rather than generalised training. In this respect, CGIAR Centres could do more in terms of distance- and e-Learning and supporting the efforts of local universities. Currently this is a priority area for some CGIAR Centres.

It is timely to undertake a systematic individual and meta-evaluation of the various approaches so that lessons can be drawn to guide future initiatives in building sustainable capacity for agricultural innovation in sub-Saharan Africa. It is important to identify the 'best fits' from the 'best practices' to suit the specific situation.

## Factors that Contribute to Sustainable AET Capacity Strengthening

The number of Latin American (Brazil, Chile) and Asian countries (India, Malaysia) that have achieved significant economic growth since the late 1990s clearly demonstrates that it is possible to build productive and financially sustainable AET and research systems in developing countries. This was achieved through a consistently high level of investment in AET by the countries themselves and their development partners. A World Bank review (2007a) identified a set of critical factors that contributed to building a productive AET system. These include:

- mobilising and sustaining political support for AET investments;
- supporting public investment in capacity building to create the scientific leadership needed to implement agricultural and rural development strategies;
- building a system of core AET institutes that make a sustained commitment over multiple generations to produce returns;

- undertaking massive campaigns to develop human capital;
- establishing closer links between research and higher education institutes and recognising that the administrative separation of research and higher education in many African countries has crippled the development of NAIS; and
- creating incentives to retain the well-qualified and experienced staff in research, extension, and educational institutions.

The same study also identified a number of measures to revitalise and build the next generation of agricultural professionals. Key reforms are suggested to enhance the effectiveness of AET for agricultural innovation and development. These include aligning the mandate of AET organisations with national development aspirations (including changes in the culture of AET organisations through the introduction of educational programmes and links beyond AET) and enhancing innovative individual and organisational capacity by improving incentives to forge stronger links between AET and other stakeholders (Davis et al., 2008). The CGIAR Centres can contribute a great deal to implementing the suggested reform processes. However, the initiative should be led by African institutes—they should own the process and the products.

## CGIAR change process and capacity-strengthening

To respond to the changing context and emerging challenges, CGIAR launched its Change Management Initiative in 2008 and adopted a new business model that would enable it to do more and do better. The core elements or organs of the new CGIAR are the funds (representing the financial contributors) and the consortium of CGIAR Centres.

The vision of the remodelled CGIAR is to reduce poverty and hunger, improve human health and nutrition, and enhance ecosystem resilience through high quality international agricultural research, partnership, and leadership.

In the new structure of the CGIAR, donor contributions will be harmonised to improve the quality and quantity of funding available, engender greater financial stability, and reverse the trend towards restricted funding. The consortium will unite the Centres under one legal entity and take the lead in developing the strategy and results framework and mega programmes in partnership with key stakeholders. The consortium and the funders (those who contribute to and manage funds) will enter into programme performance agreements for the delivery of research through the portfolio of CGIAR Research Programmes (CRPs) and two platforms. The current list of research programmes includes dry-land cereals; grain legumes; livestock and fish; maize; rice; roots, tubers, and bananas; wheat; agriculture and food security; climate change; trees and agro-forestry; water, land, and ecosystems; integrated systems for humid tropics; aquatic agricultural systems; dry-land systems; policies, institutions, and markets; agriculture for nutrition and health; and managing and sustaining crop collections. In addition, CGIAR implements several innovative 'challenge programmes' designed to confront global or regional issues of vital importance through broad-based research partnerships. Recognising the importance of cross cutting issues, two platforms—a gender platform and a capacity-strengthening platform—are created to support the CRPs.

It is envisaged that the capacity-strengthening platform will serve as a support unit to serve CRPs, Centres, and partners in collaboration with the Global Forum for Agricultural Research (GFAR) and others. This platform is expected to contribute to:

- enhanced participation of national scientists in global innovation systems;
- more effective participation by NARS in mega programmes;
- development and use by NARS of new Information Communication Technologies (ICTs) and knowledge management tools for innovation; and
- greater engagement of universities.

The details of the priority activities and modalities of operation of this platform are still being developed. The underlying assumption is that each of the CRPs will have its own capacity-strengthening activities. It is also interesting to note that this is the first time that universities are explicitly considered as partners or clients of the CGIAR Centres. Specific activities and performance indicators of the capacity-strengthening platform will be identified in the results framework. The CGIAR capacity-strengthening platform is still in the process of developing new strategies.

In further developing the CRPs and the activities of the capacity-strengthening platform, several points are worth considering.

- Each CRP should have an explicit capacity-strengthening component with the strategy, activities, deliverables, and expected outcomes spelled out. Priorities should be based on context-specific training needs of CGIAR partners. Adequate resources should be clearly earmarked and sourced for these activities.
- Coaching and mentoring should be given due consideration in programme planning and implementation.

- Contributions to capacity-strengthening should be used explicitly as one of the criteria for staff performance assessment.
- Sustainability of initiatives beyond the programme period should be given high priority in design and implementation. Building the capacity of partner institutes should be an essential component.
- In order to avoid duplication of efforts and exploit the complementarities and synergies, common capacity needs of the CRPs should be addressed by the platform. Developing the much needed soft skills should be an important component of this agenda.
- An open and competitive graduate fellowship programme similar to the Commonwealth Scholarships should be designed and implemented by the CGIAR system. This programme could also address the training needs of smaller countries and NARS, as well as issues not adequately covered in the CRPs. This fellowship programme should be implemented through the platform in close collaboration with sub-regional organisations and universities.
- CGIAR does not yet have a systematic approach to monitoring, tracking, and reporting capacity-strengthening activities. There is a need to devise and implement a common monitoring and evaluation system to systematically assess the effectiveness, quality, efficiency, relevance, sustainability, and impact of CGIAR's capacity-strengthening activities, which should also assist in providing feedback and documenting and disseminating best practices and lessons learned.
- The capacity-strengthening platform and the consortium should establish working relationships with bilateral and multilateral donors and provide input for the design and implementation of their support to capacity-strengthening activities.

## Concluding Comments

Capacity-strengthening has been and still is a priority activity of the CGIAR Centres. In line with their mandate, most of the capacity-strengthening activities of the Centres are mission-driven, research-based, and largely delivered through partnerships. Over the years, there has been an increase in informal and short-course training linked to collaborative research. Internationally recruited scientists spend on average 25 percent of their time on formal and informal training activities, and this has increased over the past five years (CGIAR Science Council, 2006). The training activities of CGIAR cover a wide range of commodities, resources, and themes.

The CGIAR Centres collectively contribute to capacity-strengthening in a number of ways. These include offering thesis research opportunities for MSc and PhD degrees, building research partnerships (South-South-North), building skills through short courses, contributing to curriculum development and review, and developing and ensuring wider dissemination of training materials and learning approaches.

Over the years CGIAR Centres have experimented and are still experimenting with a number of collaborative approaches to strengthening the capacities of their partners. These experiences will offer valuable lessons to initiate future capacity-strengthening activities in sub-Saharan Africa. Currently, the Centres are working on a number of collective action areas in order to enhance the effectiveness, efficiency, and impact of CGIAR's investments in capacity-strengthening, especially in sub-Saharan Africa.

Recognising the importance of human capital in development, the ongoing CGIAR change process is creating a capacity-strengthening platform that goes beyond simple consultation and project-based activity to identify and implement specific activities that will address the needs of the CRPs, Centres, and partners. In the author's view, each of the proposed CRPs should have an explicit capacity-strengthening strategy—in addition to overall training and capacity-building—clearly identifying objectives, activities, outputs, and outcomes. Appropriate resources (both financial and human) should be allocated to implement these initiatives. Contributions to capacity-strengthening should be used as one of the criteria for assessing staff performance at the Centre or programme level. Training-needs assessments should guide priorities at all levels. The CGIAR should also consider an open, competitive fellowship programme to support long-term education. It can also be argued that if capacity-strengthening is to live up to its mandate to support Agricultural Research for Development (AR4D), CGIAR should regard it as an important complement to research and create the organisational and management structures that will allow the design, implementation, monitoring, and evaluation of appropriate capacity development interventions.

To remain relevant, the AET system should transform itself to address the developmental needs of sub-Saharan African countries. A transformation is required in the training and education system to enable agricultural professionals to become co-learners and reflective practitioners in finding solutions to problems that exist or will emerge in society. In addition to building the capacity to become technical experts, agricultural professionals need to become agents of change. Technical competency is only one of the necessary conditions for the next generation of the R&D community. Additional competencies are needed if technical knowledge and disciplinary know-how are to have any impact. The CGIAR system can contribute a great deal to this transformation process. Without question, AET and capacity-strengthening are now back on the development agenda. The efforts and outcomes of the national and regional organisations can be enhanced and complemented by the CGIAR system, both in the area of knowledge generation and of capacity-strengthening. To achieve this, it is crucial for CGIAR Centres to have a capacity-strengthening strategy, as well as effective links and close working relationships with the AET system at large.

## References

Anandajayasekeram, P. 1993. *On-Farm Research (OFR): Impacts on Training in Southern Africa.* In Paul, H. and W. Stephen, eds. Impacts of On-Farm Research: Proceedings of a Net Workshop on Impacts of On-Farm Research in Eastern and Southern Africa, June 23–26, Harare, Zimbabwe.

Babu, S. 1997. Facing Donor Missions with Informed Policy Decisions: Lessons from CS for Policy Analysis in Malawi. *African Development 22: 2, 5–24.*

Bawden, R. 1998. *Agricultural Education and Training: Future Perspectives.* Paper prepared for the Agricultural Knowledge and Information System (AKIS) Thematic Team in the Rural Family, for the World Bank.

CGIAR. 2005. *System Priorities for CGIAR Research 2005–2015*. CGIAR Science Council.

CGIAR Science Council. 2006a. *Evaluation and Impact of Training in the CGIAR,* Science Council Secretariat, Rome, Italy

—. 2006b. *Impact Assessment of Policy-Oriented Research in the CGIAR: A Scoping Study Report.* Science Council Secretariat, Rome, Italy.

Davis, K., J. Ekboir, and D. Spielman. 2008. Strengthening Agricultural Education and Training in sub-Saharan Africa from an Innovation Systems Perspective: A Case Study of Mozambique. *Journal of Agricultural Education and Extension* 14: 1, 35–51.

Hansen, G. 1990. *Beyond the Neoclassical University: Agricultural Higher Education in the Developing World: An Interpretive Essay*. A.I.D. Programme Evaluation Report No. 20. Washington, DC. http://alliance.cgxchange.org/strategy-and-results-framework-and-mega-programs

ILRI. 2008. ILRI's Capacity-strengthening Strategy. ILRI, Nairobi, Kenya.

Kibwika, P. 2008. *Descending the Ivory Tower and re-making higher education in the era of (un)sustainability.* Knowledge for Development: Observatory on Science, Technology and innovation for ACP Agricultural and Rural Development.
http://knowledge.cta.int/Dossiers/S-T-Policy/Reshaping-tertiary-agricultural-education/Feature-articles/Descending-the-Ivory-Tower-and-re-making-higher-education-in-the-era-of-un-sustainability

Maguire, C. 2000. *Agricultural Education in Africa: Managing Change.* Paper prepared for a workshop organised by the Sasakawa Africa Association, Accra and Cape Coast, September 4–6, Ghana.

Morgan, P. 1997. *The design and use of capacity development indicators.* Paper prepared for the Political and Social Policies Division, Policy Branch, CIDA. Hull: Canadian International Development Agency (CIDA).

New Partnership for Africa's Development (NEPAD). 2002. *Comprehensive Africa Agriculture Development Programme.* Midrand, South Africa.

Rukuni, M. 1987. *The University of Zimbabwe/CIMMYT Collaborative Training Programme for East and Southern Africa*. Paper presented at the Fourth Senior Research and Extension Administrators Workshop, May 5–8, Capital Hotel, Lilongwe, Malawi.

Technical Advisory Committee (TAC). 1986. *Training in the CGIAR System: Building Human Resources for Research to Improve Food Production in Developing Countries*. A Report Prepared by the TAC Secretariat and Food and Agricultural Organization of the United Nations, Rome, Italy.

Tripp, R., P. Anandajayasekeram, and G. Sain. 1990. *The Design and Management of Call System Training in On-Farm Research.* CIMMYT Economics Working Paper 90/01, Mexico, D. F., Mexico.

—. 2007a. *Cultivating Knowledge and Skills to Grow African Agriculture: A Synthesis of an Institutional, Regional and International Review.* World Bank, Washington, DC.

World Bank. 2007b. *World Development Report 2008: Agriculture for Development*. World Bank, Washington, DC.

# PART THREE

# BUILDING AFRICAN CAPACITIES

# CHAPTER NINE

# BUILDING AFRICAN CAPACITY IN AGRICULTURE, FOOD, AND NUTRITION

CARL EICHER AND ALDO STROEBEL

This chapter draws inspiration from the role of human capital, education, and training in fueling Asia's Green Revolution in the 1960s (Eicher and Witt, 1964) and from progress made in Latin America (De Schutter and Vanloqueren, 2011).

In the 1960s, the vision of agriculture in Latin America was pessimistic. In 1965, T.W. Schultz surveyed United States (U.S.) foreign aid in Latin America over the period 1945 to 1965 and concluded: "the sad truth is that not a single first-class agricultural research centre has been developed (in Latin America) as a consequence of these (U.S.) activities." Yet today, Brazil's Agricultural Research Corporation, EMBRAPA (*Empresa Brasileira du Pesquisa Agropecuária*) is an agricultural powerhouse, and Brazil's agricultural research budget is forty times the size of Nigeria's (World Bank, 2009).

Appropriate and adequate education and training at all levels has been essential to bringing about this change and underpins all other development efforts (Haggblade, 2013). Yet the current approaches to agricultural education and training (AET) in Africa are largely outdated and out of touch with reality. A radical refocusing centred on building appropriate indigenous capacities—of people and institutions able to work from within African contexts and with African experiences—is needed to create a uniquely African Green Revolution and to address the challenges facing African agriculture today (FARA, 2006). Such a strategy would give large- and small-scale African agricultural enterprises the capacity to exploit opportunities for access to the global agrifood business system, to effectively address food security and vulnerability through environmentally and commercially sustainable agricultural practices, and to improve agricultural governance and leadership (including enabling farm producers and women) through an interactive and responsive AET system (Juma, 2011).

***Key Messages***

- *A radical refocusing of Africa's outdated approaches to AET* is needed to address the challenges facing African agriculture. At the centre of this is the need to build appropriate indigenous capacities for a uniquely African Green Revolution—people and institutions able to work from within African contexts, and with African experiences, to exploit opportunities for large and smallholder African agriculture to gain access to the growing global agri-food business system, to effectively address food security and vulnerability through sustainable agricultural practices, and to improve agricultural governance through an interactive and responsive AET system.
- Despite the "brain drain" and cutbacks in aid for overseas training, African universities must rapidly expand postgraduate programmes and research output. Success in this regard requires transforming Africa's approach to AET—a process that should be informed by the valuable lessons learned from the experiences of *Latin America and Asia.*
- Human capital, technology, institutions, and biological and physical capital are *four main drivers of agricultural development.*
- The *critical elements of a transformed AET system* include focused support and financing; relevance of training abroad; the returnee challenge and the role of incentives; and linkages between key stakeholders.
  - There is a renewed interest in investment in African AET. However, a lack of coordination across initiatives and countries results in *fragmented donor funding,* which dilutes investments and prevents achieving a critical mass for impact. The recently-established TEAM-Africa (chapter 1) will play a critical role in coordinating AET investments and enhancing ownership by African partners.
  - Long-term *overseas training* in and of itself is a narrow and incomplete approach to capacity building. Two specific challenges result from such training: 1) a high percentage of African students remain overseas after degree completion; and 2) the applied disciplines of AET are not always appropriately contextualised for African needs.
  - Careful in-country selection, mentoring while overseas, small grants for use following return, and formal links between long-term overseas training and national [African] research programmes all *increase returnee rates* for African students studying abroad. Monetary incentives have also been used successfully by developing countries such as Malaysia, Chile, Brazil, and Cameroon to develop and retain human capacity for AET.
  - *Offering applied disciplines at African institutions* rather than abroad will ensure appropriate contextualisation. This requires productive and sustainable departments of agriculture, which in turn require a multi-generational time frame and an indigenous and cost-effective human capital renewal system. Unique models for AET, such as India's state agricultural universities and the American land grant model should also be considered.
  - Rwanda's development of a long-term national information strategy makes it is one of the leading countries on the continent in terms of the rapid *increase in the use of ICTs* for development in Africa.

As noted in Chapter 1, Africa faces a crisis related to the under-performance of its vital agriculture sector.

- Population growth in sub-Saharan Africa is expected to more than double, from 809 million in 2008 to 1.7 billion in 2050 (Thompson, 2009).
- During the past five decades, per capita production of cereals has fallen by approximately 15 percent. Cereal yields have risen less than 50 percent compared to rises of 180–300 percent in Asia (FAOSTAT, 2011).
- Africa's share of world agricultural trade declined from 6 percent during the 1970s to 2 percent in 2000, before recovering somewhat to 3.5 percent in 2009 (FAOSTAT, 2011).
- Genetically modified (GM) crops are produced commercially by only three countries in Africa: South Africa, Uganda and Burkina Faso, with negative consequences for increased cereal production (Eicher et al., 2006).

It is in this context that this chapter reflects on the crisis in AET in Africa. It draws insights from the experience of Africa's faculties and schools of agriculture and natural resources and advances an agenda to reform and revitalise AET over the next twenty to twenty-five years (in other words, between now and about 2040). This agenda for reform is informed by the recently-established Tertiary Education for Agriculture Mechanism in Africa (TEAM-Africa) initiative, described in Chapter 1.

## Donor Support for AET in Africa

During the 1960s and 1970s, there was optimism about Africa's ability to catch up with Asia by importing agricultural technology and new crop varieties. For example, the International Crops Research Institute for the Semi-Arid Tropics (ICRISAT) exported sorghum varieties from India in the 1970s and, after a decade, reported that "the results were generally disappointing. Very little of the introduced material was adapted to West African conditions. It was obvious that the project had to go back to basics to plan a longer term programme" (Swindale, 1984).

During the 1960s and 1970s, many African political leaders and intellectuals shared dreams of becoming industrial nations by the year 2000. Under the leadership of Kwame Nkrumah, the government of Ghana abolished its national extension service for smallholder farmers and built state farms modelled after those in the Soviet Union. Today, Ghana has a struggling agriculture sector and a per capita gross national income of US$520 (World Bank, 2007). What happened? Why have Ghana and virtually every country in sub-

Saharan Africa (hereafter Africa) failed to achieve an agricultural transformation after a half century of development projects and billions of dollars of foreign aid?

The role of science and technology in promoting African agricultural development has been given boosts of promise and dreams of payoff during Africa's fifty years of independence. The first call to action came when Kenya's Minister of Economic Planning, Thomas Mboya, addressed the opening meeting of the Economic Commission for Africa and articulated a vision of Africa becoming a major food exporter:

> No matter how successful our efforts are to industrialise, Africa will be for many generations primarily a producer of agricultural and other primary products. We must learn to do it well and on a rapidly growing scale. This will require a massive frontal attack, not only on the research needs to which I have already referred, but also on the practical problems of production, storage and marketing (Mboya, 1967).

However, twenty-five years later Vernon Ruttan (1991) reported:

> The thing that bothers me is that the donors have consistently tried to avoid the issue of institution-building in Africa. In South and Southeast Asia in the 1950s, the donors were building the institutional capacity it took to create the growth that began in the 1960s. In the 1970s, we didn't do it in Africa because we were on the basic needs and rural development kick. An agronomist was viewed as doing elite stuff. A plant breeder was even more elite. I think it's time that the donors begin to take the issue of institution-building seriously or in 2010 we are going to be having this same conversation.

In the 1960s, Africa's new universities focused on training BSc students to replace colonial research officers and teachers (Odhiambo, 1967). The new universities were set up primarily for teaching purposes, while the ministries of agriculture continued to carry out research and extension programmes. Most new governments invested in industrialisation as a means of bypassing the agrarian stage of development. However, this industry-first strategy failed throughout Africa.

Since then, the experiences of Asia and Latin America have demonstrated that there are four main drivers of agricultural development: human capital, technology, institutions, and biological and physical capital. The coordination and sequencing of these interlinked investments is a complicated political and technical exercise at both the micro and macro levels (Bonnen, 1998). Lele (1991) studied the agricultural development experience of six countries in Africa in the 1980s and concluded that "the World Bank has not had a long-term strategy for broadly-based growth in Africa, nor has it fully appreciated the need for a balanced accumulation of human, institutional, and technological capacity for an appropriate sequencing and phasing of investments."

During the 1980s and 1990s there was widespread belief that investment in primary and secondary education would generate higher rates of return than investments in higher education. The World Bank and many donors accepted this belief and reduced their investments in higher education during this period. But Schultz (1981), Bloom et al. (2006), and others have shown that higher education can produce public as well as private benefits. Many donors now accept this updated view of the payoff to investment in AET and recognise the importance of these programmes in Africa, because universities (Eicher, 2006, Rivera, 2006):

- often have the potential to support agricultural research and extension programmes using existing staff and facilities (e.g. libraries, laboratories, demonstration farms) at little extra cost;
- can access global research findings and share this information with academic staff and students, as well as researchers in National Agricultural Research and Innovation (NARIs) and instructors in extension training programmes; and
- are the principal means for replenishing the stock of human capital in research, extension and agribusiness organisations in Africa.

In addition, agriculture is highly location-specific. Appropriate AET requires a detailed and intimate knowledge of local farming systems.

It is generally accepted that AET is a critical component for agricultural development, confirmed by examples from Vietnam, Malaysia, Japan, India, and the U.S. (Eicher, 2006; Swanepoel and Stroebel, 2012). Moreover, the importance of addressing capacity gaps to enable Africa to achieve a sustainable growth of 6 percent in the agriculture sector has been widely articulated within Africa and was explicitly addressed in the Communiqué from the Ministerial Conference on Higher Education in Agriculture in Africa (CHEA, 2010) held in Uganda in 2010. As a consequence, role players in the sector have come together in a series of workshops and discussions involving academic experts, development partners, and African AET leaders, which has resulted in the establishment of TEAM-Africa. This Mechanism has been established to facilitate the integration of AET into the Comprehensive African Agriculture Development Plan (CAADP) processes and plans to catalyse change in these institutions. The goal should be to produce graduates able to identify, analyse and resolve real-world problems and to mobilise and harmonise investments to achieve the critical mass needed to meet these challenges.

The third AET Partnership meeting, hosted by Stellenbosch University, South Africa in November, 2012, was the first of these meetings to be held in Africa. The meeting highlighted the importance of a shared vision and cohesive strategy supported by the various African role players, from senior political leaders at continental and regional levels, to professional networks and institutions involved in AET (Swanepoel and Stroebel, 2012).

## Higher Education in Africa

When sixteen African colonies won their independence in 1960, there were a few high quality African universities with an average enrollment of several thousand students. In 1961, the United Nations Educational, Scientific and Cultural Organisation (UNESCO) passed a motion that all African States should "aim at achieving universal primary education within a minimum of twenty years" (UNESCO, 1961). In 1962, "Development of Higher Education in Africa," a conference convened by UNESCO in Tananarive, Madagascar, endorsed an academic staff-to-student ratio of one to ten and a minimum enrolment of five thousand students per university (UNESCO, 1963). Looking back, the 1960s and 1970s were marked by a dramatic increase in student numbers in African education. For example, the University of Ibadan in Nigeria doubled its enrolment to more than fourteen thousand with no significant expansion in the physical facilities that had earlier accommodated only five thousand (Ekong, 1993).

By the mid-1980s, higher education in Africa was in a deep financial crisis (Saint, 1992; Ajayi et al., 1996). During the 1980s donors sharply reduced support for AET, while the number and size of universities kept increasing. Today, the University of Bamako in Mali has around fifty thousand students, followed by forty thousand at Makerere University in Uganda and thirty eight thousand at Egerton in Kenya. There are today an estimated two hundred public universities in Africa, with agriculture taught at roughly one hundred of these.

The current AET crisis is thus part of the broader challenge in the African higher education system.

## Elements of a Transformed and Revised AET System

International evidence suggests that the following dimensions have a critical role to play in moving towards a transformed and revised AET system in Africa that can meet the continent's need for skilled people to spearhead agricultural renewal.

## Focused support

As role players in the sector respond to this challenge in reforming and revitalising their AET programmes, they can gain some insights from Asia's institution-building experience. In the 1950s, when the Rockefeller Foundation decided to shift its programme support from Europe and the U.S. to developing countries, it was faced with a critical problem of choice. Trustees recalled previous periods when "scattering" prevailed. As donors struggled to make choices, fragmented efforts diluted investments, preventing a critical mass for impact. In one case, after grants were made to fifty separate institutions in a single year in Latin America, an evaluation concluded that the impact of the grants was minimal. As a result, the Rockefeller Foundation Trustees endorsed the principal of concentration and decided to assist only one university per country that had "the potential of serving national or regional needs" (Thompson, 1972).

To aid the selection of the universities in Asia, the Foundation met with prospective candidates and asked difficult questions such as:

- What is to be your university's role in the wider geographic region?
- How do you weigh numerical growth against the pursuit of excellence?
- What are some examples of your university serving regional needs?
- Has the leadership made a fresh and self-critical review of its strengths and weaknesses?
- How far has leadership gone in establishing priorities for determining points of emphasis next year, and in three-to-five years hence (Thompson, 1972)?

The Asian experience is valuable today for stakeholders and donors considering proposals from African universities. For example, if a country has four or five public faculties of agriculture and natural resources, a planning grant could provide time (say a year) to develop a joint business plan that eliminates duplicate courses, improves quality, and sets strategies to increase graduate training and research output.

## The changing demand for AET

Rural Africa is changing rapidly. This raises a number of questions about the skills mix needed to transform African agriculture:

- What is the changing demand for education and training at various levels, including primary and secondary levels, schools of agriculture, and universities (Johanson and Adams 2004)?

- What is the demand for education and training in new fields such as food security, marketing, food science, supply chain management, and finance?
- Should students be trained on the job or through distance education?
- Should donors be encouraged to contribute to the development of entire faculties of agriculture or to filling academic silos, discipline-by-discipline, such as agricultural economics?

As the rural world in Africa changes, young people need to be prepared to capitalise on new opportunities. However, the World Bank (2007) contends that education and training institutions in Africa are not adequately prepared to meet this challenge. Without question, improving the quality of basic education is a high priority for African governments. Industry and smallholder farmers' needs have to be determined to guide the development of AET programmes offered by African governments and the private sector (Kingslow, 2007; World Bank, 2008). In Namibia, the government has set up seven community skills centres to impart basic skills through wage or self-employment, while local experts carry out demand studies to identify learning skills required in these programmes to meet the needs of youth and employers.

Demand studies need to be carried out to document what kinds of foundational skills such as literacy, numeracy, and basic science knowledge are needed by children who are likely to become farmers and by those who are likely to migrate to urban centres. In many countries there is undue emphasis on training students to be farmers, even though a high percentage will most likely spend their lives in nearby market towns and cities.

In higher education in Africa, there is an urgent need to upgrade the curricula to meet the needs of modern agriculture, with emphasis on national and global issues such as food security, oil, water, marketing, and climate change. There is an urgent need to conduct demand-driven studies of the skills needed by university graduates to address these new challenges (Blackie et al., 2009). A growing number of faculties of agriculture have developed new and relevant programmes in marketing, agribusiness, and food science. For example, Egerton University in Kenya is currently receiving more than one hundred applications a year for its BSc in Food Science programme, but only has capacity to admit fifty new students. Without question, both African governments and donors need updated demand studies to guide the allocation of African and donor support for AET.

In this context a significant initiative, TEAM-Africa, has been established by a number of Africa's development partners to support the emerging African vision for transformation of AET on the continent. The process that led to the establishment of TEAM-Africa is elucidated in this chapter.

## Relevance of overseas education and training

In European industrialised countries, the agriculture sector directly contributes less than 2 percent to gross domestic product (GDP) and employment (Beintema and Elliott, 2009). This reality is reflected in the fact that some European faculties of agriculture have changed their titles to faculty of bioscience, or bioscience and engineering, or life sciences. For example, Wageningen Agricultural University in the Netherlands changed its name to Wageningen University and Research Centre (WUR). KVL, the well-known Agricultural and Veterinary University of Copenhagen, has merged with the Danish University of Sciences and the University of Copenhagen. It is now called the Faculty of Life Sciences of the University of Copenhagen, the largest university in Scandinavia. Neither agriculture nor livestock is mentioned on the KVL web site because of student concern in Europe over the environment, health (animal diseases), and animal welfare. In the Netherlands, food and livestock production matters form only part of the agenda of the Ministry of Agriculture, Nature Management and Fisheries.

With this broadening of the agricultural curriculum in European universities, tropical agriculture is being relegated to a minor specialisation, taught as a part of rural development in the tropics. The legendary Wye College in the United Kingdom (UK) closed its doors after 125 years of operation. With the population expected to double in many countries in Africa in twenty-five to thirty years, food production must be increased to help meet food security goals. African scientists require education and training in the traditional applied disciplines such as soil fertility management, plant breeding and plant protection, post-harvest management, livestock production, and animal nutrition, contextualised for the realities of smallholder farming systems and tropical agricultural systems. In order to ensure that this contextualisation is practically demonstrated, there is a need for these applied courses to be offered by African institutions. Where in Europe or North America could an African agronomist receive practical exposure in the agronomy of palm oil, cassava or rubber production?

## The returnee challenge

Despite the well-known benefits of long-term overseas training of postgraduate students, the percentage of students who remain overseas after degree completion is considered high (Eicher, 2006). This topic has generated numerous debates among African educators and other stakeholders. Data on this issue is sparse and anecdotal. However, it is possible to draw on two sources: the Asian experience of the 1960s (Stevenson and Locke, 1989) and a study of MSc and PhD students supported in various agricultural disciplines in universities in Africa and Latin America (Jamora, 2007).

The Agricultural Development Council (ADC) of New York was set up by John D. Rockefeller in 1953 to support agricultural development in Asia. During its thirty-two-year life, the ADC provided fellowships for 532 Asian students to study overseas, with an impressive 91 percent of the students returning home (Stevenson and Locke, 1989). This high rate was credited to institutional innovations and incentives students received after returning home, such as funding to attend professional meetings, small research grants and a supportive environment for young returning scholars to build productive careers. African governments and donors should study the history of the ADC in Asia, which is regarded as an important success story. Its total expenditure over its three-decade existence was US$100 million (adjusted to 1985 rates).

A study by Jamora (2007) tracked ninety-seven MSc and PhD graduates enrolled in Collaborative Research Support Programmes (CRSPs) financed by the U.S. Agency for International Development (USAID), in fifteen countries in Latin America and Africa between 1980 and 2005. It found that 84 percent of the graduates had returned home, 12 percent remained overseas, and around 4 percent were still studying. The question is: why were the returnee rates so high in the ADC study in Asia and in the recent CRSP study in Africa and Latin America? From these studies, it is clear that there are several reasons for the high returnee rate. Students were carefully observed in their home country by local and visiting U.S. researchers before being awarded a scholarship, they were mentored in overseas classrooms and labs, and they secured small research and travel grants from the Foundation and CRSPs for a period of time after they returned home.

From the analysis of the above two case studies it is clear that long-term training overseas should be coupled with support given to national research programmes and faculties of agriculture. This is to provide young researchers with access to laboratories and professional incentives once they return, so they can become researchers, teachers and agribusiness managers. Long-term overseas training alone is a narrow and incomplete approach to building capacity and developing agricultural skills for both public and private sectors.

## Small countries and regional networks

Small countries present special problems for development planners and educators. Using the World Bank Institute definition of a small country— one that has fewer than 1.5 million people (Venner, 2009) —eleven of Africa's fifty-one countries would fall into this category. Most of them do not have schools and faculties of agriculture because the small number of students interested in agriculture can pursue their education in neighbouring countries. Therefore, around forty African countries would need support for their AET. Donor aid is still distributed to some extent on a fragmented basis to universities. Many of these countries have difficulty supporting a critical mass of educators and scientists, and therefore increasingly work in partnership with other institutions and regional networks (Eyzaguirre, 1996; CHEA, 2010, Ekwamu, 2012).

However, there are risks associated with regional institutions. For example, civil unrest in the Ivory Coast in the 1970s and in Zimbabwe over the past two decades highlights the danger of concentrating African capacity-building programmes in a few regional centres of excellence. Experience has shown that even though donors have underwritten regional institutions of higher education for a decade or longer, it often has been difficult to get African political leaders to make financial commitments to continue programmes after foreign aid is finished (Mrema et al., 2004). An evaluation of a GIZ (*Gesellschaft für Internationale Zusammenarbeit*) financed MSc project of thirteen years in southern Africa concluded that this regional Masters programme was highly successful in building capacity in the focused disciplinary areas hosted by regional partners (Wollny et al., 2002). In addition, a recent review of the Regional Universities' Forum for Capacity Building in Africa (RUFORUM) concluded that the regional Masters and PhD programmes offered under the auspices of RUFORUM have been successful. These successes include strengthened capacity of universities to foster innovations and being responsive to the demands of smallholder farmers through transformed and relevant AET programmes that help meet the continent's need for skilled people to spearhead agricultural renewal (Tusiime et al., 2012).

## The financing of AET

Beyond improving on the disjointed support to returning students and the uncoordinated efforts of regional programmes, there is a need for increased investment in AET to flow through much less fragmented mechanisms than has been the case in the past (FARA, 2011). While there is a renewed interest in investing in African AET, there has been little if any coordination across initiatives and, at times, little coordination among institutions in the countries receiving investments. The limited attention given to AET within the CAADP framework—which presents Africa's primary mechanism for coordinating investment in the agriculture sector across donors, governments, and the private sector—is both a symptom of and a contributor to this lack of coordination. African governments and development partners recognize these challenges, and funding for the sector has begun to increase along with interest in greater coordination. From about 2012 onwards there is a real possibility of much greater funding from development partners, from African governments (with increased fiscal resources due to rapid economic growth), and potentially from new large-scale private investors in Africa's rural sector. A key question is how to coordinate new investments to achieve the maximum positive impact. To this end, TEAM-Africa will play a critical role in facilitating better coordination for new and increased investments aimed at maximum positive impact.

## The critical role of incentives

Many commentators have criticised African students who do not return home, while failing to mention that many African countries treat their scientists like "high class clerks in the bureaucracy of their government" (Schultz, 1984). Today, a large number of commentators are promoting the idea that African scientists in the diaspora should return home to pursue their professional careers and assist their countries. Others argue that it is important to nurture and care for the brains remaining in the country.

The role of incentives should be given more attention in institution-building programmes (Rukuni et al., 1998; Jamora 2007). For example, Malaysia and several Latin American countries have skilfully used incentives to develop and retain human capital in universities and National Agricultural Research Systems (NARS). In Malaysia, overseas training is considered an investment in the national research system. Malaysia has used monetary incentives such as subsidized automobile and housing loans, and non-monetary incentives such as a free trip to Mecca to reward and retain scientists. New research officers in Malaysia have an opportunity for overseas study after one to three years on the job (Eicher, 2006). Chile has developed world-class PhD programmes, and its universities are an extremely important part of NARS. Competitive research grant programmes represent a large share of research funds for universities (Byerlee and Echeverria, 2002). Today, Chile has sixteen universities with agricultural research capacity. In short, Chile has developed a system of institutions and incentives to reward and retain scientists for decades.

Brazil has undertaken a human capital improvement programme and, at the same time

- expanded the number of graduate programmes;
- increased the percentage of faculty members with advanced degrees;

- purchased books and journals for the libraries, and equipment for the laboratories;
- created global human capital chains; and
- generated a demand for university graduates, research, and faculty consultancies in response to the growth of private input and marketing firms in the 1970s and 1980s (World Bank, 2012).

In Cameroon, a positive example of rewarding scientists in Africa began unfolding in 2009 when the government made an important political decision to increase salaries at the University of Yaoundé and set up a central fund of US$9.5 million to increase the number of scientists and stabilise the research environment. Prior to 2009, the top salary for professors was US$500 per month. The new salary for a professor is US$1850 per month, and lecturers are now paid US$1100, up from US$490. The President has vowed to increase salaries every year because he contends that "universities are at the forefront of development" (SciDev, 2010).

## Crafting linkages between faculties of agriculture, AGRA and NARS

The decision of the Bill and Melinda Gates Foundation (BMGF) to invest in African agriculture and support institutions has encouraged other donors to increase their support for agricultural development programmes in Africa (Lele et al., 2012). The BMGF's commitment to agriculture is an admission that massive investments in health and education in Africa are unlikely to be financially sustainable unless they are financed by rising rural incomes. In addition, the BMGF and the Rockefeller Foundation pooled their resources to develop the Alliance for a Green Revolution in Africa (AGRA) with a primary goal of developing high-yielding food crop varieties by drawing on the experience of the Rockefeller partnership with the Government of Mexico that began in 1943. Under this programme, the late Norman Borlaug devoted his career to the development of high-yielding Mexican wheat varieties that were transferred to India in the 1960s and served as the foundation of the Asian Green Revolution (Lele and Goldsmith, 1989).

If AGRA's five-year budget of US$150 million were equally distributed among the 50 countries in sub-Saharan Africa, it would represent only US$3 million per country for five years. The AGRA model focuses on plant breeding of a few food security staples such as maize, complemented by investments in seed production, credit, graduate training, and pilot extension models. AGRA has broadened and operationalised the African Knowledge Information System (AKIS) paradigm of interlinked investments in education, research and extension. It has added input and product markets to ensure that non-governmental organisations (NGOs), producer groups, and private input suppliers are linked with public sector plant breeders. This initiative represents a turning point in thinking about what needs to be done to generate an African green revolution (Ejeta, 2010). Because of the failure of the single input (fertiliser, seed) and single factor (extension, credit) models to achieve African green revolutions, AGRA is combining plant breeding with research on pilot seed, fertiliser, and credit programmes.

India's Green Revolution experience offers some important insights for the implementation of AGRA's strategy for Africa. In the early 1960s, USAID, the Rockefeller Foundation, and the Ford Foundation assisted India's Ministry of Agriculture in building the scientific foundation for training scientists and boosting cereal yields in India. India reached a consensus on how to tackle its agricultural education problems. It set up commissions, such as Indo-American Study Teams, that held meetings throughout India and visited a large number of overseas countries to observe different agricultural institutions in practice (Goldsmith, 1990). These meetings helped nurture political and scientific agreements to develop a new type of agricultural university—a state agricultural university (SAU). To train Indian scientists at the postgraduate level, USAID helped to fund graduate studies for one thousand Indian scientists studying agriculture in the U.S. and mobilised five U.S. universities to help develop nine new SAUs and supply three hundred professors on two- or three-year assignments. Today, there are forty-one SAUs in India, with at least one in every state (Herdt, 2010).

Furthermore, the unique and internationally recognised American Land Grant Model of AET, whereby a college of agriculture at a university manages the three arms of the agricultural knowledge triangle (agricultural education, research, and extension), also offers particular insights for Africa. Many African countries attempted to replicate the three components of the model without success. The main reasons for this failure include path-dependency (i.e. the historical development process of agriculture in different countries) and entrenched government research and extension departments (Lynam and Blackie, 1994). A detailed account of the Land Grant Model and it implications for Africa can be found in Chapter 5 of this volume.

## Increasing the use of ICT

Over the past decade, there has been a rapid increase in the use of information and communication technology (ICT) to spur development, open new markets, and support teaching and research. There is abundant evidence that IC technologies such as radio and television, mobile phones, and the Internet can leapfrog poor infrastructure, thus bypassing drawbacks such as distance from markets. Moreover, mobile phones have expanded without the use of government intervention because the mobile phone providers are private companies and economically self-supporting. All governments have to do is establish a clear and transparent regulatory framework. Currently, there are around 400 million mobile phones in Africa, of which about 30 percent are in rural areas (Chimwaza, 2010).

Mobile phones and community radio programmes are especially popular in Africa because they can provide farm news and supply farmers with information on market prices and trends. In addition, they can provide intelligence on new market opportunities and technical information on how to garner seed and credit to tap these new opportunities.

ICT has also increased Africa's access to global scientific literature. Scientific isolation is a common problem for researchers and academic staff who do not have access to the latest scientific literature or the funds to purchase scientific journals. In 1990, the Rockefeller Foundation addressed this problem by supporting the work of librarians at Cornell University developing The Essential Electronic Agricultural Library (TEEAL), an electronic library of 140 scientific journals from 60 major publishers that are made available free of charge to users in poor countries. The TEEAL system is available on CD or a Local Area Network (LAN). African institutions appreciate TEEAL because of its high quality and the fact that it does not rely on phone, Internet connectivity or bandwidth, making it a workable solution for research stations and agricultural schools that cannot yet offer reliable and affordable Internet access to the majority of their academic staff and students.

TEEAL represents a brilliant institutional innovation, but numerous donors and publishers have raised questions about its coverage and use. These questions were addressed by RUFORUM, which carried out a survey in 2009 and found that TEEAL was available at an impressive 80 percent of the then twenty-five RUFORUM universities (RUFORUM, 2009). However, the report also found that most ICT use by researchers and teachers in RUFORUM universities was for personal use and recommended that e-learning classes should be used to expand the use of ICT in teaching and research.

Community radio is popular in Africa. For many smallholder farmers it has been an important way to obtain agricultural information. However, radio was historically a "one-way technology" because of the lack of means for listeners to contact the radio stations. Now mobile phones permit a two-way flow of information. In Mali, community radio delivers market news to farmers, but the large number of local languages (up to seven in parts of the country), makes it expensive to operate a radio service that can serve such a diversity of languages.

Rwanda is in the forefront in Africa in terms of developing a long-term national information strategy. In the mid-1990s, President Kagame made a commitment to a twenty-year development strategy that encompasses ICT at its core. Rwanda plans to supply information to the agriculture sector and use the expanded agricultural production and farm incomes to move from a predominately agricultural economy to a predominately information and knowledge economy (Torero and Von Braun, 2006; Centre of Educational Technology, 2007).

### Increasing PhD and research output from faculties of agriculture

Since academics in African universities typically spend about 25 percent of their time on research, it is important that their research address high priority national development problems rather than turning out what one university's vice-chancellor called a "thick sludge of consultancy reports" for donors. The key to greater research output is to invest public resources in strengthening graduate programmes and research facilities in order to increase funding for African graduate students who can work in research partnerships with their academic advisors. For example, the Federal University of Ceará in Brazil produced 335 MSc theses between 1973 and 1989 (Fox, 2006). This is a low-cost, proven model to increase the research output of universities. At the same time, because of heavy teaching loads on staff, only a trickle of research output can be expected from African universities that only offer BSc programmes.

## Internationalisation of Agricultural Education

Learning from international experiences is important for AET in Africa. In this context, consider what India learned from the Japanese experience in designing its higher education system. Japan's Meiji Revolution of 1868 launched a quest for gaining scientific knowledge and new technology from abroad. Japan sent a large mission to the U.S. and Europe that brought home a number of mechanical innovations, new crops, and agronomic practices. Japan also imported teachers from the U.S. and Europe and sent over six hundred Japanese students overseas to study agriculture. India and Brazil followed the Japanese model and each sent around a thousand students overseas for training in agriculture in the 1960s and 1970s.

**India** has one hundred years of experience in agricultural education programmes and is aggressively pursuing the internationalisation of its higher education. Currently, several million Indian students are studying abroad.

India has been especially adroit in developing regional and global partnerships in food and agriculture. India has forty-one SAUs, many of which have developing partnerships with overseas universities and research institutes (World Bank, 2012). The SAUs are being charged with playing a major role in training a new category of students to address problems of modern agriculture such as climate change, globalisation, access to water, and value-added market chains. Market-driven curriculum reforms in the broadest sense are being pursued at both undergraduate and graduate levels. India has the capacity to educate and train African postgraduate students in fields such as agribusiness, irrigation, marketing, and food security.

**Brazil** endows its National Agricultural Research System, EMBRAPA, with a budget of US$300 million to support over two thousand scientists. In addition, Brazil has opened an office in Accra, Ghana, with the aim of promoting scientific cooperation between Brazilian and African universities and NARS, especially in Mozambique and Angola.

**China** is an example of how a country lifted more than 600 million people out of poverty within a generation (Ravallion, 2009; World Bank, 2013). China reformed its higher agricultural education system in 2004 and is aggressively building human capital chains with African teachers and researchers (Broadman et al., 2007). Some African countries are currently receiving more graduate scholarships from China than from any other country. The Chinese Academy of Sciences is attracting researchers from throughout the world to come to China and work with their senior scientists, promising foreign researchers more research funding than they were receiving at home. China also has an impressive plan to collaborate with Africa in developing its human capital, including in agriculture (Taylor, 2010).

## Lessons Learned

Africa has a weak scientific foundation for developing a modern agriculture. Long-term public investments are required to build the scientific and institutional foundations for productive agriculture. This is difficult to achieve, but Japan accomplished this goal between 1868 and 1914, and Malaysia, Brazil, Argentina, and Chile have achieved it from a beginning in the 1970s.

Africa has around two hundred public universities today, compared to twenty in 1960. Of the current universities, around one hundred offer courses in agriculture and natural resource management. Africa also has an unknown number of agricultural training colleges and farm institutes that can produce the middle-level human resources and skills critically needed for the development of the agriculture sector.

Building productive and sustainable faculties of agriculture requires a multi-generational time frame. International experiences in capacity-building have shown that the development of an indigenous and cost-effective human capital renewal system is central to the long-term sustainability of universities and research-and-extension institutions. However, due to cutbacks in foreign aid for overseas training as well as the problem of brain drain, African universities need to expand graduate programmes and research outputs.

Accelerating agricultural development often depends on institutions and incentives as much as policy reforms. For example, at the University of Pretoria in South Africa, economists who studied the introduction of GM cotton in South Africa over the past decade described the venture as a "technological triumph and an institutional failure" (Gouse, 2009).

Africa has had decades of experience in building and operating schools of agriculture to train extension workers. Many of these schools have been upgraded to faculties and universities of agriculture. Sicherman (2005) traced the history of the Uganda School of Agriculture from 1924 to its makeover as Makerere University in 1970, and Oniang'o and Eicher (2004) studied agricultural education in Kenya. These and other studies have documented the chronic lack of funding for most agricultural schools and the frequent lack of cooperation between ministries of agriculture and ministries of education over AET matters.

## Concluding Comments

Most donor aid to Africa is distributed on a piecemeal basis to various ministries and a vast array of university partnerships, NGO training programmes, and regional programmes such as the Collaborative Master's in Agricultural and Applied Economics (CMAAE). However, some regional PhD programmes are being launched by African universities without an assurance of funding and adequate academic staff. There is a need to move from a donor-led piecemeal approach to a national policy dialogue on how to reform and revitalise AET in Africa.

African nations, universities, and donor organisations should come together and agree that the AET crisis will require the "massive frontal attack" recommended by Kenya's former Minister of Economic Planning, Thomas Mboya, in 1967 (Eicher, 1999). As a leading example of this, there are high expectations by all stakeholders that TEAM-Africa will continue to promote debate on activities within the accepted action plan framework.

It is important to persevere through the challenges of the start-up phase of AET initiatives by continuing the practice of open and regular communication and by continuing to build trust.

The financial mechanism to support TEAM-Africa as coordinator in developing AET systems, central in meeting labor market needs for agriculture, rural development, and natural resource management, is crucial. Aligning support around a broader African Strategic Plan for AET in Africa that leverages the abilities of existing networks would allow development plans to ensure harmonisation within the framework of CAADP and TEAM-Africa. Hopefully, this will eliminate further fragmentation of financial investments for African AET while enhancing ownership by the African partners. Various potential donors have expressed interest in channeling their funds through TEAM-Africa to create an environment of sustained growth and support.

## References

Ajayi, J., L. Goma, and G. Johnson. 1996. *The African Experience with Higher Education*. Ohio University Press: Athens, Ohio.

Beintema, N. and H. Elliott. 2009. Setting Meaningful Investment Targets in Agricultural Research and Development: Challenges, Opportunities, and Fiscal Realities. Paper prepared for the expert meeting "How to Feed the World in 2050." October 12–13, Rome, Italy.

Blackie, M., M. Mutema, and A. Ward. 2009. *A study of Agricultural Graduates in Eastern, Central and Southern Africa: Demand, Quality and Job Performance Issues*. A report to ASARECA and RUFORUM, RUFORUM, Kampala.

Bloom, D., D. Canning, and K. Chan. 2006. *Higher Education and Economic Development in Africa*. World Bank. Washington, DC.

Bonnen, J. 1998. Agricultural Development: Transforming Human Capital, Technology, and Institutions. In C.K. Eicher and J.M. Staatz, Eds. *International Agricultural Development, Third Edition.* Johns Hopkins University Press: Baltimore, USA.

Broadman, H, G. Isik, S. Plaza, X. Ye, and Y. Yoshino. 2007. *Africa's Silk Road: China and India's New Economic Frontier*. World Bank. Washington, DC.

Byerlee, D. and R. Echeverria, Eds. 2002. *Agricultural Research Policy in an Era of Privatization.* CABI Publishing: New York, USA.

Centre of Educational Technology. 2007. *ICTs and Higher Education in Africa.* University of Cape Town, Cape Town, South Africa.

CHEA. 2010. *Ministerial Communiqué issued at the Ministerial Conference on Higher Education in Agriculture in Africa (CHEA).* November 15–19, Kampala, Uganda.

Chimwaza, G. 2010. *E-mail*. March 9.

De Schutter, O. and G. Vanloqueren. 2011. The New Green Revolution: How Twenty-First-Century Science Can Feed the World. *Solutions Journal* 2: 4, 33–44.

Eicher, C. 1999. *Institutions and the African Farmer.* Consultative Group on International Agricultural Research. Washington, DC.

—. 2006. *The Evolution of Agricultural Education and Training: Global Insights of Relevance for Africa.* Staff Paper 2006–26. East Lansing, Michigan, Department of Agricultural Economics, Michigan State University.

Eicher, C., K. Maredia, and I. Sithole-Niang. 2006. Crop Biotechnology and the African Farmer. *Food Policy* 31:6, 504–527.

Eicher, C. and L. Witt, Eds. 1964. *Agriculture in Economic Development.* McGraw-Hill: New York, USA.

Ejeta, G. 2010. African Green Revolution Needn't Be a Mirage. *Science* 327: 831–832.

Ekong, D. 1993. *Challenges Facing African Universities: Notes of an Address to the Royal African Society.* Association of African Universities. Accra, Ghana.

Ekwamu, A. 2012. Higher Education Regional Initiatives and Experiences in sub-Saharan Africa. Africa-U.S. Higher Education Initiative Meeting. February 13–15, Addis Ababa, Ethiopia.

Eyzaguirre, P. 1996. *Agriculture and Environmental Research in Small Countries: Innovative Approaches to Strategic Planning.* John Wiley and Sons: New York, NY.

FAOSTAT. 2011. *Food and Agricultural Production.* Food and Agriculture Organization of the United Nations. Rome, Italy.

FARA. 2006. *Framework for African Agricultural Productivity (2006).* Forum for Agricultural Research in Africa. Accra, Ghana.

FARA. 2011. *Transforming African agriculture through research, advisory services, education and training.* Forum for Agricultural Research in Africa. Accra, Ghana.

Fox, R. 2006. E-mail 27 February, 2006. University of Arizona.

Goldsmith, A. 1990. *Building Agricultural Institutions: Transferring the Land-Grant Model to India and Nigeria.* Westview Press: Boulder, Colorado.

Gouse, M. 2009. Ten Years of Bt Cotton in South Africa: Putting the Smallholder Experience into Context. In Tripp, R., Ed. *Biotechnology and Agricultural Development: Transgenic Cotton, Rural Institutions and Resource-poor Farmers.* Routledge: London.

Haggblade, S. 2013. Unscrambling Africa: Regional Requirements for Achieving Food Security. *Development Policy Review* 31, 149–176.

Herdt, R. 2010. Development Aid and Agriculture. In Pingali, P. and R. Evenson, Eds. *Handbook of Agricultural Economics, Vol 4*. Elsevier: London.

Jamora, N. 2007. *Assessment of the Impact of CRSP Graduate Training in the United States on Capacity Building in Latin America, Africa and the United States.* Michigan State University (Master's thesis), East Lansing, Michigan.

Juma, C. 2011. *The New Harvest: Agricultural Innovations in Africa.* Oxford University Press: New York.

Johanson, R. and A. Adams. 2004. *Skills Development in Sub-Saharan Africa.* World Bank. Washington D.C.

Kingslow, M. 2007. *Agricultural Education and Training in Africa: A Survey of Programmes.* A Report prepared for the W.K. Kellogg Foundation. Kingslow Associates LLC. Annapolis, Maryland.

Lele, U., Ed. 1991. *Aid to African Agriculture: Lessons from two Decades of Donors' Experience.* Johns Hopkins University Press: Baltimore, Maryland.

Lele, U. and A. Goldsmith. 1989. Development of National Agricultural Research Capability: India's Experience with the Rockefeller Foundation and its Significance for Africa. *Economic Development and Cultural Change* 37, 305–44.

Lele, U., M. Agarwal, P. Timmer, and S. Goswami. 2012. *Patterns of Structural Transformation and Agricultural Productivity Growth with Special Focus on Brazil, China, Indonesia and India.* Keynote Address at the Global Conference on Agricultural Research for Development, October, Punta del Esta, Uruguay.

Lynam, J.K. and M.J. Blackie. 1994. Building Effective Agricultural Research Capacity: The African Challenge. In Anderson, J.R., Ed. *Agricultural Technology: Policy Issues for the International Community.* CAB International: Wallingford, UK.

Mrema, G., M. Noor, P. Eyzaguirre, and M. Toure. 2004. "Economies of Scale through Regional Cooperation in Agricultural Research in Africa." In Ndiritu, C., J. Lyman and A. Mbabu, Eds. *Transformation of Agricultural Research Systems in Africa.* Michigan State University Press: East Lansing, Michigan.

Mboya, T. 1967. *A Development Strategy for Africa: Problems & Proposals.* Ministry of Economic Planning and Development, Nairobi.

Odhiambo, T. 1967. East Africa: Science for Development. *Science* 158, 876–881.

Oniang'o, R. and C. Eicher. 2004. Universities and Agricultural Development in Africa: Insights from Kenya. In Ndiritu, C., J. Lynam, and A. Mbabu, Eds. *Transformation of Agricultural Research Systems in Kenya: Lessons from Kenya.* Michigan State University Press: East Lansing, Michigan.

Ravallion, M. 2009. Are there Lessons for Africa from China's Success Against Poverty? *World Development* 36: 11, 303–313.

Rivera, W. 2006. Transforming Post-Secondary Agricultural Education and Training by Design: Solutions for Sub-Saharan Africa. Unpublished paper. World Bank. Washington, DC.

RUFORUM. 2009. *Situation Analysis of ICT Capacity and Infrastructure in RUFORUM Universities: Final RUFORUM Report.* November 6, RUFORUM, Kampala.

Rukuni, M., M. Blackie, and C. Eicher. 1998. Crafting Smallholder-driven Agricultural Research Systems in Southern Africa. *World Development* 26: 6, 1073–1087.

Ruttan, V. 1991. Statement Presented at a Seminar on African Development: Lessons From Asia. Winrock International: Morrilton, Arkansas.

Saint, W. 1992. *Universities in Africa: Strategies for Stabilization and Revitalization.* Technical Paper No. 194. World Bank. Washington D.C.

SciDev. 2010. *Cameroon Puts Brain Drain into Reverse.* Science and Development Network, November 1.

Schultz, T. 1965. *Economic Crises in World Agriculture.* University of Michigan Press: Ann Arbor, Michigan.

—. 1981. *Investing in People: The Economics of Population Quality.* University of California Press: Berkeley, California.

Schultz, T. 1984. The Economics of Agricultural Research. In Eicher, C. and J. Staatz, Eds. *Agricultural Development in the Third World.* Johns Hopkins University Press: Baltimore. Maryland.

Sicherman, C. 2005. *Becoming an African University: Makerere, 1922–2000.* Africa World Press, Trenton, New Jersey.

Stevenson, R., and V.O. Locke. 1989. The Agricultural Development Council: A History. Winrock International Institute for Agricultural Development, Morrilton, AR.

Swanepoel, F. and A. Stroebel. 2012. *Capacity Building for African Agricultural Development: the Need for Transformation in Tertiary Agricultural Education.* Keynote address at the Third TAE-Partnership Workshop on Tertiary Agricultural Education in Africa. November 13–15, Stellenbosch, South Africa.

Swindale, L. 1984. *ICRISAT: Its Programmes and Its Work in Africa.* SADCC Agricultural Research Conference, Department of Agricultural Research. Sebele, Botswana.

Taylor, I. 2010. *China's New Role in Africa.* Lynne Rienner: Boulder, CO.

Thompson, K. 1972. *Higher Education and National Development: One Model for Technical Assistance.* Paper Prepared for a Conference at the Villa Serbelloni, May 3–5. Sponsored by the Rockefeller Foundation and the Ford Foundation. Belagio, Italy.

Thompson, R. 2009. *Long Term Trends in Agriculture Supply and Demand: Implications for U.S. Development Assistance to Sub-Saharan Africa.* Paper Presented at a Conference on Global Agriculture. Chicago Council on Foreign Affairs. Chicago, Illinois.

Torero, M. and J. von Braun. 2006. *Information and Communication Technologies for Development and Poverty Reduction: The Potential of Telecommunications.* The John Hopkins University Press: Baltimore, Maryland.

Tusiime, G., J. Majaliwa Mwanjololo, P. Nampala, and E. Adipala, Eds. 2012. *Proceedings of the Third RUFORUM Biennial Regional Conference on Partnerships and Networking for Strengthening Agricultural Innovation and Higher Education in Africa.* September 24–28. Working Document Series No. 7. RUFORUM, Entebbe, Uganda.

UNESCO. 1961. *Outline of a Plan for African Educational Development.* UNESCO, Paris.

—. 1963. *The Development of Higher Education in Africa*. UNESCO, Paris.

USAID. 1985. *Plan for Supporting Agricultural Research and Faculties of Agriculture in Africa.* Washington D.C.

Venner, K. 2009. *Small States Face Big Challenges.* Development Outreach. World Bank. Washington D.C.

Wollny, C., M. Chagunda, K. Molapong, W. Richter, and P. Mumba. 2002. *African-German Cooperation in Higher Education Supported by the German Technical Cooperation Agency (GTZ): Experiences and Lessons Learned.* DIES I. South-North Dialogue on Innovative Strategies in International Postgraduate Education: Challenges, Obstacles and Best Practices. February 25, German Academic Exchange Service (DAAD), Bonn, Bonn/Berlin.

World Bank. 2007. *World Development Report 2008: Agriculture for Development.* World Bank. Washington, DC.

—. 2008. *Agriculture for Development.* World Bank. Washington, DC.

—. 2009. *Awakening Africa's Sleeping Giant: Prospects for Commercial Agriculture in the Guinea Savannah Zone and Beyond.* World Bank. Washington, DC.

—. 2012. *Agricultural Innovation Systems: An Investment Sourcebook.* World Bank. Washington, DC.

—. 2013. *China Overview*. World Bank. Washington, DC. http://www.worldbank.org/en/country/china/overview

# CHAPTER TEN

# AGRICULTURAL SMALLHOLDER FARMER LEARNING STRATEGIES IN SUB-SAHARAN AFRICA

MARCUS HAKUTANGWI, JOHN MAKINA AND RANGA TARUVINGA

In order to establish an appropriate and practical farmer education framework in sub-Saharan Africa (SSA) and elsewhere, it is important to first understand how farmers learn and how farmer education in its general sense is given. Therefore, this chapter focuses on growing and increasing understanding of farmers' learning strategies, approaches, and methods.

For this purpose, information has been gathered and consolidated from development projects and initiatives in which innovative farmer education and learning has led to improved food production and enhanced value chains. The initiatives included in this analysis cut across crop and livestock production, agro-ecological regions, socio-political and land tenure systems, as well as different implementing agencies such as multi-lateral and bilateral agencies, and international, regional and national non-governmental organisations (NGOs).

Since colonial times, efforts have been made to change rural agriculture from subsistence to commercialised production in order to improve food security and increase income. Various methods and approaches have been used and applied by governments, NGOs, parastatals, bilateral organisations, the private sector and entrepreneurs. This plethora of actors has in most instances gone by with little coordination, leading to duplication and in some instances confusion on the part of the smallholder farmers. The situation has been exacerbated by varying levels of qualification and experience among field and extension agents.

Farmer education and learning strategies are further complicated by the wide variety of terms and concepts used to describe projects, initiatives and programmes. Most of these terms come from the fields of education and agricultural science and have been adapted to suit farmer education. Therefore, it is important to explore in detail some of the terms and concepts commonly used.

***Key messages***

- AET in sub-Saharan Africa must take into account African farmers' learning strategies, approaches, and methods. This can be achieved by considering practical examples of innovative farmer education with proven impact on the quality of food production and value chains.
- Traditionally, there are three components of farmer education and learning systems: the farmer as the learner or main beneficiary of the transfer of knowledge and skills; the extension services or change agents who must recognise how farmers accesses that knowledge; and the knowledge-generating systems, especially agricultural research systems.
- The principles that guide farmer education are closely related to classical adult education principles. Generally, adult learners:
    - learn only what they want to learn;
    - bring their own experiences, which they are willing to share with others, to the learning situation;
    - expect respect and not to be treated like children;
    - bring some solutions and benefits to the problem-solving process;
    - expect rewards and recognition in the learning process;
    - appreciate having an opportunity to practise what they are learning;
    - have critical time constraints; and
    - need to be motivated.
- The landscape of agricultural learning systems in sub-Saharan African countries shows a wide and varied learning terrain consisting of different approaches, execution systems, participation levels, and outcomes. Farmer study groups and learning circles represent one valuable learning approach, as these settings allow farmers to be active participants and to set their own agendas for knowledge acquisition, knowledge management, and the embedding of knowledge in their communities. Other approaches include farmer-to-farmer strategies, farmer field schools, master farmer training, training and visit (T&V) settings, training in environmental and agricultural management (TEAM), community livestock auxiliaries (CLAs), and young commercial farmers' education programmes.
- Knowledge management underpins all of these learning approaches and strategies. It is the interface between rural farmer education and formal education institutions.
- Improving food production and food security in the sub-Saharan region will require a new system of farmer education. The system must take into account farmers, extension agents, and research scientists; it must include nonformal and informal education; and it must emphasise investment in value-added studies.

## Clarifying Basic Concepts

The main concepts explored are education, training, and extension. Traditionally, farmer education and learning systems have three components: 1) the farmer as the learner or main beneficiary of the transfer of knowledge and skills exchange; 2) the extension services or change agents who must recognise how the farmer accesses that knowledge; and 3) the knowledge-generating systems, especially National Agricultural Research Systems (NARS). Farmers' needs are quite specific and personal, yet extension agents often treat them as a homogeneous group.

Knowledge transfer is about communication—the ability to reach farmers in the appropriate way at suitable levels. NARS faces the dual challenge of developing relevant technologies to meet the needs of farmers and transferring that knowledge to them.

### Farmers' needs

The principles that guide farmer education are closely related to classical adult education principles. Alan Rogers (2004) has pointed out that as adult learners, farmers exhibit maturity. This means that they have the ability to expand and utilise their personal talents, and they have the attributes needed to make sound judgment about themselves and others. They appreciate being given autonomy; they assume responsibility for decision making; and in most cases they learn voluntarily.

Adults have some consciousness of what they can and cannot do and have their own learning patterns, which vary considerably. Knowledge needs of farmers can also be differentiated according to "group needs," related, for example, to their farming systems and related production needs, or to their economic status or level of experience.

Adults learn voluntary and intentionally, are independent and self-directed, are problem-oriented and practical. As such, adult education helps the person to learn; the teacher becomes a facilitator and the learner a participant (Rogers, 2004). In some cases, according to Muller (1993), adult learning has a participatory and collaborative element, because adults prefer to meet as equals in small groups to explore issues and concerns and to take common action as a result of dialogue and inter-learning by discourse. The group becomes a "learning co-operative," providing the opportunity for inter-learning. Within the group, the teacher as well as other group members play the role of facilitators.

Muller (1993) identified a "dialectical process of needs negotiation" during which it is important to bring together the needs as felt by the learners and the needs as seen by the adult educators in order to reach consensus on the "real" needs. These real needs must correspond to the experience of adult learners. If adults feel their experience is not being valued, they feel personally rejected. New learning takes on meaning as adults are able to relate new ideas to their own life experiences. Experienced adult educators allow learners to plan and rehearse how they are going to apply their new learning in their day-to-day lives or duties, and to combine training with transfer and application.

To summarise, adults:

- learn only what they want to learn;
- bring their experiences, which they are willing to share with others, to the learning situation;
- expect respect and not to be treated like children;
- bring some solutions and benefits if the learning is about problem solving;
- expect rewards and recognition in the learning process;
- appreciate having an opportunity to practise what they are learning;
- have critical time constraints; and
- need to be motivated.

Linking farmer education and adult education requires recognition of rural farmers as adult learners and extension agents as adult educators. This leads to functional and successful farmer education and extension services.

## Agricultural Extension

Since the 1950s the accumulation of knowledge by agricultural researchers and the dissemination of relevant information to farmers through extension have become formalised and have developed into the discipline of agricultural extension—"assistance to farmers to help them identify and analyse their production problems and become aware of their opportunities for improvement" (Adams, 1982). This includes the provision of information to farmers in order to achieve objectives such as increased agricultural production, adoption of new technologies, and the protection of natural resources and the environment. Other scholars have defined extension, especially as related to agriculture.

Leeuwis and van den Ban (2004) defined extension as a series of embedded communicative interventions that are meant "to develop and/or induce innovations which supposedly help to resolve (usually multi-actor) problematic situations." On the other hand, Maunder (1973) defined extension as "a service or system which assists farm people, through educational procedures, in improving farming methods and techniques, increasing production efficiency and income, bettering their levels of living, and lifting social and educational standards."

Extension is based on various forms of adult education. It is a voluntary educational programme that employs teaching and learning principles, leads to changes in the lives of farmers, and is generally carried out in an atmosphere of mutual trust and respect between the agricultural agents and their clients. Extension education is concerned with three basic educational tasks: 1) the dissemination of useful and practical information related to agriculture and home economics; 2) practical application of such knowledge to help people analyse their problems; and 3) assisting learners in using this technical knowledge to solve problems on their own (Williams, 1981).

Once known as the application of scientific research and new knowledge to agricultural practices through farmer education, agricultural extension encompasses a wider range of communication and learning activities organised for rural people by professionals from different disciplines, including agriculture, health, and business studies. Therefore, agricultural and rural extension has been expanded to include non-farm rural micro-enterprise development, technical extension, marketing extension, and farmers' associations (Rivera et al., 2001).

One of the roles of extension is to help farmers improve production, thereby increasing the value of the goods they produce. However, increasing production without improving downstream activities such as processing and marketing still leaves farmers impoverished due to the cost of increased inputs required to produce more crops. This indicates that, to be successful, extension efforts should include areas such as marketing, financial management, pricing, negotiation strategies, beneficiation of farm produce, local processing, and other agribusiness subjects.

Farmer learning starts with farmer-to-farmer contact, while farmer education and support (as described in Table 10.1) begin with extension auxiliaries at the lower level and end with research officers at the higher level. Operationally, extension auxiliaries and assistants are firmly rooted in the rural small farming sector, while extension specialists and research officers operate in the commercial farming sector. The extension officer who operates in both sectors provides the bridge between the two. Academic institutions and universities continue to increase their participation in extension education through outreach programmes and research in extension itself. The premise is that agricultural extension involves human resource development, which includes education and training. Extension service also includes aspects of technology adoption and use, communication, and information dissemination. The communication component is based on three basic models, namely linear (in conventional extension practices), interactional, and transactional (replacing the linear model in modern participatory extension). A combination of all the players in the development, promotion, and consumption of agricultural extension efforts forms a country's NARS.

| Type /level of Extension worker | Description | Type of institution | Farmer category served |
|---|---|---|---|
| Extension auxiliaries | Community based, with basic training usually in just one aspect of farming, such as livestock auxiliaries. | NGOs, multi- and bilateral agencies | Rural livestock family farmers |
| Extension assistants | Broad training background. Link between farmers and Extension Officers | Government agricultural colleges | Small-scale rural farmers |
| Extension officers | Highly trained. Perform primary, secondary, and referral duties. | Government agricultural colleges | All farmers, rural and commercial |
| Research extension officers and extension specialists | Highly trained in research | Universities | Commercial farmers and extension officers |
| Researchers | Highly trained in research | Universities and research organizations | All farmers, through the extension channels |

**Table 10.1. Functions of the extension cadre**

## National Agricultural Research Systems

In its broadest context, a NARS comprises all organisations in a country that have the principal mission of executing research in the different fields of agricultural sciences, primarily the mainstream agricultural research institutions and universities. Their aim is to generate new knowledge and technology to improve agricultural productivity. Although farmers stand to benefit directly from this new knowledge, there are problems with ensuring that the research is relevant and with establishing practical farmer-research-extension linkages.

Traditionally, research institutions tend to focus on well-supported commodities such as coffee and tea at the expense of subsistence crops. For example, post-independence agricultural research in Zimbabwe focused on the needs of commercial farmers. In most countries, the dominant model for the generation and transfer/exchange of agricultural technology is based on systems of breeding, testing, and distributing crop varieties (Ewel, 1998).

Universities have moved from treating smallholder farmers as objects of research to participatory on-farm research extension. Ewel (1998) identified the separation between research institutions and extension as one of the major limiting factors in developing relevant technology and suggested research extension linkages could be improved through establishing avenues for informal cooperation, setting up research extension coordinating committees including extension field staff in research, and integrating research and extension programmes. Michelsen et al. (2003) recommended similar strategies for universities to respond to the needs of agricultural technology development.

The higher education institutions in Africa continue to increase direct linkages with smallholders and their environments to make some of the technologies generated at these institutions more appropriate for smallholder farmers. However, smallholder farmers must organise themselves to demand appropriate research support.

With regard to farmer education through short courses, many colleges and universities are still described as detached ivory tower institutions. To be more relevant to smallholder farmers, universities could offer short courses for farmers and extension staff and provide accreditation for the farmer-learning that eventually takes place, thus recognising the continuing education of farmers.

## Farmer Education

Adult education contrasts with formal education, which is defined by Coombs and Manzoor (1974) as a highly institutionalised, chronologically graded, and hierarchically structured education system, spanning from lower primary school to the upper reaches of the university. Most agricultural education consists of formal courses in a highly structured system.

However, there is also the lesser known nonformal education, which Coombs and Manzoor (1974) describe as any organised, systematic educational activity carried on outside the framework of the formal system to provide selected types of learning to particular subgroups in the population, adults as well as children. Nonformal education's ability to cater to a wide range of clientele has been used to transform societies, and its variety of programmes has uplifted rural and urban communities alike, from the slums of

Nairobi and the farming villages of India to the housing estates of the UK (Fordham et al., 1979). Because of their relaxed entry requirements, nonformal education programmes suit rural farmers and provide chances to enter formalised education.

Informal education is described by Coombs and Manzoor (1974) as "the lifelong process by which every person acquires and accumulates knowledge, skills, attitudes and insights from daily experiences and exposure to the environment—at home, at work, at play; from the example and attitudes of family and friends; from travel, reading newspapers and books; or by listening to the radio or viewing films or television." Informal education is the learning, knowledge, skills and attitude acquisition that goes on all the time—generally referred to in education circles as 'the school of hard knocks.'

For those who have completed formal courses and are practising in the field, there are learning opportunities through continuing education—the formalised instructional programmes provided for adults after they have left formal education. The programmes bring participants up to date in particular areas of knowledge and skills through such diverse methods as independent study; broadcast, video recordings, online instruction, and other forms of distance learning; group discussion; study circles; conferences, seminars, and workshops; and part-time classroom study.

The relationship between the various components of agricultural education is illustrated in Figure 10.1.

Agricultural education thus can be formal, nonformal, informal, or lifelong/continuing in nature and can manifest itself wherever cultivation, livestock rearing, or harvesting take place. Conservation of natural resources is not a new science. On the contrary, valuable indigenous knowledge about the wise use of resources has been passed from one generation to the other for millennia; thus agricultural education improves quality of life by helping farmers increase production, conserve resources, and provide nutritious food.

## Farmer Learning Strategies

A learning strategy can be described as a group or combination of interrelated, interdependent activities and methods through which individuals acquire knowledge, skills, and attitudes that lead to the permanent alteration of their behaviour. These learning strategies can be elaborate—as in national formal systems, traditional initiation (learning) systems, and agricultural extension systems, or simple—as in a competence learning system.

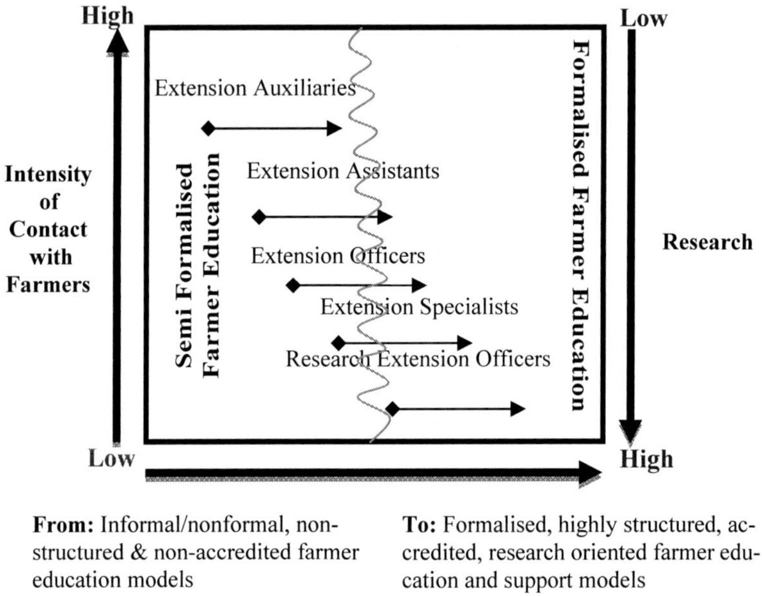


Figure 10.1. Provision of farmer education

The learning strategies explored in this section are mostly subsystems of the traditional agricultural extension learning that has been in use in SSA since the latter half of the last century. Thirty-eight projects in eleven countries were reviewed to identify the farmer learning strategies analysed in this chapter. No particular criteria were applied to select the projects except for the assumption that their existence implied that farmer learning strategies had been employed in production of main food crops (maize, rice, cassava, vegetables) and livestock.

Most projects aim to reorganise and coordinate rural farming communities into learning groups, leading to improved food production and value chain activities. Successful projects have a high level of sustainability mechanisms that include education, integrated project practices, respect for and use of indigenous knowledge and practices, and types and levels of downstream value chain activities (i.e. the extent to which these activities elevate the farmer from primary to secondary and tertiary levels of the value chain). An understanding of the nature and structure of the farmer education systems used in these projects is required.

Projects have a defined geographical area and tend to concentrate and confine activities in this defined area. This means that similar areas in the same country do not benefit from project intervention. The short duration of most project interventions is not sufficient for innovation to take root in the community. Even those projects deemed successful do not continue and are not taken up by the countries' governments. However, farmer learning systems have the potential to be applied to different countries and situations. In the following sections, short descriptions of the various strategies in the eleven African countries are analysed. The learning methods are not necessarily mutually exclusive and may even overlap in some areas.

## Farmer-to-Farmer Learning

This is a common method of learning, especially among communal farmers. It simply involves farmers learning from their fellow farmers and has no formal structures. To capitalise on this informal learning, structured extension systems usually organise these farmers into groups or "clubs" and use demonstration plots as a learning tool. Local farmers are chosen and trained to act as facilitators. The clubs usually meet weekly for training, which includes a visit to the demonstration plot where farmers observe and discuss a specific topic, for example crop management requirements. Demonstrations have a high level of adoptability and afford learning from and teaching of peers. The Garden Africa Project in Lesotho, run by the NGO Maluti Crop Share (MCS) is an example of a successful demonstration for smallholder farmers (Makina and Taruvinga, 2010).

Farmer-to-farmer learning can also be implemented and conducted by the farmers themselves. For example, the "Zaï Field School" approach was initiated in 1992 by the farmers of Somyanga village in the Yatenga Region of Burkina Faso, under the leadership of a local farmer, Ousséni Zoromé. *Zaï* is a planting system that requires planting crops in small holes filled with compost. In an effort to improve farmers' *zaï* skills, the farmers who initiated this farmer-to-farmer extension became service providers without remuneration for their time. This programme was implemented without a donor or implementing organisation other than an innovative farmer.

The scope of learning in this project includes rehabilitating degraded land, improving soil fertility for growing millet and sorghum, and re-establishing woodlands. Each farmer group pays a contribution of 5,000 CFA (US$8.00) to become a member of a regional union. Formal researchers and extension officials who have witnessed the success of this initiative recognise that it is a practical model for participatory innovation development that puts the farmers at the centre of a self-directed process of learning and sharing. Farmers in

Yatenga Region and in other parts of the densely populated Central Plateau of Burkina Faso are becoming increasingly interested in *zaï* (Hakutangwi and Mongbo, 1997).

Four other projects—Training in Environmental and Agricultural Management (TEAM), Lesotho; the Sustainable Agriculture and Natural Resource Management Programme; Garden Africa: Maluti Crop Share (MCS); and Keyhole Gardens, Lesotho (IFAD, 2009)—also illustrate the farmer-to-farmer learning strategy (Makina and Taruvinga, 2010).

Farmer-to-farmer extension is a good example of community learning. It affords peer teaching and learning, which contributes to building strong communities, promotes the use of indigenous knowledge, has a high level of adoptability, and is strongly rooted in basic adult education principles. Emphasis of farmer-to-farmer learning is largely on production skills, although it can also incorporate aspects of the value chain such as inputs acquisition and distribution, purchase and maintenance of farming implements, marketing and sale of surplus produce, processing of grain through large or small milling, packaging, transportation, wholesaling, and retail sales.

Extension agents act as catalysts in setting up farmer-to-farmer learning networks but are generally absent on a day to day basis.

## Farmer Field Schools (FFS)

The Farmer Field School (FFS) emerged in Indonesia in 1988, growing out of the Training and Visit (T&V) initiative, an extension method that emphasises crop management practices and timely interventions by the extension workers. At that time, Indonesia was threatened by a rice insect outbreak that required individual farmers to make complex decisions in their fields. Thus, the FFS form and structure grew from the need for practical field observation and decision-making practices.

In general, Farmer Field Schools consist of groups of farmers with a common interest who get together on a regular basis to study the "how and why" of a particular topic. They are supported by a trained facilitator (extension agent). The topics covered can vary considerably—from Integrated Pest Management (IPM), organic agriculture, animal husbandry, and soil fertility to income-generating activities such as handicrafts. Learning can thus cover various aspects of the value chain and supply chain. FFSs, however, are particularly well suited for field studies, which require specific hands-on management skills and conceptual understanding based on nonformal adult education principles.

The FFS curricula follows the natural cycle of its subject, be it crops, animals, soil, or handicrafts. For example, the cycle may be "seed-to-seed" or "egg-to-egg" approaches that allow all aspects of the subject to be covered. Most FFS programmes exist within larger programmes run by government or civil society organizations and have a good programme leader who can support the training of facilitators, organise materials for the field, solve problems in participatory ways, and nurture field staff facilitators.

In FFSs, extension agent members are the trained facilitators. However, FFSs also promote farmer facilitators. Farmer facilitators are often better than external extension agents because they know the participants and the local terminology, e.g. for insects and pathogens. The hands-on nature of FFS experiential learning represents the good educational practice of using most senses in a learning process. Meetings are scheduled according to crop cycles but also consider farmers' schedules; this means, for example, that FFS meetings are usually held weekly for annual crops and livestock, bi-weekly for long-term crops, and monthly for perennial crops. Functional literacy, children's primary health care, and other practices can simultaneously be included under an adult education team leader.

After a brief introduction in Sudan in 1993 and Kenya in 1995, a larger scale launch of the FFS approach in Africa took place in Zimbabwe in 1997. Today, FFSs are conducted by a wide range of institutions in Africa, such as the Food and Agriculture Organization of the United Nations (FAO), the Danish International Development Agency (DANIDA), many national governments, and numerous NGOs. In Africa, FFSs have provided guidance about overuse or misuse of pesticides, but have also been used to inform farmers on special topics such as health and nutrition in response to farmers' low level of awareness of the dynamics of diseases such as HIV/AIDS and malaria, which are crippling many rural communities (Hakutangwi and Mongbo, 1997).

As a spin-off of FFS, women's groups in Western Kenya have developed "commercial plots," with proceeds shared among the group, adjacent to the FFS learning plots. The International Fund for Agricultural Development (IFAD) funded a four-country effort to develop the methodology by working with these innovative FFS groups. Many other exciting developments, such as Farmer Life Schools (FLSs) and Junior Farmer Field and Life Schools (JFFLSs), which serve thousands of farmers in various parts of Africa, have been supported by international organisations and institutions.

## Master Farmer (MF) Programme

The Master Farmer (MF) programme in Zimbabwe is another example of semi-formalised farmer training for rural smallholder farmers. Under this scheme, agricultural extension workers conduct a two-year training programme for a group of 25 to 30 farmers in crop production, farm machinery, farm management, and livestock production. Farmers have to practice what they learn as a group in their individual fields. The number of sessions attended and what the farmers practice in their individual plots forms the basis for qualifying for a Master Farmer Certificate. Literate farmers can proceed to a second Advanced Master Farmer stage, which has a written examination at the end of a year and offers an Advanced Master Farmer Certificate. The group meets twice a month for four hours. The approach follows training modules and, where possible, the theoretical training is supplemented by demonstrations (Hakutangwi, 2008a; 2008b).

The MF training scheme is centred on a well-resourced farmer training institute with a well-developed curriculum. These are often established by governments or commodity producer associations. Groups of farmers identified in the community come for training. Extension support and follow-up are guaranteed after the formal training. The fact that training takes place at training centres and follows a formalised curriculum is one of the main strengths of this certified farmer training-continuing education strategy.

## Community Livestock Auxiliaries (CLAs)

The Community Livestock Auxiliaries (CLAs) are an example of continuing education programmes. The CLAs teach clients as part of their service, use the revolving fund concept for farm inputs, have a high level of community participation, and offer opportunities for certified nonformal training projects. This strategy was successfully used by the CARE trained community livestock auxiliaries ("barefoot vets") in Zambia (USAID, 2002).

Trained community CLAs were part of a certified, nonformal training and continuing education strategy and provided a scope for strong community participation. Trained CLA members were deployed in the field to teach clients as part of their service. The CLAs used the revolving funds concept to supply veterinary products.

In the Zambian case, training was provided to CLAs by the Veterinary Department. The CLAs were deployed in the livestock-rearing districts, where they undertook routine samplings of animals to monitor disease incidence. They collected blood and lymph node smears from animals and sent the samples to the Veterinary Department for analysis. Where animals tested positive with East Coast fever, red water sickness, and gall sickness, the veterinary services prescribed drugs accordingly, which farmers accessed through the CARE-facilitated drug loan scheme. In addition to treating animals, CLAs also reported to the Veterinary Department on movements of animals as a way of containing the spread of East Coast Fever.

## Young Commercial Farmers Education Programme (YCFP)

The Young Commercial Farmers Education Programme (YCFP) was built around the same principles as the Green Certificate in the UK; the training was a prerequisite to owning farm land and undertaking farming. This approach assumes that commercial farmer education and training is an essential step in ensuring that farmers appreciate and practice farming as a business. The three-year programme, developed by the Danish International Development Agency and based on Danish experience, was piloted in Zimbabwe at Kushinga Phikelela Agricultural College and National Farmer Training Centre, but has since been introduced and expanded upon in Uganda.

YCFP is designed to equip youth with the competence to become full-time commercial farmers through a programme that is 20 percent theoretical and 80 percent practical education and training. The farmers are trained in specific agricultural skills such as growing vegetables and field crops and livestock production.

However, the strategy is not suited for out-scaling. The training focuses on developing self-learning materials on a block release system around competency-based learning. It costs US$4,200 to put one student through the conventional certificate and diploma system, yet YCFP students can expect to earn US$12,500 income per year upon completion of the program, which indicates that the programme could be self-sustaining, with a surplus. Decentralisation into satellite training centres would increase the annual intake of students tenfold. While the group learning component is not obvious, the participants learn from each other and grouping becomes spontaneous.

## Farmer Study Group Strategy - Study Circles

The farmer study group strategy helps communities to organise themselves into study-circle learning groups. Each circle usually has between five and fifteen participants who meet periodically to carry out planned studies under an accepted, trained leader. The leader is responsible for setting up a study plan and producing related study materials based on the participants' specific experiences and skills.

The unique pedagogical method of the study circle builds on a participatory, dynamic process that incorporates the experiences of the participants, the skills of the study circle leader, and the quality of the study material. As opposed to conventional education based on a hierarchical model in which the teacher disseminates knowledge from the top down, a study circle involves interaction in which all participants contribute.

Study circles are free and voluntary studies of conditions and opportunities for humanity within the society, working life, nature, and culture of the surroundings. They lead to understanding the strength, power, and pleasure of creating something together and of working with others to acquire new knowledge and to scrutinise conditions with a view of development in individual communities.

In the study circle, work is built around the participants' search for knowledge according to individual needs and interests. Work is characterised by democratic values, with individuals' exchanges of experience and analysis guiding the process. The studies depend on active contribution from the participants in the planning and implementation of the work.

The work in a study circle is a vivid, active process that relates knowledge to action, social change, development, and creativity. The method can be described in terms of the basic principles of equality and democracy, experience and cooperation, freedom and the right to set objectives, continuity, and planning and active participation.

Although implementation of farmer study circles varies from country to country, all can provide opportunities for people to learn together—as in the common saying "if you want to move forward, go into a circle," to strengthen their opportunities for influencing their own life situations, and to fill society's knowledge and information gaps. Through regular meetings to study a certain subject or theme or to take part in a practical activity, participants take responsibility for basing their circle's work on their own needs and interests. An important part of the study circle methodology is the exchange of experiences and ideas among participants—they make their own analyses of the subject studied and, therefore, learning is based on the collected wisdom of the group. By managing the process themselves and coming to conclusions or making decisions through discussion, listening, and understanding, participants are also practicing democracy and systems of shared power.

Unlike the traditional classroom in which inequality defines the student-teacher relationship, the technical expert in a study circle does not have higher status than other group members. Often, there is a facilitator—a study group leader—whose responsibility is not to master the subject matter but to keep the learning process moving forward. Study circles focus on self-realisation and social transformation, encouraging participants to blossom as individuals and to bring about change in society. For participants, the ultimate outcomes of working in study circles include systematic knowledge and skills acquisition through participatory education, learning, and action, and knowledge sharing which can lead to substantive community empowerment, community development, and improved livelihoods.

Women and youth have active roles in study groups, often working as organisers and ward facilitators, or in leadership and community documentation roles. Community documentation is a means for communities to write their own histories during the process of community and societal development. Scientists have been challenged by research and questions emerging from the experimentation of participants in study circles, indicating that educated minds start by experimenting, become inquisitive and question, and then demand services. This learning method certainly has the power to unlock wisdom in the villages. Study circles have no literacy bounds because every village has someone who can read. The essence of this approach is not about reading but about sharing ideas, learning, and growing together (The MDG Steering Group, 2008).

## Other Learning Strategies

Many other learning strategies that have a bearing on rural farmer education may be classified as variants of the above, but are equally important. The following examples illustrate the variations that exist.

The *Farmer Extension Facilitators' strategy (FEFs)* has been used to promote participatory learning opportunities, use of indigenous knowledge and social systems, and peer teaching and learning among farmers. The FEF strategy serves to advance integrated learning systems and is a suitable candidate to be considered a nonformal education programme.

The *Training and Visit (T&V)* learning approach combines formal education and training with informal education and networking. It also promotes nonformal education for the farmers and continuing education for the extension workers. Traditional T&V is good for demonstrations and also promotes community cohesion. The method has been extensively tried in Africa but proved expensive, and a substantive number of countries have dropped the method or modified it.

*Training in environmental and agricultural management (TEAM)* is strong in farmer-to-farmer extension methodology. It follows a unified extension system model, which promotes a more participatory, client-centred approach to agricultural extension. Farmer extension facilitators, who form the core of the TEAM approach, provide training and capacity building for community institutions and groups to address some of their developmental constraints, facilitate the transformation of rural farmers into small commercial farmers, and train locals to train others with the support of readily available manuals (Makina and Taruvinga, 2010).

## Comparison of Smallholder Farmer Learning Strategies

Table 10.2 compares farmer learning strategies. The landscape of agricultural learning systems in SSA countries shows a wide and varied learning terrain consisting of different approaches, execution systems, participation levels, and final results.

| **Learning Strategy** | **Farmers' needs** | **Extension Services** | **Knowledge sources** |
|---|---|---|---|
| Farmer-to-farmer strategy | Farmer-driven and easily accessible | No extension role, Informal learning | Based on experience of other farmers – Indigenous knowledge |
| Farmer field schools | May not meet day-to-day needs of the farmer but demonstrations meet holistic community needs | Extension active in establishing demonstration plots. Informal but structured learning | National Agriculture research systems |
| Master farmer training | Helps create career farmers and meets long-term farming needs | Farmer training centres with extension follow- up. Formal System | NARS |
| Training & visit (T&V) | Meets day-to-day needs of the farmer but is resource extensive | Extension active in establishing demonstration plots. Informal but structured learning | National Agriculture extension and research systems |
| Training in environmental and agricultural management (TEAM) | Active participation by farmers and helps to build community cohesion | Informal and semi- structured learning | Development Partners and National Agriculture research systems |
| Community Livestock Auxiliaries (CLAs) | Based on, and meets farmers' immediate needs. Based on Just in Time (JiT) approach | Continuing Education after formalised instructional programs. Combines various strategies such as T&V, Farmer to Farmer, Agricultural colleges Extension support on request | National Agriculture extension, research systems and Development Partners |

| | | | |
|---|---|---|---|
| Young commercial farmers | Helps create career farmers and meets long-term farming needs | Agricultural colleges Extension support on request. Formal system | Agricultural colleges |
| Farmer study groups and learning circle | Active participation by farmer in needs and learning | The programmes bring participants up to date in particular areas of knowledge and skills through such diverse methods as independent study; broadcast, video recordings, online instruction, and other forms of distance learning; group discussion; study circles; conferences, seminars, workshops; and part-time classroom study. | NARS information packets designed for use by farmers. Indigenous Knowledge |

**Table 10.2. Summary profile of farmer learning strategies**

## Farmer Knowledge Needs and Related Learning Strategies

Farmers are practical people who seek knowledge to solve their individual problems. Most farmers would prefer initially to contact their neighbours when they need support, which indicates the importance of farmer-to-farmer strategies for access to knowledge and learning. However, learning is limited by the knowledge level of the neighbour. There is clearly a reliance on indigenous knowledge passed on from generation to generation and among neighbours. Thus there is a need to recognise the importance of traditional knowledge systems, especially in exploring various activities and understanding cultural values that impact the success of new initiatives (Metcalfe, 1992).

This knowledge base can further be strengthened by appropriate extension services. Thus the creation of farmer-to-farmer learning, supported by mechanisms for infusing new knowledge, makes the farmer learning circles a very powerful method for organising and increasing knowledge and technology at the grassroots level. The ability of various role players to articulate a powerful vision, to manage information, and to encourage investment in new capabilities is the source of agricultural success.

The extension worker is the catalyst needed to both initiate and continue learning and is also the conduit through which knowledge resources are channelled and administered. Initial training of extension workers is carried out by government and higher education institutions, while private institutions are active in the nonformal and continuing education sectors. Agricultural colleges have been active in producing extension agents. Special colleges are being set up to train the professional and commercial African farmer.

There is limited extension support for farmer-to-farmer learning strategies. However, farmer education in general is mostly provided by the state with some support from NGOs, multilateral and private organisations. Because of the rigid nature of state educational provision, coordination of the new farmer education programmes is better done by non-governmental institutions. Successful examples abound, most visibly the Communal to Commercial Farmer projects of Humana People-to-People in Zimbabwe, and the Heifer International Livestock project in South Africa (Makina and Taruvinga, 2010).Strong extension services are however essential if farmers are to gain any meaningful new knowledge for their operations.

Commercial farmers in the formal colonial era became more successful because they had better state, financial, resource, and extension support than smallholder peasant farmers. Farmer education for the latter was provided mostly by less trained personnel, while more highly trained personnel serviced the large commercial farmers. This is a legacy of colonial land tenure and farming systems where emphasis was placed on vast land holdings and commercialisation for the production of food and cash crops. Research benefits flowed to these establishments first before trickling down to the rural smallholder farmers. However, this setup is changing. In some countries, smallholder farmers have slowly begun to play an important role in food security, eclipsing the commercial farmers' production. This change in production has not necessarily been matched by similar changes in farmer educational support. In terms of numbers, rural and smallholder farmers continue to outnumber commercial farmers, yet in terms of land holding, the commercial farmers fare much better.

Farmer education tends to be informal but effective, using community networks that pass on knowledge and skills informally, encouraging non-threatening participation, while formality may encourage better discipline and recognition of effort. Increasingly, farmers want recognition for their achievements, which would be based on a structured formal learning system. There is a tendency to recognise formal training at the expense of informal and nonformal farmer education and training. However, learning circles have a level of informality with the potential to organise formalised learning systems for lifelong education of the farmers. Master Farmer training and young commercial farmer methods, with their highly formalised structures, have the capacity to develop professional farmers.

## Research Systems and Knowledge Generation

National Agricultural Research Systems (NARS), of which universities are a part, play a significant role in generating knowledge and new technologies. The key question is whether these research systems are focusing on the needs of farmers or are simply doing research that may not be readily applicable by farmers. This is in contrast to agricultural research institutions that tend to be focused on application of research results. To their credit, universities also get involved in researching results and impact after project implementation and so complete the cycle and knowledge generating loop. Thus, farmers are being viewed less as objects for observation and more as partners in exploring problems and issues. Consultation among key players such as subject matter specialists, researchers, and technical advisory committees as well as the farmers themselves leads to generation of knowledge that is appropriate and relevant to the farmer.

As indicated above, farmer-to-farmer learning relies heavily on indigenous knowledge. The farmer field schools and Master Farmer training are well structured around formal agriculture curricula. Formal research derived through NARS can therefore be readily integrated into these learning strategies. Good learning systems encourage research and use different approaches to training and extension; give support to group learning and post-training implementation; and use methods that encourage continuous interaction and learning.

Smallholder farmers have suffered marginalisation in areas of knowledge, information, education, and research largely because they have been treated as consumers of education and information and have not participated in determining what should be taught. Yet, education has been identified as one of the key means of securing equal access to knowledge, skills, jobs, and participation in decisions affecting one's life. The value of the farmer study groups and learning circles is that farmers can be very active participants and set agendas on their knowledge needs and on how knowledge is managed and embedded into their communities. This is perhaps the most important empowerment and emancipation tool for equipping individuals with an understanding of their communities and unique needs. The centrality of education in the development of any society and in the well-being of the individual cannot be over-emphasised.

## Concluding Comments

Small local changes in the production and processing of agricultural value chains can assist in decreasing costs and improving productivity in rural farming systems through building trust, which is core to success; creating positive power relationships among role players; developing viable information and resources banks; and articulating powerful visions to manage information and encourage other chain partners to invest in new capabilities. Underpinning all these is knowledge management—the way in which knowledge is sourced, generated, disseminated and stored.

## References

Adams, M. 1982. *Agricultural Extension in Developing Countries*. Longman: United Kingdom.

Care SA Lesotho. 2009.

Coombs, P. and A. Manzoor. 1974. *Attacking Rural Poverty: How Nonformal Education Can Help*. A Research Report for the World Bank. The Johns Hopkins University Press: Maryland, USA.

Ewel, P. 1998. *Linkages between on-Farm Research and Extension in Nine Countries*. OFCOR Comparative Study No. 4. ISNAR.

Fordham P., G. Poulton, and L. Randle. 1979. *Learning Networks in Adult Education*. Routledge and Kegan Paul: London.

Hakutangwi, M. and L. Mongbo. 1997. *Support to Training of Agricultural Service Staff and Managers in Africa: Identification Mission Constraints Analysis Study of Extension Personnel Training in Africa and Identification of Case Studies of Innovative Training Programmes for Wider Application in Africa.* Paper Presented at Workshop on Extension Personnel Training in Africa Third Informal Consultation of International Supporters of Agricultural Extension Systems in Africa, October, Cape Coast University, Ghana.

Hakutangwi, M. 2008a. *Establishing Learning Circles in Community Sites*. Paper presented at the Kellogg Foundation Regional Workshop on the Art and Skill of Building Vibrant African Rural Communities. Aloe Ridge Hotel, Johannesburg.

—. 2008b. *Towards the Establishment of an Agricultural Education Business School, Manica Province, Mozambique, the Barefoot University and Study Circles*. Paper presented at Workshop on Agricultural Education and Training Initiatives in Africa, a Southern African Networking and Clarification, Manzini, Swaziland.

IFAD. 2009. *Sustainable Agriculture and Natural Resource Management Programme in Lesotho*. International Fund for Agricultural Development (IFAD).

Leeuwis, C. and A. Van den Ban. 2004. *Communication for Rural Innovation: Rethinking Agricultural Extension (Third Edition)*. Blackwell Publishing: New Jersey, USA.

Makina, C. and R. Taruvinga. 2010. *Agricultural Education Initiatives (AEI) Mapping Paper*. Unpublished conference paper presented at the WK Kellogg Regional Workshop. Mananga, Swaziland. Mananga Centre for Regional Integration and Management Development.

Maunder, A. 1973. *Agricultural Extension: A Reference Manual First Edition)*. FAO: Rome.

Metcalfe, S. 1992. *Communal Areas Management Programme for Indigenous Resources (CAMPFIRE) Perspectives*. In Western, D., M. Wright, and S. Strum, Eds. Communal-Based Context. Island Press: Washington, DC.

Michelsen H., L. Zuidema, C. Hoste, and D. Shapiro. 2003. *Improving Agricultural Research at Universities in sub-Saharan Africa: A Study Guide*. ISNAR Research Management Guidelines. No. 6. International Service for National Agricultural Research, The Hague.

Muller, J. 1993. *The Action Training Model and its Educational Foundations*. Adult Education and Development 41: 239–253.

Rogers, A. 2004. Looking again at Nonformal and Informal Education: Towards a New Paradigm. *The Encyclopaedia of Informal Education*. www.infed.org/biblio/non_formal_paradigm.htm.

Rivera, W., M. Kalim Qamar, and L. van Crowder. 2001. Extension, Education and Communication. Service, Research, Extension, and Training Division, FAO Sustainable Development Department, FAO, Rome. November.

USAID. 2002. *Assessment and programmatic recommendations: addressing pastoralist conflict*. United States Agency for International Development, Washington, DC.

The MDG Steering Group. 2008. *Achieving the Millennium Development Goals in Africa: Recommendations of the MDG Africa Steering Group*. United Nations EUROPEAN, June, Training in Environmental and Agricultural Management (TEAM) Project, Lesotho. http://africanagriculture.blogspot.com/search/label/Lesotho

Williams, T. 1981. *Agriculture and Adult Education. In Bown, L. and J. Okedara, Eds*. An Introduction to the Study of Adult Education: A Multi-Disciplinary and Cross-Cultural Approach for Developing Countries. University Press Limited: Ibadan.

# CHAPTER ELEVEN

# CAPACITY BUILDING IN AGRIBUSINESS EDUCATION AND TRAINING FOR SUB-SAHARAN AFRICA

EDWARD MABAYA, RALPH CHRISTY AND MAUREEN BANAMA

The central role of agriculture in economic growth and development in Africa has long been widely recognised. To spur rural development and food security, agricultural education and training (AET) in Africa has traditionally focused on increasing agricultural productivity on farms. More recently, development practitioners and policy makers have broadened their attention to include agribusiness or agro-industries—the post-harvest activities involved in the transformation, preservation, and preparation of agricultural production for intermediary or final consumption (Wilkinson and Rocha, 2008).

Because agro-industries are uniquely situated between natural sources of supply and the demand for food and fibre, promotion of agribusiness development can have positive impacts on employment in both rural and urban areas. Furthermore, agribusiness development can increase market access to smallholder agriculture, present business linkages to small and medium-sized enterprises, enhance food security by reducing post-harvest losses, and extend the shelf-life of food and fibres for the rapidly increasing number of urban poor. The combined effects of employment gains and food security through improved agro-industry competitiveness can serve as an important strategy for reducing overall poverty in developing countries. Yet, as noted in a recent World Bank report on Africa's agribusiness potential, the continent lacks "well-trained university and diploma graduates with skills in communication and teamwork and practical skills in business management, marketing, and finance" (World Bank, 2013).

Developing strong and viable agro-industries requires a combination of skills, policies, and institutions different from the traditional, farmer-focused framework. Agribusinesses have a different objective—maximising profits—and require a specific enabling environment to thrive. For this purpose a multitude of new policies, initiatives and institutions have emerged in developing countries since the 1990s.

These interventions, usually aimed at facilitating the participation of small and medium-sized enterprises in agribusiness, tend to address issues such as warehouse receipts, business clusters, microfinance institutions, technology parks, business development services, contract farming, and public investment in transport and infrastructure.

Much has been written about both the theoretical basis and empirical evidence for these interventions. Yet the role of AET in fostering agribusiness in developing countries remains underexplored. That exploration is the focus of this chapter.

## Scope and Outline of the Chapter

While the status of AET in Africa has been analysed since the start of the millennium, there has been a limited focus on agribusiness. The job market for students graduating from agricultural science departments has changed significantly over the years, from mainly government sector to private agribusiness sector and non-profit organisations. Historically, African countries based their AET systems on meeting the human resource requirements of public service or parastatal agencies. Most graduates found employment in government civil service.

Today, economic crises and structural adjustment programmes have in most countries minimised civil service recruitment. The private sector and entrepreneurship have become increasingly important within globalised, newly liberalised, and increasingly technologically advanced economies. Demand for agribusiness leadership and management has come from an industry characterised by evolving structures and business models (De Lange and Van Rooyen, 2002). Academic institutions and non-governmental organisations (NGOs) that provide education and training are struggling to meet the new demand for skills from the emerging agribusiness sector in terms of the number of people educated or trained, the quality of programmes and the relevance of skills offered. Further complicating the matter, very little research has been done to increase understanding of the current dynamics in academic as well as executive training for agribusiness in Africa. Consequently, investors, non-governmental organisations, policy makers, and leaders of academic institutions often make key decisions on how to adapt to the changing agribusiness environment, yet they have limited information or expertise on which to base these decisions.

**Key Messages**

- Improving agribusiness, with its combined effects of employment gains and food security, can be an important strategy for reducing poverty within developing countries. The objective of agribusiness—'maximizing profits'—often requires an enabling environment to thrive. To address this need, a multitude of new policies, initiatives, and institutions—such as interventions to facilitate participation of small and medium enterprises—have emerged in developing countries. While much has been written about these interventions, *agribusiness education and training to foster agribusiness growth* in developing countries remains relatively under-explored. There are several regional initiatives such as CAADP, NEPAD, and AGRA promoting agribusiness development and helping to shape agribusiness development in Africa.
- Academic institutions and NGOs that provide education and training are *struggling to meet the demand for new skills from the emerging agribusiness sector* in terms of number of people trained, quality of training, and relevance of skills offered. Investors, NGOs, policy makers, and leaders of academic institutions often make key decisions on how to adapt to the changing agribusiness environment with limited information and expertise on the subject. While several African universities have responded to market demand by including specialization in their agricultural economics departments, the quality of graduates remains questionable.
- Agribusiness managers have to contend with a broad range of socio-economic issues such as gender, HIV and AIDS, conflict resolution, technological advancements, environmental sustainability, natural disasters, the global economic crisis, and ever-changing socio-political dynamics that are part of the agribusiness operating environment. *Adaptive graduates who can tackle today's challenges as well as future unknowns are needed.* The new agricultural professional must be market oriented, able to learn, innovate, and solve problems, and able to listen and communicate well. The target market is more diverse than traditional AET and includes students, farmers, small and medium enterprises, and agribusiness executives.
- Agribusiness education should aim to provide diversity and exposure to diversity (thought, race, gender, and culture).
- *Key recommendations for improving AET for agribusiness* in SSA include:
  - A paradigm shift—from agriculture to agribusiness
  - Gender inclusivity
  - Balance quantity with quality
  - Engage the private sector
  - Use and develop local case studies
  - Adapt to changing environments
- The *implication for AET is that institutions need to re-invent* themselves continuously in order to keep pace with contemporary and future challenges. Better yet, AET needs to produce postgraduates who are able to meet future challenges without the need for further training—the *"adaptive graduate."*

This chapter analyses the current state of capacity-building in agribusiness education and training for sub-Saharan Africa (SSA). It is organised in five sections:

- Characterising the macroeconomic environment in which agribusiness development is taking place in Africa;
- Exploring the implications of a growing and evolving agribusiness sector on AET;
- Assessing the current status of agribusiness education and training in African academic institutions;
- Benchmarking new models of executive training being offered by non-governmental organisations, academic institutions, and the private sector;
- Recommending strategies for developing agribusiness education and training initiatives that bridge the gap between current offerings and industry needs.

Throughout the chapter the term "agribusiness education and training" is used instead of the more general notion of AET that includes, but is not limited to, agribusiness.

## Agribusiness in the Economic Development of sub-Saharan Africa

Before considering the supply and demand dynamics of agribusiness education and training, it is important to understand the macroeconomic environment in which agribusiness development is taking place in sub-Saharan Africa. In addition to the general contributions of agribusiness to economic development mentioned in the introduction to this chapter, urgent efforts to improve agro-industries are further justified by the potential long-term negative impacts of the 2008 financial crisis on African economies, as well as the need to satisfy the food security demands of a growing population.

In many countries the financial crisis led to a slowdown in economic growth, reduction in trade, and recession. The crisis had and will continue to have an impact on African agribusiness and on the pace of agro-industrialisation. The rising anti-globalisation movement calling for protectionism and a resurgence of economic nationalism may also play a role in altering the policy approaches in certain countries. While the prospects for Africa's integration into the global economy are likely to improve as the world economy recovers, it is at this juncture important to understand the implications of the crisis so that African agro-industries will be in a better position to compete in national, regional, and international markets.

## Regional initiatives promoting agribusiness development

In addition to numerous country-level initiatives designed to accelerate agribusiness growth, several regional initiatives have recognised agribusinesses as important economic engines in poverty reduction. These include the Comprehensive Africa Agriculture Development Programme (CAADP) of the New Partnership for Africa's Development (NEPAD), the Alliance for a Green Revolution in Africa (AGRA), and the United Nations (UN) Comprehensive Framework for Action.

NEPAD's CAADP was designed to sensitise African leaders to the need for action along four key "pillars": land and water management, market access, food security and hunger, and agricultural research. CAADP is directed by African policy makers in NEPAD's own structures, national policy makers in public and private sectors, non-governmental organisations, academic institutions, think tanks concerned with Africa's development, and officials in the development cooperation agencies of donor and multilateral bodies.

Pillar two of CAAPD—increasing market access through improved infrastructure and trade-related interventions—is directly related to a policy agenda and programme initiatives for accelerating the development of agribusiness and agro-industries in Africa. The objective of this pillar is to "accelerate growth in the agriculture sector by raising the capacities of private entrepreneurs, including commercial and smallholder farmers, to meet the increasingly complex quality and logistics requirements of domestic, regional, and international markets" (NEPAD, 2010). This objective will be met through a focus on four strategic areas:

- increasing competitiveness and taking advantage of opportunities in domestic, regional, and international markets;
- investing in commercial and trade infrastructure to enhance the cost effectiveness of reaching those markets;
- developing value-chain and financial services; and
- strengthening the commercial and technical capacities of farmer organisations and trade associations.

These strategic areas, especially improvement in infrastructure and market access to local, regional and extra-regional supply chains, are essential to agribusiness and agro-industry development, which in turn are key to the sector's contribution to economic growth and poverty reduction goals.

AGRA was established by the Rockefeller Foundation and the Bill and Melinda Gates Foundation in 2006 as a comprehensive approach to raising the productivity and incomes of African smallholder farmers. While its programme initiatives focus on smallholders through its work on seed systems, soil nutrients and water, AGRA also has broader interests in off-farm programmatic activities that include improving access to input and output markets and encouraging government policies that support small-scale farmers. To this end, several investments have been made in developing agribusinesses, such as fertiliser and seed companies, along the smallholder farmer input supply chain.

The UN Comprehensive Framework for Action was developed by the high level task force on the global food crisis established in April 2008 by the UN, World Bank and International Monetary Fund. The task force is chaired by the UN Secretary-General. In July 2008, eight factors that support smallholder food production were identified in the Comprehensive Framework for Action. Based on these factors, a series of short-, medium- and long-term actions, such as supporting the sustainable increase of food production by smallholder farmers, was proposed. Agribusiness and agro-industries will play a critical role in addressing five of the eight factors, namely: 1) stimulating private and public investment in agriculture; 2) improving rural infrastructure; 3) ensuring sustained access to competitive and transparent private sector-led markets for food produce and quality inputs; 4) supporting development of producer organisations; and 5) strengthening the access of smallholders and other food chain actors to financial and risk management instruments.

The World Bank has also been responsive to an agenda that accelerates the development of African agribusiness and agro-industries. Its report, *Modernizing Africa's Agro-Food Systems: Analytical Framework and Implications for Operations*, recorded a strategy for World Bank Group assistance in support of agro-enterprise development and the more general modernisation of agro-food systems in Africa. An important contribution of this work was identifying five "priority themes and leverage points for activity"—food security, food safety, competitiveness, smallholder and small- and medium-sized enterprises market integration, and risk management—that contribute to the stated goal of fostering the competitiveness of African client agro-food systems in a manner that both accelerates growth and reduces poverty (Jaffee et al., 2003).

## Forces driving agribusiness development in Africa

If agribusiness capacity-building and training is to make meaningful contributions and remain relevant, the major forces shaping development of this sector must be identified and assessed. Examining the forces that influence agribusiness helps to provide insight into the policies and innovative strategies needed to ensure that this sector will be able to contribute significantly to economic growth and poverty reduction goals. Furthermore, these factors can be used in designing curricula for agribusiness education and training. Table 11.1 provides an overview of many of the key forces that influence the structure and performance of agribusiness and agro-industries in Africa. As a basis for further discussion, the main opportunities and threats shaping African agribusiness are identified using a policy, economic, social, and technology (PEST) framework.

Today's agribusiness managers have to contend with a broad range of socio-economic issues such as gender, HIV and AIDS, conflict resolution, technological advancements, environmental sustainability, natural disasters, the global economic crisis, and the ever-changing socio-political dynamics that are part of the broad agribusiness operating environment. It is the responsibility of academic institutions and other capacity-building organisations to deliver the type and number of graduates that are able to tackle not only today's challenges, but also future challenges that are as yet unknown. This will require great vision, careful planning and possibly institutional innovations that depart from traditional paradigms.

## Implications for AET

Having recognised the forces shaping agribusiness development in Africa, the next step is to explore the implications for AET. Significant human capital development is required to fulfil the growing need for skills and expertise to exploit the opportunities and mitigate the threats mentioned in Table 11.1. The implications of these changes for AET can be divided into three groups: changes for service providers, changes in target markets, and changes in curriculum design.

Unlike other traditional fields of AET, the responsibility for building skills and expertise in agribusiness does not lie exclusively in the public domain. Academic institutions are no longer the "sole guardians of knowledge." In addition to traditional institutions such as universities, colleges, technical colleges, vocational schools, and extension agencies, new models are emerging from non-governmental organisation entrepreneurship and executive training programmes. Both academic institutions and non-governmental organisations have the opportunity to go beyond window-dressing existing programmes to design and implement more effective, successful agribusiness programmes that remain relevant well into the future.

| Factor | Opportunities | Threats |
|---|---|---|
| **Political Legal** | • Political stability in most countries<br>• Regional harmonisation of policies<br>• Preferential trade agreements, trade policy | • Political conflict in some countries<br>• Corrupt government regimes<br>• Limited enforcement of property rights<br>• Poor business climate<br>• Limited law enforcement |
| **Economic** | • Economic liberalisation well underway<br>• High economic growth rates in recent years<br>• Increasing direct foreign investments<br>• Central role of agriculture in most economies | • Global recession<br>• Global financial crisis<br>• Poor transport infrastructure<br>• Legal barriers to trade<br>• High interest rates<br>• High inflation and unemployment<br>• No macroeconomic stability |
| **Social Cultural Human Demographic** | • Growing population – increasing market<br>• Rapid urbanisation<br>• Improved access to education and health<br>• Gender—growing recognition of women's rights | • HIV and AIDS<br>• Limited adoption of new technology<br>• Cultural preferences for specific staples |
| **Technological** | • Improved access to information (mobile phone and Internet)<br>• Advances in biotechnology<br>• Technology leap-frogging<br>• Investment in public research through National Agricultural Research Systems (NARS) and Consultative Group on International Agricultural Research (CGIAR) | • Diverse agro-ecologic conditions limit technology diffusion<br>• Weak national scientific and research institutions and universities<br>• Limited enforcement of intellectual property rights |

**Table 11.1. Key forces influencing agribusiness in Africa**

The envisaged result is that institutions produce the kind of graduates who meet the agribusiness needs of today and are prepared and able to tackle challenges yet to come. Figure 11.1 illustrates the stylised approaches to capacity-building and training for agribusiness. On one end are the traditional AET options offered in African universities and colleges. Some of these academic institutions have added agribusiness management either as special elective courses under the traditional disciplines of the faculties of agriculture (animal science, crop science, agricultural economics, soil science, or agricultural engineering courses) or as a separate field. On the other end are entrepreneurship development initiatives, mostly offered by donor-supported non-governmental organisations or government initiatives. Several of these initiatives have segmented their business development services into sectors that are popular for small and medium-sized enterprises, such as tourism, retailing, catering, and agribusiness.

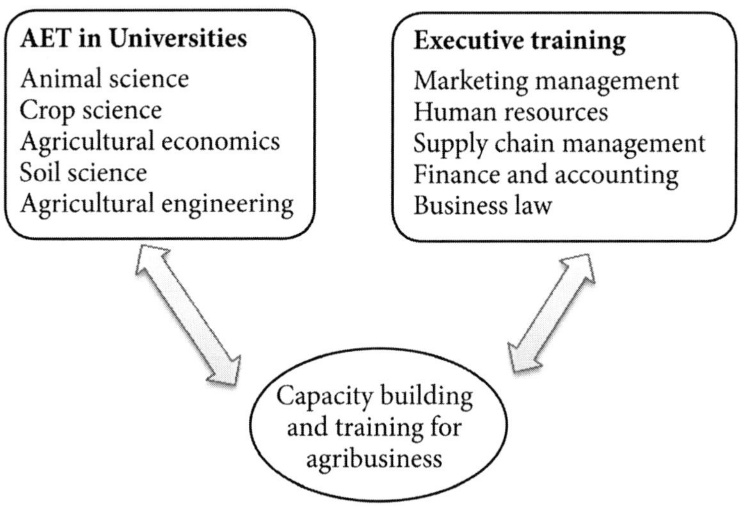


Figure 11.1. Approaches to capacity building and training for agribusiness

The target market or beneficiaries of agribusiness education and training are far more diverse than those of traditional AET. Beneficiaries include students, farmers, and small and medium-sized enterprise and agribusiness executives. The rest of this chapter focuses on those higher-end agribusiness skills aimed at students and executives from established agribusinesses.

The highly diverse target market and wide range of service providers lead to a plethora of curricula that appear to satisfy unique niches. They vary in length, format, and level of rigour. They include short courses (one day to week long), seminars and conferences, and full- and part-time diploma, certificate, and degree programmes. Figure 11.2 highlights the current offerings of agribusiness education and training in Africa. Along the vertical axis lie the different targeted beneficiaries. The horizontal axis distinguishes between the practical and academic orientation of a curriculum.

## Academic Programmes

Although the term "agribusiness" was coined at Harvard University in 1957, it took decades before the area developed into a field of academic specialisation. "Traditionally, agricultural economics departments offered this to an individual with a degree in agricultural economics who had taken a management course or two during their graduate program" (Akridge et al., 1994). Pioneers in the graduate agribusiness education market include Harvard University, Santa Clara University, and Iowa State University, all in the U.S. As noted by Akridge et al. (1994), "these agribusiness programmes are orthodox business school programmes that have agribusiness components."

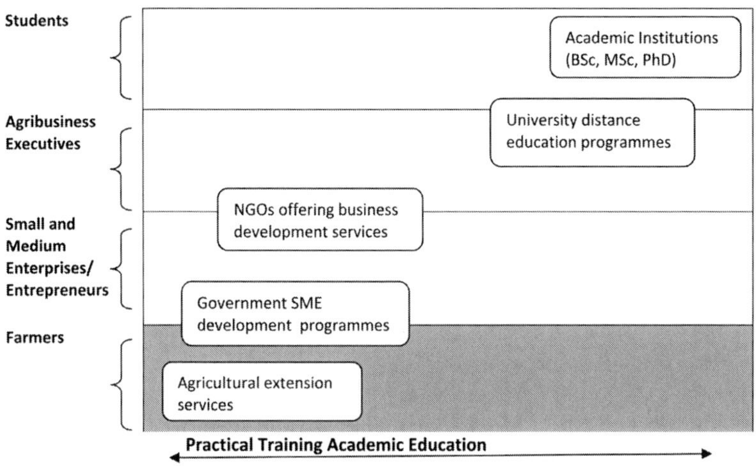


Figure 11.2. Benchmark of agribusiness education and training programmes in Africa

As agriculture became more industrialised, research needs within agricultural economics changed, and a greater emphasis was placed on agribusiness. There is an ever-increasing need for research on the operations of the agribusiness sector as supply chains become more tightly aligned; businesses become more consolidated with mergers, acquisitions, joint ventures, and strategic alliances; and a general industrialisation mentality unfolds (Dooley and Fulton, 1999). Agricultural economics departments have increasingly taken note of these trends and new realities, and have started responding to these demands. Norman and Obwona (2001) noted how institutions had moved from an initial response of merely changing department names to reflect agribusiness to having programmes that place more explicit emphasis on agribusiness. It was found that general agriculture degrees, sometimes with an option in agricultural economics, are increasingly complemented by agribusiness and agricultural economics bachelor's degrees. Since the mid-1990s there has been a marked increase in the number of institutions offering bachelor-level agribusiness degrees in East and South Africa. The University of Pretoria established the Association of African Business Schools Agribusiness Consortium (AABSA) agribusiness chair in 1999, while 2001 saw the first group of agribusiness graduates from Egerton University in Kenya (Norman and Obwona, 2001). Makerere University's Bachelor of Agribusiness Management programme produced its first graduates in the 2003/2004 academic year.

Academic institutions have a central role to play in capacity-building in agribusiness. African universities will ultimately be responsible for replenishing the stock of human capital in their research and extension services and for providing their graduates with the broader set of skills necessary to grow agriculture in the twenty-first century (World Bank, 2007). Christy (2009) has identified three roles for universities in the development of agribusiness:

- the analysis of public policies and regulations that impact, both positively and negatively, the competitiveness of agribusiness;
- the design and evaluation of alternative institutions that strengthen small and medium-sized enterprise linkages as well as smallholder access to national, regional, and global agro-industry supply chains;
- the offering of innovative capacity-building programmes to enhance the economic performance of agribusiness.

Yet key questions remain unanswered about the role of African universities in delivering the requisite skills and expertise for the emerging agribusiness sector:

- Are academic institutions producing the kind of graduates that agribusiness finds useful and attractive?
- Is the demand for agribusiness expertise being sufficiently met?
- How can current programmes be improved to better suit the needs of agribusiness—not only today, but in the future?
- How can gender diversity best be promoted and accomplished in the agribusiness sector, including raising the number and profile of female participants?

## Current offerings by African universities and colleges

Africa now has roughly 300 universities, and three-quarters of African countries offer some higher education-level training in agricultural sciences.

This means that extensive institutional infrastructure for AET is in place (World Bank, 2007). Table 11.2 shows some key institutions with programmes with an agribusiness focus (AET Africa, 2012).

How should academic programmes in agribusiness be structured to produce the type of graduate who can be at the forefront of the development agenda? A report by the World Bank established that AET supply is often out of sync with labour market demand in terms of knowledge and practical competencies, especially in agribusiness, basic management, and problem solving (World Bank, 2007). In the U.S., a number of studies on the state of agribusiness education have been commissioned over the years. Boland et al. (2001) found 115 programmes in the U.S. at undergraduate level with degrees or majors in agribusiness management, although some were lacking essential elements such as human capital, agricultural or business law, strategy, and other integrative management courses, as well as international and global dimensions of agriculture.

In 2004, the National Food and Agribusiness Management Education Commission conducted a study with the objective of describing the current state of food and agribusiness management education at two- and four-year colleges and universities in the U.S. It identified a set of key issues facing food and agribusiness management programmes, developed a set of recommendations on these issues, and then communicated the recommendations broadly through a national conference, professional meeting presentations, reports, and a web site. The study revealed that most agribusiness curricula remain rooted in agricultural economics. Business management courses were the exception rather than the rule. The undergraduate curricula for agribusiness could best be characterised as collections of courses rather than cohesive programmes (National Food and Agribusiness Management Education Commission, 2006).

According to the World Bank (2007), current AET systems suffer from five main weaknesses: declining enrolments, professional isolation, nar-row and outdated curricula, inadequate staffing, and insufficient pedagogical inputs. Surveys of agribusiness executives (Boland et al., 2001; Breazeale et al., 2004; Wachenheim and Lesch, 2004) found that the most important skills for new graduates were personal qualities—characteristics such as the ability to communicate and work well with others as well as administrative and leadership qualities.

| Programme | Institution | Country |
|---|---|---|
| BSc Agricultural Economics and Agribusiness | University of Ghana | Ghana |
| BSc in Agribusiness Management | Egerton University | Kenya |
| BSc Agribusiness Economics and Food Industry Management; | Jomo Kenyatta University of Agriculture & Technology | Kenya |
| BSc Agribusiness Management and Enterprise Development | Jomo Kenyatta University of Agriculture and Technology | Kenya |
| BSc in Agribusiness Management | University of Malawi | Malawi |
| BCom (Agribusiness Management) | University of Pretoria | South Africa |
| BSc Agricultural Economics: Agribusiness Management | University of Pretoria | South Africa |
| Bachelor of Agribusiness Management (AgricAdmin) | University of Stellenbosch | South Africa |
| BSc Agricultural Economics and Agribusiness Management | University of Swaziland | Swaziland |
| BSc in Agricultural Economics and Agribusiness | Sokoine University of Agriculture | Tanzania |
| Bachelor of Agribusiness Management | Makerere University | Uganda |
| BSc Agriculture and Natural Resources (Agribusiness) | Africa University | Zimbabwe |
| Dip. Agribusiness and Marketing | Bukura Agricultural College | Kenya |
| Postgraduate Diploma in Agribusiness Management | Busoga University | Uganda |
| Agribusiness Management | Co-operative College | Zambia |
| Cooperative Development and Agri-business | Co-operative College | Zambia |
| MSc in Collaborative Master Programme in Agriculture and Applied Economics | Egerton University | Kenya |
| Master of Science in Agribusiness Management | University of Malawi | Malawi |
| Master of Agribusiness Management | Makerere University | Uganda |
| PhD Agricultural Economics and Agribusiness | University of Ghana | Ghana |
| PhD in Agriculture and Applied Economics | University of Malawi | Malawi |

**Table 11.2. Agribusiness programmes. Source: AET Africa (2012) and University Websites**

These findings are corroborated by those of the National Food and Agribusiness Management Education Commission shown in Figure 11.3.

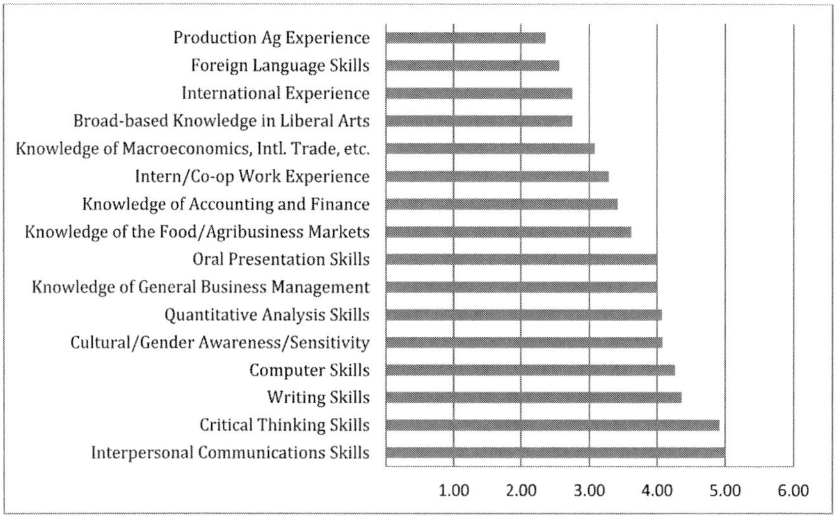


Figure 11.3. Rating skills, capabilities, and experiences sought in new hires with the potential to become future company leaders . Source: National Food and Agribusiness Management Education Commission (2006)

The "new" agricultural professional must be market oriented, able to learn and adapt, able to innovate and solve problems, and able to listen and communicate (World Bank, 2007). In a survey of agribusiness executives in the U.S., Wachenheim and Lesch (2004) found that the ability to communicate and work well with others was the most desirable skill for new graduates. In Uganda, Makerere University used the Agribusiness Management Aptitude and Skills Survey questionnaire. Its results were concurrent with results worldwide. Personal quality was rated the highest, followed by administrative and leadership qualities. Communication skills were rated third and business and economic skills fourth, followed by technical skills and computer, quantitative, and management skills. The least important category was previous employment and general experience (Breazeale et al., 2004).

Agribusiness courses also need to be reflective of diverse industry interests, as graduates find employment in agribusiness firms, non-governmental organisations, governments, research organisations, and consulting firms as well as academia. This can be done by cultivating linkages with agribusiness and recognising the qualities that agribusiness seeks in graduates. Linking academic programmes to actual agribusiness firms is useful, in particular for:

- developing programmes that not only suit current needs of firms but equip and train students for future needs;
- providing practical training opportunities and industrial experience; and
- developing soft skills in students that may be difficult or impossible to acquire in classroom environments.

The International Food and Agribusiness Management Association conducted a survey to understand preferences for components of undergraduate agribusiness education from a diverse set of firms tied by a common interest in international agribusiness (Wachenheim and Lesch, 2004). Among the most important industry executive suggestions to agribusiness programme leaders were diversity and exposure to diversity (of thought, race, gender, and culture). Agribusiness education should therefore aim to provide this kind of exposure.

The Faculty of Agriculture at Makerere University in Uganda restructured its undergraduate and graduate programmes to reflect the economic and political realities faced by African universities. Without this measure, Makerere University risked becoming less relevant to the agribusiness sector in the country. The Faculty's traditional focus on producing agricultural graduates for the public sector had left it unprepared to meet private sector needs. Its curriculum was out of date and there was an acute shortage of information and knowledge describing the training needs and qualifications demanded by the private sector.

In order to rectify this situation, the Faculty made a strategic decision to work with the private sector in updating its curriculum to produce more desirable graduates (Breazeale et al., 2004). The master's programme was divided into two distinct degrees—agricultural economics and agribusiness (Norman and Obwona, 2001). The agribusiness degree was offered on both part- and full-time bases, with evening classes as an option; the University aimed to attract private sponsorship from local agribusiness firms for the part-time degree. For the agribusiness master's degree the University set up an advisory board with representatives from the private sector. This association with the private sector:

- gives the agribusiness degree credibility and visibility;
- improves the potential for private sponsorship of students;
- nurtures linkages between the University department and private sector;
- ensures that course content remains relevant to the needs of the rapidly evolving private sector; and
- improves the likelihood of students doing theses on topics relevant to the private sector (e.g., possibly working on a topic of interest to the sponsoring institution) (Norman and Obwona, 2001).

Such efforts illustrate the level of progress that can be achieved when academic institutions work together with agribusiness in shaping curricula. Similar collaboration with agribusiness can be found elsewhere in sub-Saharan Africa. The University of Pretoria in South Africa introduced the AABSA Consortium Chair in Agribusiness Management in 1999 as part of a portfolio of courses presented by the University's Department of Agricultural Economics and Rural Development. Its goal is to supplement the capability of agribusiness to face the challenges of a sector in flux through innovative training programmes, research, and outreach. The University also offers Agribusiness Management through its Department of Agricultural Economics as one of three areas of specialisation in agricultural economics. Programmes on offer include BCom (agribusiness management) and BSc Agriculture (agribusiness management/agricultural economics) (University of Pretoria, 2013).

Critical to African development is a new generation of entrepreneurs equipped to capitalise on the vast opportunities that lie in the agribusiness sphere. The African agribusiness education system, in departing from primarily supplying the civil service with employees, is an important platform for developing students' entrepreneurial skills, enabling them to start and manage successful agribusiness enterprises of their own.

## Quality of current academic training in agribusiness

As highlighted above, several African universities have responded to market demand by including agribusiness specialisation in their agricultural economics departments. What remains unclear is the quality of the graduates emerging from these programmes. To explore this issue, a litmus test for agribusiness graduates is the response to questions such as: are holders of agribusiness qualifications able to compete with those who have business administration backgrounds? In other words, in applying for a job at an agribusiness company, how well do students with a master's degree in agribusiness compete against students with a traditional master's degree in business administration?

While data to allow rigorous analysis of this question are not available, anecdotal evidence points to a higher preference and remuneration for MBA graduates than for their MA counterparts. The unfortunate outcome of the current effort may therefore be the delivery of graduates who cannot compete in the market because they possess neither the technical training to work in other sectors of agriculture, nor enough management training to be a manager of choice over an MBA candidate. Under this scenario an agribusiness student may be viewed by employers as a "jack of all trades and master of none". "More than minor tinkering would be required to convert the master's degree in agricultural economics, with its economics underpinning, into a close substitute for an MBA degree" (Akridge *et al.*, 1994).

But it is not all doom and gloom for academic programmes in agribusiness management; there are several success stories. Sokoine University's agribusiness programme is not only generously funded by the government of Tanzania, but also receives funding from donors such as the Norwegian Agency for Development Cooperation (NORAD). Since inception of its agribusiness programme, the department has experienced capacity problems because demand has been very high (De Lange and Van Rooyen, 2002). The University of Zambia invites current agribusiness managers and leaders as guest lecturers in order to provide students with insight on how to run a successful business. Following consultations with industry and academia, De Lange and Van Rooyen's study found that agribusiness programmes were likely to be successful if they addressed at least some of the following (De Lange and Van Rooyen, 2002):

- utilised a case-study approach that focused on real-world application,
- provided more practical training and exposure for students,
- developed computer literacy,
- established strong foundations of economics and finance among agribusiness students,
- improved negotiation skills and ensured better understanding of global trade issues, procurement and other supply chain-related issues, or
- made a conscious effort to educate female students and female farmers in agribusiness.

African agricultural institutions are aware of the challenges and inadequacies of current agribusiness education and training. In response to the need to develop a more robust training platform, the African Network of Agriculture, Agroforestry and Natural Resources Education (ANAFE), a network established in 1993 and made up of over 130 agricultural universities and colleges in 35 African countries, embarked on a programme, "Strengthening Africa's Strategic Agricultural Capacity for Impact on Development" (SASACID). The programme comprises seven key components that aim to address some of the most important challenges (ANAFE, 2011).

**Box 11.1: Key components of the ANAFE SASACID Programme**

- Refocusing agricultural learning objectives and improving curricula
- Establishing the capacity of agricultural scientists to develop relevant learning resources based on African knowledge and experiences;
- Building capacity for innovation systems approach: Linking agricultural policy with research, education, industry and practice.
- Strengthening capacity for agri-business education and training, particularly strengthening the interest and capacity of women and youth to take up agricultural careers;
- Managing risk and uncertainty in agriculture, including agrochemicals, biosafety and climate change
- Strengthening methods for teaching and learning and enhancing agricultural information and knowledge management, and
- Revitalising agricultural capacity and institutions in fragile countries.

It is important to note that despite their best intentions, the host departments of agribusiness programmes in African universities are still plagued by many problems. Some transcend the university while others seem unique to the faculties of agriculture. In a comprehensive study entitled "Agricultural Education in Africa: Managing Change" by the Rural Development Department at the World Bank, Maguire (2000) identified the following as main problems plaguing African universities:

- enrolment that is greater than capacity,
- unsustainable patterns of expenditure for higher education,
- decline in quality of education,
- declining relevance to national needs,
- exodus of teaching and research staff to areas of higher pay and better conditions,
- too many universities in certain countries unsustainable with existing budgets, and
- a disconnect with the employers of university graduates (Maguire, 2000).

The same study identified the following problems specific to agricultural education:

- isolation of the agricultural university from other parts of the university system,
- lack of communication with employers of university graduates,
- poor practical skills,

- decreased funding as the urban focus gathers strength,
- weak connection with other parts of the agricultural education system such as colleges, vocational schools and farmer training networks,
- high unemployment of university graduates, often due to lack of relevance of curriculum,
- failure to attract best quality students from secondary schools,
- in-breeding, as teaching and research staff are frequently graduates of the same university or college, thereby excluding the entry of new ideas from the wider world of academia and research (Maguire, 2000).

## Distance education

Visits by two researchers from Cornell University and the International Food Policy Research Institute (IFPRI) to agricultural economics departments in African universities established that the level of interest in agribusiness education was quite high (Wyeth, 2002). Yet agribusiness education and training efforts in sub-Saharan Africa are plagued by the fact that faculties are often understaffed and lack PhD programmes. It remains a challenge to retain staff who go overseas to further their education.

There are also several serious constraints to the effective implementation of distance education. Some institutions have "technological constraints", and their information and communication technologies need to improve. The Internet is in most cases inadequate or expensive or both (Wyeth, 2002). This means, for example, that if a programme is delivered only over the Internet some countries would be unable to take full advantage immediately. Therefore all forms of media need to be utilised to make delivery more effective.

Students complete core courses in the first and second semesters and then take one of the above as an elected field of specialisation together with a course in institutional and behavioural economics. The final two semesters are spent on thesis writing and oral examinations. The programme is designed to attract a range of applicants, including women, by providing information on prerequisite courses and cost-effective means to obtain them. Participants are drawn from a range of applicants and have the opportunity to take "make-up" courses which is meant to make the programme more appealing to women who often have family responsibilities that interrupt their studies (African Agricultural Economics Education Network, 2004).

**Case Study:** *Eastern, Central and Southern Africa: Collaborative Master's Programme in Agricultural and Applied Economics*

In 2005, the demand for high-quality agricultural economists exceeded the supply. Recognizing this as an opportunity, the African Agricultural Economics Education Network launched the Collaborative Master's Programme in Agricultural and Applied Economics in Eastern, Central and Southern Africa. The programme was built on the premise that highly trained local professionals could address the challenges posed by far-reaching changes in global and local economies, technology and marketing by adapting their advanced knowledge and methods to the particular institutional, political and economic circumstances of East, Central and Southern Africa. This effort ran in parallel with the recognition that human capital at postgraduate level needs to be developed as part of capacity building for African agriculture.

In its first five years, it drew participants from institutions in Botswana, Ethiopia, Kenya, Malawi, Mozambique, Rwanda, South Africa, Swaziland, Tanzania, Uganda, Zambia and Zimbabwe. The programme is structured to reflect the changing needs of African agriculture and offers specialisation in:

- agricultural policy and trade,
- agriculture and rural development,
- environmental and natural resource management,
- agribusiness management.

It also offers elective courses in areas such as quantitative methods in agribusiness management and agribusiness management.

## Executive Training in Agribusiness

The preceding section focused on agribusiness education offered by African academic institutions, mainly departments of agricultural economics in universities and colleges. However, the market for professional agribusiness education is highly fragmented. Potential clients for such programmes range from major food companies to small, rural, owner-managed farm input firms. Even within firms, needs that might be met by professional agribusiness education programmes are diverse, ranging from professional development for the Chief Executive Officer, to sales training for field sales personnel, to an outlook seminar for the corporate planning department (Akridge *et al.,* 1994).

Continuing management education for practicing agribusiness managers in mid-size and larger firms is also in demand for executive agribusiness education and training. Agribusiness managers need regular updates on new technological developments and their management implications.

This section explores the executive training for agribusiness professionals that is currently offered by African governments, non-governmental organisations, academic institutions and the private sector.

## Government in executive agribusiness training

Most governments offer a host of business development and support services, either directly or through contracted agencies. Most of these initiatives are generic entrepreneurship development courses, but a few have divisions that are specially focused on agribusiness.

- **South Africa:** The Real Enterprise Development (RED) Door is an initiative of the Department of Economic Development and Tourism, Provincial Government of the Western Cape. Although it is not specifically aimed at agribusiness, farmers benefit from its support to new and existing entrepreneurs in starting and growing their businesses (Red Door, 2010).
- **South Africa:** The Small Enterprise Development Agency (SEDA), a government initiative that provides business development and support services to small enterprises through its national network in partnership with other actors in small enterprise support. The Small Enterprise Development Agency also implements programmes aimed at business development in areas prioritised by the government. Products and services to which entrepreneurs have access include, among others, information and advice, training and mentoring, business planning and registration, incubation and technology transfer as well as sector-specific training and development programmes. While the programme is stratified according to level of growth of the enterprise, there is no specialisation for agribusiness (SEDA, 2010).
- **Botswana:** The Local Enterprise Authority (LEA) promotes and facilitates entrepreneurship and enterprise development through targeted interventions within manufacturing, tourism and agriculture, as well as any services that support the three business sectors. In particular, the Local Enterprise Authority targets women, youth and the unemployed. It provides development and support services to the local industry needs of small, medium and micro enterprises, encompassing training, mentoring, business plan finalisation, market access facilitation, and facilitation of technology adaptation and adoption. The Local Enterprise Authority does not provide financial support to entrepreneurs, but it facilitates access to finance and provides support services to small, medium and micro enterprises, before and after funding. The intervention is also stratified accord-

ing to level of enterprise development – from pre-start-up to business growth – and further specialised by type of industry (LEA, 2013).

- **Kenya:** The Micro, Small and Medium Enterprises (MSMEs) Competitiveness Project is implemented through public and private sector partnerships. Established by the Kenyan government with assistance of the World Bank, its objective is to increase the growth and competitiveness of micro, small and medium enterprises in Kenya, primarily through facilitating access to finance, strengthening enterprise skills and market linkages and improving the business environment. It has two successful subcomponents.
  a) The Global Business School Network improves opportunities and training for entrepreneurs and micro, small and medium enterprises by strengthening business education in business schools, and deepening the pool of competent local managers by helping local business schools enhance their capabilities. Its partner schools (Jomo Kenyatta University of Agriculture and Technology, Strathmore and Kenyatta Universities) develop local case studies which are incorporated into their curricula, and support the training needs of micro, small and medium enterprises managers and owners.
  b) The Value Chain Based Matching Grant Fund addresses market failures that limit the ability of micro, small and medium enterprises to obtain the skills and business services necessary to exploit opportunities and overcome bottlenecks in the business environment. The project beneficiaries are Kenyan-based individuals and micro, small and medium enterprises operating in any one of the selected value chains, namely cotton, pyrethrum, coffee and leather sectors. Through its projects, the Fund offers farmer organisational capacity building, dissemination of marketing information (price, quality standards), recordkeeping and business management training, management and marketing skills, development of marketing strategies, market linkages, farmer training and educational tours. Teams comprising representatives from across the value chain formulate project proposals (MSME Kenya, 2013).

## Executive training by academic institutions

A few universities have developed special part-time programmes for executives to boost revenue streams. While this model is common in most business schools, it is relatively underexplored in agriculture departments. Even rarer are universities that have customised agribusiness management programmes aimed at industry executives. A handful of South African universities are pioneering in this area.

The University of the Free State Department of Agricultural Economics runs an Agribusiness Unit which organises market workshops every term with a vision of "innovating strategy and management for the future success of agribusiness." Market workshops are held in an informal setting, usually attended by about a 100 farmers, bankers or agribusiness people. They interact with guest speakers who are experts in their field, and participate in discussions on

- macro-economic issues: interest rates, exchange rate, commodity prices;
- grain: current prices, imports and exports, international and local production volumes and forecasts; and
- livestock: current prices, imports and exports, international and local production volumes and forecasts (University of the Free State, 2010).

The emphasis is on group discussions of future market trends and appropriate business strategies. Scenarios and strategy sessions are also offered to agribusinesses and farmer or industry groups on request. Each semester it offers a short course: "Introduction to grain marketing in SA", and a one-day course to assist in using the South African Futures Exchange (SAFEX) as a marketing tool, looking at the SAFEX environment and the use of options and futures for hedging (University of Free State, 2010).

The North-West University School of Agricultural Sciences hosts the Centre for Agribusiness and Entrepreneurial Training. The centre primarily services small farmers with short courses, and provides training for extension officers and for any other contract business such as land assessment and feasibility studies for resettlement (North-West University, 2010).

The Stellenbosch University Standard Bank Centre for Agribusiness Leadership and Mentorship Development was established in partnership with Standard Bank. With the premise that successful agricultural transformation and land reform require skills and capacity development to enable proactive agribusiness leadership, its programme covers effective business linkages and support systems as well as pre- and post-farm settlement support, coaching and mentoring. Stellenbosch University initiates and coordinates a multidisciplinary postgraduate-level education consisting of doctoral and master's degree programmes for knowledge development and research, offers a mentoring programme, provides practical experience through various accredited short courses, and has a database of projects and mentors from which participants can draw.

PricewaterhouseCoopers is one of the strategic partners involved in the development and presentation of the short courses. The Centre's major client groups consist of agribusiness executives responsible for development initiatives and agri-mentors and coaches. The Centre positions Stellenbosch University to engage with the agribusiness community to develop the required human capacity and knowledge systems to play a leadership role in a competitive, equitable and environmentally sustainable agriculture sector (Van Rooyen, 2010).

## NGOs in agribusiness education and training

Several non-governmental organisations have stepped into the realm of agribusiness training by offering adapted training courses, most of which are donor supported and target small, specialised client bases. Length, cost and structure of such programmes vary widely by country, funding sources and macro-economic context. Below are a few examples of such initiatives.

The African Organic Farming Foundation has developed an agribusiness model for Southern Africa in the field of organic agriculture. The Foundation's Marketing and Agro-Business Development Programme is a market-led, trade capacity-building partnership that facilitates inclusion of the resource poor in organic production and trade by linking small farmers with markets to improve nutrition, food security, incomes and community decision making (African Organic Farming Foundation, 2010).

The Swedish Cooperative Centre works directly with farmers and contributes to capacity building in agribusiness education and training within sub-Saharan Africa. Its initiatives include the Farmer Empowerment through Agribusiness Development (FETAD) project in Uganda as well as the Swedish Cooperative Centre and Agroforestry project in Rwanda. Farmer Empowerment through Agribusiness Development builds on the success of previous projects, extending organisational development support to beneficiaries by addressing outcomes related to improved agricultural productivity, improved market linkages, strengthened organisational capacity and improved accessibility to financial services. As its contribution to improving livelihood among targeted communities, the project works in the field of agroforestry and agribusiness development to address environmental problems and poverty alleviation. In promoting agribusiness development through the concept of farming as a business, it trains farmers in ways to convert from subsistence farming to a commercialised form of agricultural production, which entails aspects of enterprise selection, proper recordkeeping, utilisation of projected income statements and marketing. This approach to farming is seen as a means of increasing household incomes and improving farmers' livelihoods. The project also focuses on organising farmers into democratic, member-based organisations in order to consolidate the quantity and quality of farmers' produce to meet market standards and help farmers realise substantial profits as a result of improved bargaining power (Swedish Cooperative Centre, 2010).

The Trade and Investment Programme for Competitive Export Economy (TIPCEE) is a CARE project in Ghana that targets smallholder farmers who have already begun to integrate into commercial marketing chains. The project recognises the importance of strengthening the role of women as owners of small and medium-sized enterprises, as smallholder farmers and as productive workers in processing and marketing activities. Even though about 60 percent of Ghanaian farmers are women, the challenge lies in finding commercially oriented women farmers and including them in the target groups. A critical aspect of activities within the project is aligning large and medium-sized agricultural enterprises with overseas markets and linking smallholders to modern supply chains for agro-processing and export (CARE, 2010). The project offers various free short courses to farmers and agribusinesses, based on identified need.

The Seed Enterprise Management Institute at the College of Agriculture and Veterinary Sciences is being established in cooperation with the Iowa State University Seed Science Center, the University of Nairobi, the International Maize and Wheat Improvement Centre, AGRA and private business experts. Its goal is to eradicate food insecurity through capacity building in sub-Saharan Africa, where seed supply chains are lacking or inadequate. The Institute also provides seed training to graduate students who pursue plant breeding in African universities, supports the production of improved seed varieties and has created a Web-based network for information exchange on seed technology (Iowa State University, 2010).

The Ghana Institute of Management and Public Administration (GIMPA) and the IFPRI have partnered to launch a four-week Agribusiness Entrepreneurship Programme in Ghana. The course curriculum, developed and delivered by the University of Ghana, the Indian Institute of Management in Ahmedabad and the Indian Institute of Management in Bangalore makes use of conventional learning techniques for principles of good business practice and case studies as problem-solving tool. It also includes field trips and class projects to enable participants gain practical experience in managing agribusiness. This programme targets mid-level personnel engaged in managing the food, feed, and agriculture system in the private sector (IFPRI, 2013).

The Association of African Business Schools, established in 2005 and comprising 28 business schools, has formed an Agribusiness Consortium (AAC) to champion the cause of Agribusiness management education in Africa. The consortium consists of six schools - Sokoine University of Agriculture, Tanzania; GIMPA, Ghana; Lagos Business School, Nigeria; United States International University (USIU), Kenya; University of Stellenbosch

Business School, South Africa; and the Jomo Kenyatta University of Agriculture and Technology (JKUAT), Kenya. The aim is to strengthen the capacity of African business schools to offer agribusiness management training programs, develop the business, management and leadership skills of actors in the sector, and to strengthen linkages across the value chain for greater effectiveness of projects that integrate smallholder farmers into agribusiness value chains (AABSA, 2011).

Market Matters Inc. has offered all-encompassing agribusiness management training since 2001 including, but not limited to, the topics of agribusiness marketing, strategy, financial analysis, supply-chain management and human resource management. Its "Making Markets Matter" workshop series helps African agribusiness entrepreneurs recognise market demand and meet their customers' needs, while at the same time offering employment and promoting economic opportunities in their communities. Market Matters Inc. has organised and facilitated both large-scale (75+ participants) and smaller training programmes, including the mentorship of individual businesses. In 2010, Market Matters Inc. celebrated a decade of building capacity in small and medium-sized enterprises in emerging markets. The programme targets individuals along the agribusiness supply chain, from farmers to business owners producing value-added food products. Every year the initiative strives to increase the participation of women business owners, still underrepresented in the African agribusiness sectors. Besides providing executive management training that is otherwise difficult to access for small and medium-sized enterprises in emerging markets, "Making Markets Matter" is a venue for building business-to-business relationships. Its special luncheons and evening programmes feature speakers from large agribusinesses, multinational corporations and financial institutions with active agribusiness programme. Given its continent-wide reach in Africa, the programme is an excellent platform for larger businesses to connect with potential clients and business partners.

## Concluding Comments

Building on the preceding analysis of the current state of capacity building in agribusiness education and training for sub-Saharan Africa, this concluding section synthesises the findings into key recommendations for improving AET for agribusiness in Africa.

### A paradigm shift—from agriculture to agribusiness

The hardest aspect to change with respect to education and training for agribusiness is the overall paradigm that governs the approach to the subject.

Traditional AET has existed and been institutionalised over many decades, mostly in the same institutions that are taking on the challenge of building capacity for the agribusiness sector. This makes sense from an efficiency point of view, as it maximises economies of scope and integrates agribusinesses into traditional fields of AET.

However, this approach forces the subject matter of agribusiness to fit into the rather static infrastructure and ideologies that were developed for traditional agricultural fields – mostly farming. Doing so misses the opportunity to challenge existing paradigms for AET that have not kept pace with recent dynamics in the macro-environment. Figure 11.4 illustrates some key areas for which a new paradigm is needed in order to shift from traditional AET to agribusiness-centred education and training. *Firstly,* the approach needs to shift from producing disciplinarians with a narrow but in-depth knowledge base, to producing well rounded professionals with requisite skills to get the job done. *Secondly,* graduates of AET need to be more competitive and employable, not just in government departments but within the general market that now includes private agribusiness. Likewise, the traditional production orientation needs to be replaced by a market-oriented approach that prioritises meeting consumers' needs.

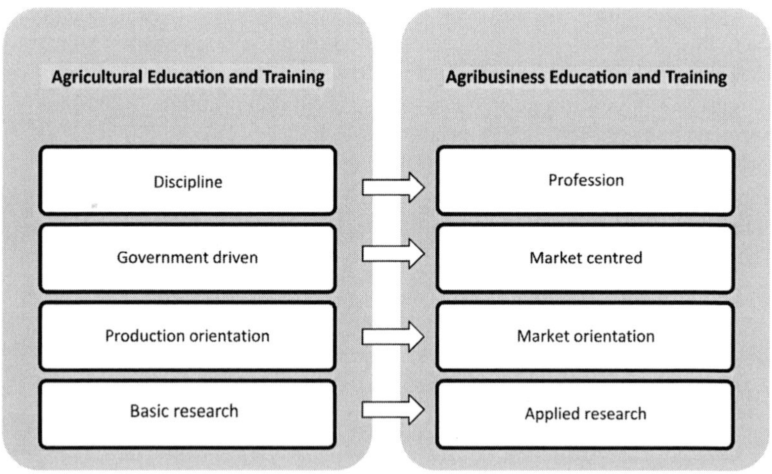


Figure 11.4. Paradigm shift from agriculture to agribusiness education and training.

Unfortunately, most institutions are designed to defend the *status quo,* and hence change might take a long time to develop and implement. An interim strategy might be to give some degree of autonomy or flexibility to agribusiness education and training programmes.

There is no need to re-invent the wheel when it comes to developing programmes and institutions for capacity building in agribusiness education and training. As has already been highlighted, the contribution of agribusiness to GDP increases as a country develops. This is in contrast with the contribution of agriculture or farming which declines with a country's growth. Consequently, for several decades, agribusiness has played a larger role in developed countries than traditional agriculture. Over the years, agribusiness education and training has evolved to meet the needs of the industries in developed countries. African countries can learn from Europe and North America about good practises for designing and implementing successful agribusiness capacity-building initiatives in both the public and private spheres.

## Learn from other developed countries and adapt for Africa

It is however important to note that Africa faces unique challenges and opportunities. Hence, good practises need to be adapted to the local context both at continent-wide and at national levels.

## Gender inclusivity

Lack of gender inclusivity is a key problem that seems to bedevil overall AET. There simply are not enough women included in the different layers of agribusiness education and training. In Africa this poses a special challenge because women play a larger role in agriculture than men; the majority of smallholder farmers are women. If the agribusiness sector is to integrate well with the smallholder farming community it needs to include more women and be cognisant of gender issues. Agribusiness management appears not to be an attractive option for female students enrolling at academic institutions; there is a significantly smaller proportion of women in agribusiness and agricultural economics departments compared to other fields and disciplines. Conscious efforts need to be made in both academic and executive training to recruit and retain female students and instructors.

## Balance quantity with quality

The first wave of response by academic institutions and non-governmental organisations to the growing need for skills and expertise in agribusiness management in Africa has been to produce as many graduates as possible. While this orientation towards an increase in numbers is to be commended, it now needs to be balanced or complimented by an orientation towards quality. Quality control measures need to be put in place to ensure the highest standards of graduates given available resources. Academic institutions need to move beyond changing department names or adding a couple of agribusiness management courses to developing comprehensive programmes that meet and keep up with industry needs.

Similarly, non-governmental organisations and private companies that offer executive training for agribusiness need to move towards accreditation to ensure sufficient rigour and enforceable quality standards. Funding institutions that are investing in agribusiness training will need to signal measures of quality in evaluating outcomes and impacts.

## Engage the private sector

The private sector needs to be engaged in agribusiness education and training. Firstly, capacity building institutions should continuously consult with private agribusiness to assess their skills requirement both in number and areas of expertise. Industry representatives should be consulted in the curriculum design process to ensure that their needs are met. Secondly, the private sector should be engage in the education and training process itself. This can be achieved through the contribution of case studies, guest presentations, student internships and even financial support in AET.

## Use and develop local case studies

Appreciation of case studies as a teaching and research tool has grown remarkably in recent years because of its applicability to real-life, contemporary, human situations and its public accessibility through written reports. Given the highly dynamic nature of agribusiness in Africa, case studies can ensure that education and training is delivered within a contemporary context. Yet there are very few documented case studies of African agribusinesses. Investments are needed for researching and documenting agribusiness case studies that highlight contemporary management issues. Further investments might be needed to training instructors on the use of case studies in teaching.

## Adapt to changing environments

Last and perhaps most importantly, the macro and micro environments facing agribusinesses in Africa will continue to change. The recommendations noted above, while relevant today, might not be applicable to tomorrow's challenges for agribusiness education and training. New issues such as the global financial crisis, the rise of genetically modified goods, environmental challenges, technological advancements and rapid urbanisation are likely to present new challenges that are impossible to anticipate.

The implication for AET is that institutions need to re-invent themselves continuously in order to stay at pace with contemporary challenges, and perhaps anticipate future ones. Better yet, AET needs to produce graduates who are able to meet future challenges without the need for further training – the adaptive graduate.

## References


AABSA. 2011. *Association of African Business Schools Agribusiness Consortium.* http://www.aagri.net/page/about_us/consortium.html

AET Africa. 2012. *Portal for Agricultural Education and Training in Africa.* http://www.aet-africa.org/

African Agricultural Economics Education Network. 2004. December – last update, *Collaborative Master of Science in Agriculture and Applied Economics.* http://www.agriculturaleconomics.net/home/home.asp?00=1

African Organic Farming Foundation. 2010. *The AOFF Agribusiness Development Model.* http://www.africanorganics.org/progsum.html/

Akridge, J., W. Dobson and M. Holschuh. 1994. Positioning Agricultural Economics Departments to Serve Agribusiness Graduate and Professional Education Markets. *American Journal of Agricultural Economics;* 76 (5): 1193–1198.

ANAFE. 2011. *Strengthening Africa's Strategic Agricultural Capacity for Impact on Development – A Transformative Programme.* ANAFE Unpublished report. http://anafe-africa.org/?q=content/strengthening-africas-strategic-agricultural-capacity-impact-development-sasacid

Boland, M., E. Lehman and J. Stroade. 2001. A Comparison of Curriculum in Baccalaureate Degree Programs in Agribusiness Management. *International Food and Agribusiness Management Review;* 4 (3): 225-235.

Breazeale, D., M. Mangheni, M. Erbaugh and S. Mbowa. 2004. *Making University Curricula and Training Programs Responsive to Employer Needs: The Experience of Makerere University's Agribusiness Education Program.* Association of International Agricultural and Extension Education 20th Annual Conference, Dublin.

CARE. 2010. *Trade and Investment Program for Competitive Export Economy.* http://www.care.org/careswork/projects/GHA032.asp

Christy, R. 2009. *The Role of Universities in Agribusiness Development*. Agribusiness and Economic Development, University of Copenhagen, Copenhagen.

De Lange, A. and J. Van Rooyen. 2002. *The Current Status in Training for Agribusiness*. In Van Rooyen, J., O. Doyer, L. D'Haese and F. Bostyn, Eds. Readings in Agribusiness: A Source Book for Agribusiness Training. University of Pretoria: Pretoria.

Dooley, F. and J. Fulton. 1999. The State of Agribusiness Teaching Research and Extension at the Turn of the Millennium. *American Journal of Agricultural Economics;* 81 (5): 1042–1049.

IFPRI. 2013. Agribusiness Entrepreneurship Programme to Strengthen Skills and Competencies of Ghanaian Agribusiness Professionals- Press Release. http://www.ifpri.org/pressrelease/agribusiness-entrepreneurship-programme-strengthen-skills-and-competencies-ghanaian-agr

Iowa State University. 2010. April—last update. *ISU Seed Center*. Homepage of Iowa State University. http://www.news.iastate.edu/news/2010/apr/semi

Jaffee, S., R. Kopicki, P. Labaste, and I. Christie. 2003. *Modernizing Africa's Agro-Food Systems: Analytical Framework and Implications for Operations*. World Bank, Washington D.C.

LEA. 2013. *Local Enterprise Authority*. http://www.lea.co.bw/

Maguire, C. 2000. *Agricultural Education in Africa: Managing Change*. Workshop 2000, Ghana, World Bank, Washington D.C.

MSME Kenya. 2013. *Micro, Small and Medium Enterprise Competitiveness Project.* http://www.msmekenya.org/downloads/MSMEBrochure.pdf

National Food and Agribusiness Management Education Commission. 2004.

—. 2006. *Agribusiness Food and Agribusiness Management Education: Future Directions.* Purdue University, Kansas State.

NEPAD. 2010. *The Comprehensive Africa Agricultural Development Programme* [Homepage of NEPAD], [Online]. http://www.nepad-caadp.net/

Norman, D. and M. Obwona. 2001. *Status of Agricultural Economics in Selected Countries in Eastern and Southern Africa.* IFPRI, Washington D.C.

North-West University. 2010. *Faculty of Agriculture, Science and Technology*. http://www.uniwest.ac.za

Red Door. 2010. *Red Door Real Enterprise Development*. http://www.reddoor.gov.za/

SEDA. 2010. 10 February 2010 – last update. *Small Enterprise Development Agency*. http://www.seda.org.za/

Swedish Cooperative Centre. 2010. *SCC Eastern and Southern Africa and Vi Agroforestry Programmes*. http://www.sccportal.org/

University of Pretoria. 2013. *Department of Agricultural Economics, Extension and Rural Development*. http://web.up.ac.za

University of the Free State. 2010. *Agribusiness Unit*. http://www.uovs.ac.za/faculties/content.php?id=6578&FCode=04&DCode=099

Van Rooyen, J. 2010. *The Standard Bank Centre for Agribusiness Leadership and Mentorship Development*. http://blogs.sun.ac.za/news/2010/04/28/standard-bank-and-su-launch-centre-to-support-transformation-in-the-agricultural-sector/

Wachenheim, C. and W. Lesch. 2004. U.S. Executives' Views on International Agribusiness Education. *International Food and Agribusiness Management Review;* 7 (1): 42–59.

Wilkinson, R. and R. Rocha. 2008. *The Agro-processing Sector: Empirical Overview, Recent Trends and Development Impacts.* Global Agro-Industries Forum.

World Bank. 2007. *Cultivating Knowledge and Skills to Grow African Agriculture: A Synthesis of Institutional, Regional and International Review.* The International Bank for Reconstruction and Development. World Bank, Washington D.C.

—. 2013. *Growing Africa: Unlocking the potential of Agribusiness.* World Bank, Washington D.C.

Wyeth, P. 2002, *Distance Delivery of Graduate Education in Agricultural Economics for Middle and Low Income Countries.* The Farm Foundation, Washington State.

# Part Four

# Critical Issues

# CHAPTER TWELVE

# MULTIDISCIPLINARY SYSTEMS APPROACHES IN AGRICULTURAL EDUCATION AND TRAINING: A LIVESTOCK PERSPECTIVE

AKKE VAN DER ZIJPP

Livestock production is an important component of African farming systems, though often undervalued by outsiders. Africans herded cattle long before they domesticated plants. According to Marshall and Hildebrand (2002), livestock has a major role in Africa because it is less risky than other endeavors; livestock are more likely to survive in arid environments than are cultivated crops.

In the past, productivity valuations indicated that, compared to crops, livestock contributed relatively little to household income, which was interpreted to mean that livestock was not a significant part of the farming system. As a result, research, extension, and education systems in the developing world too often ignored the crucial importance of livestock for the household.

However, in a study of goat keeping in tropical farming systems, Bosman et al. (1997) concluded that livestock make a significant economic contribution to the income and security of farming households. Beyond producing products such as milk, meat, hides and skins, livestock also provide power for traction, manure for fertiliser and building material, and—by serving as living banks—financial security and asset functions, all of which had been overlooked in evaluation of livestock keeping. Once these values are recognised, it is clear why farmers expend significant amounts of time and energy to keep, feed, and care for livestock (Stroebel *et al.,* 2011).

Globally, livestock production accounts for 40 percent of agricultural output. In developing countries, where agricultural output from smallholders is the backbone of economic activity, livestock production contributes up to 80 percent of total output. In emerging economies, the demand for livestock-based food is growing rapidly.

***Key Messages***

- The dynamics of livestock production, climate change and other socio-economic and environmental issues make livestock farming systems in Africa a prominent mechanism for systems science and the emerging sustainability paradigm based on **systems approaches**.
- On the one hand, livestock production contributes up to 80 percent of total output in developing countries and thus supports the livelihoods of almost a billion people. Livestock farmers in Africa as ideally positioned to benefit from the "livestock revolution" – increased demand for livestock-based foods resulting from population growth, higher incomes and urbanization – which will further contribute towards **poverty alleviation** and development outcomes on the continent. On the other hand, the contribution of intensified livestock production on **climate change** has become unacceptable in the last three decades. Developing countries in particular are often most severely affected, given the lack of mitigation strategies and limited resources. This contrast gives rise to challenging questions related to sustainability and to **trade-offs** at and between different system levels.
- The emergence of **sustainability science** (as opposed to reductionist science) supports, but also necessitates the modernisation of the FSR approach – which includes the ecological, economic and social aspects of livestock farming systems.
- A systems approach to AET will realise **a better understanding of the current system** before new technologies and new concepts for increasing productivity are introduced; educate **more competent future scientists** who can describe the context of the farming system, identify the drivers of change, frame their objectives and present relevant questions for research, design the research approach and select the methods, and analyse and report the results in the context of the livestock farming system and for different levels of society; be applied to **pull together, evaluate and optimise knowledge** form all of the different domains of livestock farming; and add a significant and effective dimension to **capacity development**, and alleviate pressure of recruitment and formal appointments through networking.
- One systems approach to solving the important issues for global livestock production development is the teaching of **T-shaped skills**: the bottom of the T represents the disciplinary sciences, while the top represents the multi- and interdisciplinary sciences.

Livestock production depends on being able to meet the fast-growing demand for feed—which may come from either local, natural resources or from imports of processed feed or maize and soybeans grown in countries endowed with highly productive land (Naylor *et al.*, 2005). By-products and residues can be used efficiently and effectively for concentrated feeds (Nonhebel, 2007), reducing the environmental impact of livestock rearing. International trade in livestock products and feed ingredients is contributing to the new demand-supply patterns for foods of animal origin, especially in emerging economies. Maintaining a balance that makes livestock rearing attractive for market-directed smallholder production systems and provides consumers with fairly priced, good quality animal foods is not easy. Designing agricultural policies that can satisfy both producers and consumers depends on local conditions and opportunities, although policy adjustment will be needed over time (World Bank, 2009).

In the past three decades, livestock has become a global focus of concern. In western Europe (Van der Zijpp, 1991), the U.S., and East and Southeast Asia (FAO, 2006a), the environmental pressure caused by intensification of livestock production became unacceptable as air quality deteriorated, waterways atrophied, acidification affected forests, and soils were overloaded with nutrients. Animal manure expressed as waste became viewed as a liability.

Africa, however, faced the reverse of these problems, incurring nutrient losses in soils from harvesting crops without sufficient nutrient replacement (Stoorvogel and Smaling, 1990). Animal manure in these systems is often the only contributor to soil fertility because chemical fertiliser is frequently unavailable or too expensive. The study *Livestock and the Environment* proposed urgent changes in techniques and policies to rescue the environment from the pollution of livestock farming systems and to preserve a productive agro-ecosystem with safeguards for food security (De Haan et al., 1997). This study identified some worrying developments that had to be addressed, such as loss of biodiversity and soil fertility and declining access to fresh water.

The publication *Livestock's Long Shadow* (FAO, 2006b) shocked the world by showing that the value chain of livestock production was responsible for 18 percent of the greenhouse gas emissions contributing to climate change, and that water and biodiversity were strongly affected by the land degradation associated with raising livestock. In a phenomenon similar to that arising from producing bio-energy from biomass, the fast changing geography of livestock production was pushing the livestock sector into competition for land, crops, water, and other natural resources. This study proposed cross-cutting policy framework solutions to these environmental problems, for example establishing correct pricing of natural resources such as land and water in order to avoid over-exploitation and pollution, removing damaging subsidies, and introducing payment for environmental services such as water, clean air, and carbon sequestration.

In 2009, with the theme "Livestock in the Balance," *The FAO State of Food and Agriculture* (FAO, 2009) reported that livestock accounted for 40 percent of global agricultural output and supported the livelihoods of almost a billion people. Yet, even though the growth of large-scale industrial producers contributes to the growing economic importance of livestock, many pastoralists and smallholders who contribute to local food security and livelihoods risk marginalisation. FAO, as well as the World Bank in its report *Minding the Stock* (2009), recommend stronger governance of the livestock sector, pointing out that it could benefit more from increasing demand if smallholders' capacities were enhanced.

The livestock sector needs to consider its role in environmental protection and environmental services. In addition, healthy animals represent a public good and deserve more public investment to prevent epidemics and animal protein losses which, in turn, affect humans through zoonotic diseases and the need for quality nutrition. Better governance would create an inclusive development process benefitting livestock farmers and consumers. The publication *Livestock in a Changing Landscape* addresses these issues with two volumes: *Drivers, Consequences, and Responses* (Steinfeld et al., 2010), and *Experiences and Regional Perspectives* (Gerber et al., 2010). This publication provides a wealth of information regarding the present social, economic, and biophysical consequences of livestock production. Appropriate land use for the increasingly predominant mono-gastric production systems based on crop production, and environmental services provided by ruminant production in marginal areas will be critical in the future. Livestock production contributes significantly to poverty reduction (FAO, 2012). Smallholders organised in cooperatives and contract farming have an opportunity to improve their livelihoods.

For many smallholders, policies supporting transition from subsistence to production for the market will be crucial, and if successful will provide people with opportunities for employment in the value chain rather than just in primary production. The requirements for food quality and safety and for on-time delivery will put even more pressure on smallholder farmers. Livestock science has to move from commodity research (Anderson, 2005) to complex systems research if it is to build the kind of sustainable livestock farming systems that will be able to feed the world's increasing population. Finding ways to address production, environment, social exclusion, and human health issues at the local level and simultaneously address global issues such as climate change and trade is a challenge that education and training in animal sciences in Africa and the world can help to meet.

## Africa's Mixed Crop/Livestock Farming System

The majority of Africans farm in mixed crop-livestock systems, based on staple crop production and livestock. Dependence on livestock is higher in arid and semi-arid conditions, where there are pastoral systems and livestock is moved to seasonally-available feed resources including crop residues and grasslands. In Africa, landless systems for poultry are also emerging, catering to consumers in urban centres. In rain-fed and irrigated areas, livestock such as dairy, pigs, sheep, goats and traction animals are part of the mixed farming systems. It can be expected that climate change will push farmers to utilise better adapted breeds or even change to species better adapted to drought and higher temperatures.

In the mixed crop-livestock farming system, livestock offers an array of goods and services, including foods such as meat, milk, and eggs; fibres such as wool; hides; power for traction; manure for bio-gas production; financial security in the form of living banks and assets; social and cultural roles for festivities; ownership status; and a means of social connectivity for security. In addition, livestock represents a means to fend off risk, be it from drought, human disease, accidents, or low market prices for crops (Stroebel et al., 2008). The diversity of functions changes with the farming system location and according to the context in which the farming system performs. With the availability of improved market outlets and access to stable banking systems, affordable health services, and education, the role of livestock changes and there is more emphasis on production for the market. The importance of traction, nutrients, and energy depends upon the livestock's interaction with crops and the availability of chemical fertilisers. Even though the multifunctional component that livestock contributes to the farming system can change according to context (Moll et al., 2007), farmers tend to maintain their traditional species because of cultural functions in family traditions, previous generations associated with the family stock, and the accumulated knowledge of traditional management in the local agro-ecology and community. Multifunctionality is not restricted to livestock. Land, crops, and ponds also have a multitude of functions, which farmers optimise to secure the continuity of their farming systems and livelihoods, to fend off risk, and to assure food security and income (Andersson et al., 2007; Thorne et al., 2003).

Human-animal relationships represent a unique aspect of farming systems with livestock. In addition to utility functions such as production and traction, animals have emotional and aesthetic values for people. The growing discomfort affluent Western societies have with the intensification and scaling up of animal production systems shows an increasing divide between consumers, citizens, and producers. The production process leads to long and complicated value chains across the world. Western demands to observe animal welfare and health regulations, originating from the Brambell report in 1965 and the Farm Animal Welfare Advisory Committee report (1979) introducing the five freedoms,[1] will ultimately impact production in developing countries.

The increasing demand for animal products coupled with higher standards of living in urban centres of emerging and developing countries will focus increased attention on animal welfare and health and food safety issues. For many smallholder farmers, these demands may appear excessive, because as farmers they have intimate relationships with livestock and a wealth of practical knowledge regarding treatment of food for consumption. For example, boiling milk before consumption is still widely practiced in Africa and, in fact, was practiced in Western Europe until the 1960s, before the advent of affordable refrigerators and pasteurised milk in the market. Becoming market oriented will mean more demands for improved product quality and animal welfare and safety in the primary production phase. These issues will become part of sustainable production systems over time. Processors and retailers will set their own standards through systems like EUROP GAP (Euro retailer produce/ Good Agricultural Practice), ISO (International Organisation for Standardisation) certification, and HACCP (Hazardous Analyses Critical Control Points).

Livestock farming systems in Africa are pressed to change for a number of reasons, starting with the so-called "livestock revolution," which refers to the increased demand for livestock-based foods resulting from population growth, higher incomes, and urbanisation (Delgado et al., 1999). Other drivers of change are the impact of intensification on the environment and emerging diseases that affect both animal and human populations (Sere et al., 2007). Although Africa has had slower income growth than Asia, its growth rates have nonetheless been substantial. In addition, African urbanisation is a continuing process, with younger and usually better educated people migrating nationally and internationally (Ruben, 2007), searching for a better life than in the countryside where some 75 percent of the population still lives in abject poverty. The human capacity that could help transform livestock farming systems and value chains may be lost in this process of migration. Africa has been importing food since the 1980s and will face higher costs of food imports to cover the growth of demand if productivity is not increased. At the same time, Asian and Middle Eastern countries, amongst others, are buying or leasing fertile land in Africa in order to meet their own demands for food, feed, and energy production because they lack adequate land or the appropri-

---

[1] 1) Freedom from Hunger and Thirst; 2) Freedom from Discomfort ; 3) Freedom from Pain, Injury or Disease; 4) Freedom to Express Normal Behaviour; 5) Freedom from Fear and Distress.

ate climate for productive agriculture (Brautigam, 2010). China, on the other hand, is introducing its own development model into Africa, implementing infrastructures, institutions, and financial agreements similar to those that drove Chinese development. These efforts could have potentially far-reaching effects and offer dynamic opportunities for Africa's development.

In short, growing demand for livestock products opens up opportunities for development of livestock farming, but environmental sustainability, capacity loss through migration, and international competition for land undermine the process of transforming the livestock farming systems for the benefit of local smallholder farmers. In areas where infrastructure such as processing facilities, roads, mobile phones and internet capabilities can contribute to meeting demand, transformation of local value chains is occurring at a fast rate. For example, dairy production in Kenya has shown how quickly mixed farming systems can adapt when marketing milk locally becomes attractive (Staal, 2002). In Mekelle, Ethiopia, increased local demand for eggs has affected the prices of hatching eggs (Aklilu, 2007), the first step in establishing larger flocks. In Mekelle, it was found that egg production was best close to town, because if production were further away, multiple traders and transporters would need to take their profits. In other words, egg production remains an activity for the local market, which allows the prices to remain low.

Ethiopia has also built internationally certified slaughter houses for export of ruminant meats, but utilisation has been difficult (Klaasen, 2005). The smallholder mixed farmers do not meet the standards of traders for live weight and quality to supply slaughter lines, so traders have reverted to fattening beef lots. This excludes the smallholders, who lose the opportunity to earn income from fattening ruminant livestock. This implies several possible problems within the value chain, including: feed availability and quality of feed, avoiding risk, unequal power relations in trading, lack of extension and information about slaughter requirements, inefficiency of small numbers, and weak farmer organisations. However, Ethiopia's donor-funded programme for Improving Productivity and Market Success (IPMS) has shown that, even with bottom-up participatory approaches, farmers can successfully identify potential for marketing of agricultural products and innovate and provide marketable, qualified products for new value chains in milk, forage seeds, and vegetables.

The complicated nature of crop-livestock systems, the dynamics of a fast-changing context, and the varied local agro-ecology (Inter Academy Council, 2004; IAASTD, 2009) and socio-economic environment of Africa indicate the need for a better understanding of the current system before introducing new technologies and new concepts for increasing productivity. It is critical to understand the complicated farming system, but even more important that farmers share their own visions for future development with relevant stakeholders.

Combining the knowledge of local and global farming systems and indigenous practices with current science and technology should bring about innovation adapted to the local situation, with an inherent bottom-up drive to adapt subsistence systems to sustainable livestock farming systems that are linked to viable markets. Farmers then become the drivers, assisted by applied scientists and extension officers who seek solutions to their problems, but always with long-term goals of maintaining a healthy environment for farming. The change will inevitably lead to loss of employment in the livestock sector, but new opportunities will emerge in the associated value chains. Input and output markets need to be developed, and once in place, these markets can easily absorb labour trained for the functions of processing, transporting, marketing, and banking. Even so, in the absence of social security programmes, social safety nets will be needed for the extremely poor and for dealing with changes in the extended family brought about through education and migration.

According to the FAO SOFA (2009), balancing societies' objectives for livestock is central to sustainable development which, by definition, refers to development that meets the needs of the present generation without compromising the ability of future generations to meet their own needs. The word balance implies trade-offs such as higher income versus stress on natural resources, or protecting the environment at the expense of increased production costs. The prioritisation will differ between countries and regions, because sustainability issues in animal production systems vary according to location and time (local resources, changing markets, food preferences, seasonal and annual climate changes, age of farmers). Even farmers in the same locations have identified different issues they think would improve the sustainability of their farming systems. Finding the way forward will require solutions from research and education as well as tailor-made extension. At the national level, there is a need to address the diversity of issues related to the context of the farming system and to identify the policies needed to support different types of farming systems. At local, national, and global levels, there is a need to increase understanding of the different impacts of livestock farming on food security, food safety, environment, and livelihoods, and provisions have to be made for international agreements recognising these differences. To improve the contribution of livestock to global society, actions are needed at all levels: local, regional, national, and international.

Thompson and Nardone (1999) identified two approaches to sustainable livestock production: resource efficiency and functional integrity. Although studying resource efficiency would implicitly assume functional integrity, the scope of resource efficiency may not cover all interactions in the system, which is the substance of functional integrity. Resource efficiency on its own may lead to irreversible losses of the economic or social components of the system. To engage in the science to support a sustainable livestock sector (complex), systems studies have the potential to provide appropriate and credible solutions (Ackoff, 1999; Collinson, 2000). Systems science recognises that multiple stakeholders benefit or suffer from livestock production.

Future scientists have to be educated so that they can describe the context of the farming system, identify the drivers of change, frame their objectives and present relevant questions for research, design the research approach and select the methods, and analyse and report the results in the context of the livestock farming system for different levels of society. In the end, the stakeholders decide among the different choices of sustainable development scenarios. The selected scenario should be evaluated in a sustainability assessment exercise by the stakeholders, leading to a cycle of continuing improvements in line with the contextual changes of the livestock farming system (Cornelissen, 2003; Mollenhorst, 2005). Change should become an inherent characteristic of the farming system and value chain, enabling it to adapt to its context and make its optimal contribution (i.e. by trade-off selection) to society.

Complex systems science starts with an analysis of the problem of a farming system or a value chain. Once the problem has been defined, which usually requires participatory approaches, the research process is designed to study the problem and potential solutions. This distinguishes systems research from disciplinary approaches, which often start with a research question that researchers assume is relevant for science or society. Systems science approaches problem formulation and designs the research process so that the research leads potentially to problem solving with inclusive outcomes on location and in time. Systems research is also powerful in identifying drivers and consequences at local, national and global levels, leading to holistic reporting such as FAO's *Livestock's Long Shadow* (2006) and *SOFA Livestock in the Balance* (2009).

## Systems Science and Livestock Farming

In the late 1960s and 1970s, systems thinking moved into agriculture. It started with field practitioners such as agronomists and economists who were dissatisfied with the success levels associated with introducing Western technologies into the field. These practitioners identified a gap between what poor, small-scale, and often illiterate farmers needed and the new technologies emerging from research stations, which were then still thriving in Africa.

Today, the link between farmers and researchers remains challenging. In Ghana, the *Convergence of Sciences Programme* (COS1) found that a majority of the scientists at research institutes had never been involved in technology transfer (Van Huis et al., 2007). How can extension officers provide appropriate information to farmers if they are not interacting with researchers? Increasingly, farmers become better informed through modern media or communication channels, for example, radio. The need for Farming Systems Research (FSR) has probably increased with the dynamic changes in the context of livestock farming. Initially, farming systems research focused almost exclusively on cropping systems. The multifunctionality of livestock was not valued until the early 1990s, when Bosman et al. (1997) explained the financial logic of keeping goats for financial security—goats could cover the costs of health, school fees, and social activities, serving as live savings banks in locations that had no reliable banking. Livestock farmers were critical of the adoption of new technologies; livestock's mitigating role against risk— of disease, environmental disasters, conflicts, or drought—discourages smallholder farmers from participating in experiments. Farmers will not comply with researchers and extension officers when such an important component of the farming system is at risk. This may also explain why livestock farming has not been a focus of farming systems research (Collinson, 2000; Mutsaers, 2007) despite its complexity and the disappointments in technology adoption.

If smallholder mixed crop-livestock and pastoral farmers, as well as consumers of livestock products, are to benefit from the livestock revolution, a farming systems approach is required. Livestock farming is about production, environmental integrity, social benefits, animal welfare, and consumer requirements such as health and food safety extending into the value chain. Knowledge of each of these domains has to be pulled together, evaluated with relevant reference criteria, and optimised through a systems approach (Lenne and Thomas, 2006).

Since the 1980s, many tools have been developed to assist in formulating research objectives and questions. Sere and Steinfeld's (1996) classification of world livestock systems according to agro-ecological and socio-economic principles was a significant milestone. Dixon et al. (2001) presented a detailed treatise of the classified systems and devoted a chapter to sub-Saharan African farming systems. Scoping studies (e.g. the Interdisciplinary Research and Education Fund [INREF] at Wageningen University) have been introduced to ensure that local and regional problems are addressed. Scoping studies make use of common tools such as transects, rapid rural appraisal, focus groups, interviews, expert knowledge, farmers' participation, and local data sources (Collinson, 2000). These tools are used to define the problem, develop associated research questions, define the value chain with its system boundaries, and describe the context of the system. Objectives and research questions are addressed by a large diversity of methods, including those listed above as well as through literature studies, modelling, and experimentation in laboratories, farms, and fields.

The common features of good systems science studies are making intelligent combinations of soft and hard systems approaches in the design, analysing results, and framing results in the appropriate context. Ideally, stakeholders should be involved in the initial prioritisation of the problem under study, and when presented with a number of choices or scenarios, they will review the usefulness of the results. The choices may involve technology, organisation, geography, communication, economics, pollution, and other aspects of the farming system or value chain that require change to become more sustainable. The evaluation should include the relevance of the system definition with its boundaries as well. Livestock (sub) systems may be very productive and effective, but externalities in the mixed crop-livestock system itself or in the specialised (industrialised) livestock system could produce imbalances that prove detrimental to soil fertility or air quality, animal welfare, or health of employees; in addition, societal issues such as gender equity, public health (nutrition/obesity), labour requirements, and environmental quality or climate change may be adversely affected. Systems studies always operate at different levels as determined by the problem being addressed (Aklilu, 2007; Mekoya, 2008; Bebe, 2003; Phong, 2010).

## Systems Science in Education and Training for Livestock Production

Systems science in livestock research and education, although not identified as such, has been around since the Second World War. Researchers such as Hoekstra (quoted by Schiere, 1995), who worked in the Netherlands, the East Indies, and recently Indonesia, were aware that knowledge of the tropical farming system and its cultural context were essential for identification of research problems. Moreover, these researchers knew that the opinions of the farmers themselves were vital for progress. Farming Systems Research (FSR), which sought basic systems information and data concerning field situations, became prominent in the 1980s. The need for FSR still applies today, because regional knowledge of farming systems is often unavailable, meaning that policy making is often disconnected from the reality of farming.

FSR develops systems concepts, research methodologies, and software (Schiere, 1995) and can link to development practices and extension; further, it can link to technology testing and evaluation, design of farming systems, on-station research and demonstrations, on-farm experiments and farm synthesis, and many other activities that enhance participation in extension and development. These activities may yield interesting baseline information, but the design of FSR is based on linkage to the objective (a farmer-relevant problem) and requires strict choices, sampling procedures, and analysis of farming systems in the appropriate context for the problem identified.

This means that FSR is not a linear animal nutrition, reproduction, or even a housing experiment that compares different rations, treatments, or arrangements with statistical inference. Instead, it entails getting information in order to identify a problem and set an objective with consecutive, relevant research questions. The process can move forward only when the farming systems researcher has designed the experiment (case study or comparison), identified the locations, and chosen the sampling system, methods, and combinations that will lead to answers that can be validated with triangulation.

FSR is a reflective science in all phases of its research process that:

- starts with a relatively abstract concept (a farming system is a dynamic mix of subsystems with physical and institutional components), and
- develops into reality through experience in the research process including experiments, observations, and interactions with feedback.

Solutions for the problem are more likely to be options for development, and each farmer may achieve improvements in a different way. Study of these farmer-based actions represents the next cycle in the process, and the cycles may continue indefinitely (Cornelissen, 2003; Bell and Morse, 2003). Provided a positive understanding between farmers and researchers is maintained, the process of evaluation and adaptation to the changing, dynamic context and introduction of new research topics will continue. The application of FSR then turns into agricultural innovation systems with a moving body of expertise and actors appropriate for what the system needs to become more sustainable.

As FSR has changed over time, it has naturally incorporated many questions related to sustainability and to trade-offs between system levels. For livestock farming systems this is a major assignment. For example, greenhouse gas emissions are a global problem, while nutrient loading— which can cause eutrophication and acidification—is a local problem. Yet the intensification of the system may be the major driver for both processes, leading to trade-offs where nutrient loading has to be stopped while reduction of climate change might still be possible. Trade-off can also refer to the issue of equity—increased demand provides socio-economic opportunities for smallholder farmers. But large-scale investors in animal production farms could negotiate exclusivity with the retailers, preventing smallholders from gaining market access. Farmers have become more dependent on the structure of value chains, where swapping of environmental impact between actors in the chain may force decision-making for the chain as a whole. For example, feed impacts may swap from feed industry processing to on-farm production and *vice versa* (Thomassen, 2008).

The important issues for global livestock production development as presented in SOFA (FAO, 2009) can only be addressed with a holistic approach. This does not exclude the value of discipline-based research, but it calls for a strong connection between disciplinary and more holistic research systems. Wageningen University has exploited the so-called T-shaped skills, presented in Figure 12.1. The bottom of the 'T' represents the disciplinary, here depicted as specialist, sciences. The top of the 'T' represents the multi- and interdisciplinary sciences. Disciplinary and multidisciplinary sciences meet in the heart of the 'T'. Communication that shares knowledge, i.e. overlaps among different researchers and partners, is essential; otherwise the expertise of the disciplinary researcher will never reach the setting of the farming system or society that is the domain of the interdisciplinary researcher or team members of different disciplines.

The importance of T-shaped skills has been growing with the changes in science. The curriculum of the animal science student is invariably restricted to animal-based biology and technology, with very limited basic knowledge of the social, plant, and environmental sciences. Moreover, most students have urban backgrounds, with interests focused on the animal and less on the farming system and sector.

## FSR Students and Courses of Study

Compared to the 1960s, student backgrounds have changed from rural (80 percent) to urban (90 percent) and the curriculum from broad-based (generalist) to specialised. An MSc in farming systems science is therefore an essential component of the curriculum, despite challenges that some students experience with the FSR approach. Personal interest at the animal level and a dominant hard systems tradition in the basic curriculum sometimes hamper students' abilities to observe and reflect, to accept farmer participation, and to consider social aspects and relevance of perceptions.

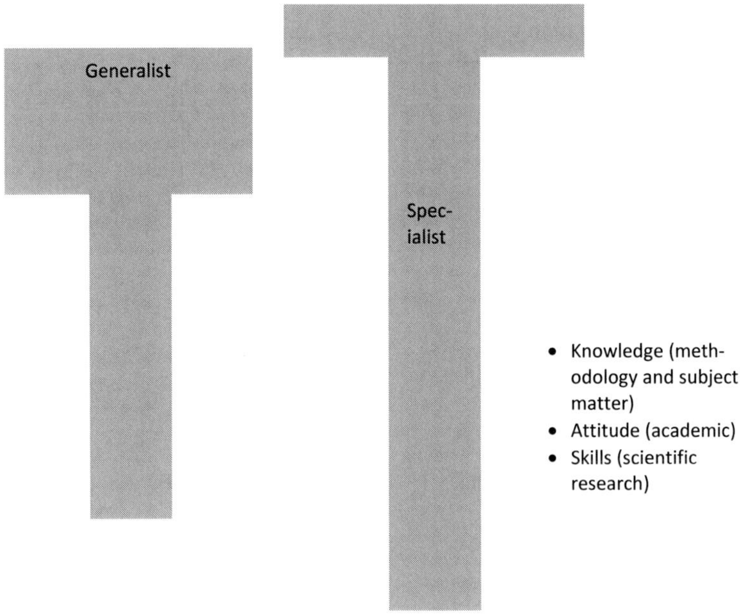


Figure 12.1. Competencies: T-shaped skills

At Wageningen University, the first FSR-based course, Systems Approach to Animal Sciences, is presented as part of the third-year curriculum of the BSc to all students of animal science and to students who select this course as the basis for a minor MSc research project. The BSc is actually part of MSc coursework (based on three years BSc and two years MSc) according to the Bologna agreement of European higher education.

These students may have a farm technology, biology, organic farming systems, or other major background. The course is also presented to international MSc students, either in the classroom or as an Internet-based course, because few students have been taught the basics of systems science in agriculture at their own universities. A selection of the courses taught at Wageningen University is summarised in Box 1 (Study Handbook, 2009/2010).

| **Box 12.1. Courses Sustainability Science for Animal Production** | |
|---|---|
| BSc Animal Science | *Systems Approach to Animal Sciences*<br>The course teaches the principles of systems science applied to animal production systems. Students are introduced to the H-R-H concept, in which "H" is holistic and "R" is the reductionist approach. First, problems in the systems context are reviewed. Each week begins with theory lectures, followed by individual exercises where staff provide feedback and facilitate joint dis- discussions. |
| MSc Speciali- Specialisation, Animal Production Systems, Ani- Animal Science | *Sustainable Development of Animal Systems: Issues and Op- Options*<br>The ecological, economic, and social aspects of sustainability, including stakeholder analysis, issues, indicators and validation, reference values, assessment, and proposal writing are incorpo-rated. The course provides students with tools and methods to experiment with, and to use later in researching livestock farm-ing systems and value chains (Bell and Morse, 1999). |
| MSc Speciali- Specialisation, Animal Production Systems, Ani- Animal Science | *Future Livestock Systems*<br>The role of systems and their drivers are explored. Students set up a logical framework related to values and functions of ani- animals and resource use, judge and evaluate farming systems, and design livelihood strategies for future development. |
| MSc across Wageningen University | *Agricultural Consultancy Training (ACT)*<br>This component is required for all MSc students and sensitises them to the interdisciplinary nature of problems in the field. Project objectives for consultancies are supplied by external societal organisations, industry, and government. An interdisci- interdisciplinary and multicultural team of MSc students prepares a project proposal including budget and timeline. The eight-week course includes project management, communication, and team building, with initial presentation modules. The coach, expert, and client evaluate the student's contributions, final report, and presentation. |
| MSc Speciali- Specialisation Animal Production Systems, Ani- Animal Science | *Thesis*<br>Students who have completed both MSc courses are entitled to start a major thesis research project specialising in Animal Pro- Production Systems. Students with one MSc level course are limited to a minor thesis. Overall the teaching emphasises analysis, observation, and critical thinking rather than descriptive content. |

## Modernising and teaching FSR

*Practical significance.* Many African universities introduce aspects of FSR in their discipline-based curricula. However, modernisation of the approach now seems necessary, in light of the emerging sustainability paradigm and the recent focus by FAO and the World Bank on intensification of livestock farming. In actuality, teaching the approach—which includes analysing the ecological, economic, and social aspects of livestock farming systems, incorporating the perceptions of livestock farmers, and designing more sustainable systems in a series of scenarios with feedback to the farmers for evaluation—is a valuable teaching tool and also elicits information that is valuable for farmers, extension officers, and policy makers. If validated, the information can be used by local governments, NGOs, and farmer associations for their own strategic prioritisation. Students become very motivated when they realise that their work has practical applications in the field.

*Teachers and researchers.* Now that sustainability science has gained prominence at some universities—e.g. Arizona State University, USA, and EARTH University, Costa Rica (see chapter 6)—training of teachers and researchers in FSR will be accelerated. Sustainability science is included in top scientific journals such as the *Proceedings of the National Academy of Science* (Clark, 2007) and new scientific journals such as *Sustainability Science,* launched by Springer publishers in 2006, and the *International Journal of Agricultural Sustainability,* by Earthscan. Systems scientists frequently encounter resistance from editors of the traditional agricultural disciplinary science journals (e.g. animal science), who are bound to reductionist science. Unfortunately, this impedes the sharing of results of systems science research, as well as the search for major grants in the field of sustainable development. Calder and Clugston (2004) provide an interesting account of the results of South Carolina's sustainable universities initiative. The study concludes that a major stumbling block for FSR (also in European universities) is recognition, difficulties in competing for major grants, and lack of prestige in the context of interdisciplinary science. To quote the authors: "This usually means narrow, disciplinary research" is needed to be successful. Offsetting these negative attitudes to sustainability science requires strong support of administrators, recognition of champions, and focus on FSR projects to leverage sustainability research and education.

After much debate, the European Association of Animal Production (EAAP) added a new Commission on Livestock Farming Systems (LFS) to its disciplinary and species-based commissions. Its role is to integrate expertise of EAAP Commissions into systems-based approaches and exchange views with all EAAP scientists. The LFS Commission provides a platform to develop systems science in animal agriculture. This is significant for European animal scientists as it is a place to start collaboration and to exchange and organise research, education, and training programmes for funding by the European Union. EAAP provides links to the disciplinary animal sciences, while the International Farming Systems Association (IFSA), originating from the FSR initiators, provides better links to the social sciences.

## Issues with Implementation in University Curricula

University curricula tend to be highly regulated, and changes are a matter of discussion for all teaching staff. In presenting an updated review of a classical undergraduate curriculum for livestock production based on the natural sciences, Casey (2010) discussed several societal reasons for extending the animal science curriculum to include communication and environmental science courses, but advised inter-university exchange as a means to address human interaction and enhance intellectual capital in South Africa.

At Wageningen University, the introduction of systems science was preceded by many years of teaching systems-based approaches, including modelling in tropical animal production. The Wageningen Animal Production Systems Group, which began in the mid-1990s, expanded the application to Western systems and introduced a framework for assessment of sustainability and many new research methods applicable to the ecological and social aspects of animal production systems. For development-oriented research, Wageningen University initiated its Interdisciplinary Research and Education Fund (INREF), which provides an on-going forum on development of interdisciplinary science and experimentation with international PhD students (INREF, 2010). Other donors, such as the Rockefeller Foundation, Winrock, and the Ministry of International Cooperation (DGIS) in The Netherlands, have supported similar programmes. The Dutch National Science Foundation (NWO) recently decided to follow a similar interdisciplinary programme approach. Even with these funding opportunities, identifying financial support is difficult, and reviews of project proposals may still be biased through disciplinary notions and lack of knowledge regarding livestock revolution opportunities. The experiences at Hawkesbury Agricultural College in Australia are a case in point (Bawden, 2003). A basic funding mechanism of university departments (independent of scientific disciplines), stamina, and a strong belief in the strength of the systems approach are therefore required to be successful.

The Hawkesbury learning approach was based on experiential learning, systems thinking, and action research. Staff and students both were part of the learning mode, experiencing the different assumptions and paradigms underlying agricultural science; the interaction of reductionist science and the hard and soft systems; and the learning cycle emphasising experience, analysis, observation, and reflection. The agro-ecosystem was seen as lying at the interface between natural and social systems. This 'learning by doing' approach led to participation of stakeholders and to a transformation involving technical experts taking on the role of facilitators.

The evolution of FSR in livestock systems research and education is moving towards studies of sustainability based on Complex Systems Research, with ever increasing numbers of methods and tools available for a diverse set of problems and research questions. Although FSR is needed more than ever, changing curricula to specialised fields like genomics to meet the interest of urban students presents some obstacles. Just as ministries need policy makers, farmers' associations as well as NGOs need FSR expertise, and as value chains grow in importance, industry also needs more insight. The demand for FSR-trained graduates is great. In Africa these graduates can provide farming systems expertise that often is not available either in the ministries or from extension and veterinary services. Farmers are now better educated and becoming better informed through traditional media such as radio and television, Internet and mobile phones. The question is not the demand, but the supply of graduates, and in turn, the adaptability of curricula designers at universities to fill the gap quickly.

Disciplinary scientists will not automatically support this movement toward FSR education and research. Pressure from university leadership, farmers' associations, ministries, and input and output suppliers in the value chain, as well as NGOs will be needed to make this change happen. For university leadership the unavailability of high quality scientific journals is no longer an excuse not to act. Action to recruit experienced scientists should become the next agenda item for collaborative action of the regional agricultural research organisations in West, South, and East Africa. These include the West- and Central-African Council for Agricultural Research and Development (WECARD), the Association for Strengthening Agricultural Research in Eastern and Central Africa (ASARECA), and the Southern African Development Community (SADC) Food, Agriculture and Natural Resource Directorate (SADC-FANR). In addition, commendable initiatives such as the Regional University Forum for Capacity Building in Agriculture (RUFORUM) and the Forum for Agricultural Research in Africa's (FARA) capacity building initiatives should be capitalised upon to strengthen FSR approaches. The recently established Tertiary Education for Agriculture Mechanism (TEAM-Africa) emphasises the importance of farming systems research to advance farming systems research in Africa (Swanepoel and Stroebel, 2012).

FSR was started by disciplinary researchers who understood farming systems and were willing to assist in improving African farmers' livelihoods. Among the present generation of discipline-trained staff there is a potential resource of interested staff with experience in extension and on-farm research who can continue learning in systems science and link to a network of graduates who have received formal degrees in animal production systems and FSR specialisation.

Networking can add a significant and effective dimension to capacity development and alleviate the pressure of recruitment and formal appointments. Examples of institutions in this context are Aarhus University in Denmark, Michigan State University and Harvard University in the USA, Wageningen University in The Netherlands, and the International Livestock Research Institute (ILRI). The collaboration should include the content of FSR in terms of sustainability science expertise and knowledge as well as the teaching skills of staff to be trained. Unfortunately, education in teaching skills does not have a prominent place in the curriculum of international students, nor in training of staff. This training should be incorporated into programmes to develop FSR education and training in Africa. This would probably add more to experiential learning and capacity building of teaching staff than would separate knowledge-focused courses (Mugabi and Gulikers, 2009).

The dynamics of livestock production, climate change, and many other socio-economic and environmental issues make livestock farming systems in Africa a prominent mechanism for systems science and the emerging sustainability science based on systems approaches. Amekawa et al. (2010) identify three areas that must be considered: inclusive policy making involving all stakeholders; learning in public institutions; and interdisciplinary science—through which the linkage between agro-ecology and social processes has to be strengthened. They argue for holistic management of sustainable, multi-functional agriculture as opposed to the highly compartmentalised organisation of modern agriculture into specialisations such as animal science. Effecting change in attitude, values and skills, as presented in the T-shaped skills illustration, takes time and perseverance. In reviewing FSR experiences in Kenya, Tanzania, and Zambia, Collinson (1988) concluded that the success of FSR was dependent on changes in institutions and professional paternalism; under the current system FSR suffers from pitiful rewards, donor interference, and impatience. The professional consensus that FSR is appropriate for development of smallholder farmers was still far away in 1988, but even in 2010, sophisticated Western agriculture is still foremost in the minds of policy makers, teachers, and extension officers.

However, learning from practice combined with innovative research that can lead to new paradigms in this dynamic environment could be very productive for science and for farming systems that fit the local and global conditions. New universities not bound to traditional curricula are emerging in Africa, and many are growing. This makes innovative curricula and expansion, as opposed to conventional curriculum building, a real possibility. The rural dynamics also challenge new study opportunities. Africa lacks leaders who grasp the potential of FSR and who can focus on sustainability for the future of rural communities (Meridian Institute, 2013). Apart from leadership for research and education in FSR, scientific communication and publication excellence have to be incorporated (Van der Zijpp, 2011). FSR has been successfully promoted in the sustainability-science section of the U.S. National Academy of Sciences Proceedings; advocates for teaching sustainability science in Africa can emulate this success. Despite all the pitfalls, Africa's dynamic livestock farming systems offer a unique entry into sustainability science.

## Concluding Comments

Farming Systems Research emerged approximately forty years ago from practitioners in the field who were disappointed by limited production increases due to low uptake of technology in developing countries. Animal scientists were not prominent among these early practitioners. Livestock farming systems are multifunctional, and farmers adjust their goals depending on the contextual dynamics of the farming system and the perceived risks.

During the past twenty years, livestock has become associated with negative environmental impacts such nutrient loading, climate change, socio-economic marginalisation, and issues of equity. Livestock farming systems have become systems in need of sustainable development to balance the benefits of high quality protein foods for consumers with negative impacts on the environment and the potential bypassing of smallholder livestock farmers.

With increasing numbers of students and the dynamics of rural development, African universities and colleges offer new opportunities for innovative curricula like FSR and sustainability science. Livestock plays a prominent role as demand for livestock foods is increasing with rising standards of living. Communication about FSR has to be addressed to students and staff who may have no concept of the complexity of the science and what its benefits can be. FSR is important for livestock development. To be successful in education and science for animal production, higher education administrators must champion interdisciplinary science as opposed to the norm of specialised disciplinary sciences. Collaboration within Africa, and with outside development partners interested in and committed to African development, could strengthen and accelerate the training of FSR scientists and teachers.

Given the present dynamics calling for economic, ecological, and social change, Africa's livestock farming systems offer a unique opportunity for growth. A continuing participatory linkage of the university with the farming communities could ensure that the innovation cycle proceeds towards higher production, a better environment, more income, and improved social welfare of the farming communities.

## References

Ackoff, R. 1999. Disciplines, the Two Cultures, and the Scianities. *Systems Research and Behavioural Science* 16: 533–537.

Aklilu, H. 2007. *Village Poultry in Ethiopia: Socio-Technical Analysis and Earning with Farmers.* Thesis, Wageningen University.

Amekawa, Y., H. Ssegya, S. Onzere, and I. Carranza. 2010. Delineating the Multifunctional Role of Agro-ecological Practices: Toward Sustainable Livelihoods for Smallholder Farmers in Developing Countries. *Journal of Sustainable Agriculture* 34:2, 202–228.

Andersson, J., A. Twine, A. Murwira, K. Giller, A. Mashingaidze, and M. Slingerland. 2007. *Land, Productivity and Agricultural Research: Towards an Understanding of the Multiple Meanings of Land.* FARA-SADC regional dialogue, Forum for Agricultural Research in Africa (FARA), Sandton, Johannesburg.

Anderson, P. 2005. *The Ecological Dimension of ARD: From Commodity Research to Complex Systems Research.* Abstracts Plenary Session 5, EFARD, Zurich, Switzerland.

Bawden, R., 2003. *The Hawkesbury Experience: Tales from a Road Less Travelled.* In Bawden, R., Ed. Coping with Complexity, a Collection of Papers on Systems Thinking and Learning. Reader produced by International Agricultural Centre, Wageningen, The Netherlands.

Bebe, B. 2003. *Herd Dynamics of Smallholder Dairy in the Kenyan Highlands*. Thesis, Wageningen University.

Bell, S. and S. Morse. 1999. *Sustainability Indicators*. Earthscan: London.

—. 2003. *Measuring Sustainability: Learning from Doing*. Earthscan: London.

Bosman, H., H. Moll, and H. Udo. 1997. Measuring and Interpreting the Benefits of Goat Keeping in Tropical Farm Systems. *Agricultural Systems* 53: 349–372.

Brambell Committee. 1965. Command Paper 2836.

Brautigam, D. 2010. *Africa's Eastern Promise.* www.foreignaffairs.com/articles/65916/deborah-brautigam/africa%E2%80%99s-eastern-promise

Calder, W. and R. Clugston. 2004. Lighting Many Fires: South Carolina's Sustainable Universities Initiative. In Corcoran, P. and A. Wals. *Higher Education and the Challenge of Sustainability: Problematics, Promise, and Practice.* Kluwer Academic Publishers: Dordrecht, the Netherlands.

Casey, N. 2010. Integrated Higher Learning: An Investment in Intellectual Capital for Livestock Production. *Livestock Science* 130: 83–94.

Clark, W. 2007. Sustainability Science: A Room of its Own. *Proceedings of the National Academy of Science* 104: 1737–1738.

Collinson, M. 1988. The Development of African Farming Systems: Some Personal Views. *Agricultural Administration and Extension* 29: 7–22.

Collinson, M., ed. 2000. *History of Farming Systems Research.* FAO and CABI Publishing: Wallingford, UK.

Cornelissen, A. 2003. *The Two Faces of Sustainability: Fuzzy Evaluation of Sustainable Development.* Thesis, Wageningen University.

De Haan, C., H. Steinfeld, and H. Blackburn. 1997. *Livestock and the Environment: Finding a Balance.* Report of a study supported by the Commission of the European Communities; The World Bank; and the Governments of Denmark, France, Germany, the Netherlands, the UK and the U.S. European Commission Directorate-General for Development, Brussels.

Delgado, C., M. Rosegrant, H. Steinfeld, S. Ehui, and C. Courbois. 1999. *Livestock to 2020: The Next Food Revolution.* Food, Agriculture and the Environment. Discussion Paper 28, International Food Policy Research Institute, Washington, DC.

Dixon, J., A. Gulliver, and D. Gibbons. 2001. *Farming Systems and Poverty: Improving Farmers' Livelihoods in a Changing World.* FAO, Rome, Italy.

FAO. 2006a. Livestock Policy Brief 02. *Pollution from Industrialized Livestock Production.* Livestock Information, Sector Analysis and Policy Branch, Animal Production and Health Division, FAO, Rome.

—. 2006b. *Livestock's Long Shadow: Environmental Issues and Options*. FAO, Rome, Italy.

—. 2009. *The State of Food and Agriculture: Livestock in the Balance.* FAO, Rome, Italy.

—. 2012. *Livestock Sector Development for Poverty Reduction: An Economic and Policy Perspective – Livestock's Many Virtues*. FAO, Rome, Italy.

Farm Animal Welfare Advisory Commission. 1979. "Five Freedoms of Farm Animals." Advice given considering the outcome of the technical committee. The Brambell Report, 1965. www.fawc.org.uk/freedoms.htm

Gerber, P., H. Mooney, J. Dijkman, S. Tarawali, and C. de Haan. 2010. *Livestock in a Changing Landscape: Experiences and Regional Perspectives*. Island Press: Washington, DC.

IFSA. www.ifsa.bokuac.at/cms/index.php?id=workshops

INREF. www.inref.wur.nl/UK/

—. 2010. *Journey into Interdisciplinarity*. Ten Years of INREF Experience. Wageningen UR, Wageningen International, Wageningen, the Netherlands.

Inter Academy Council (IAC). 2004. *Realising the Promise and Potential of African Agriculture.* Science and Technology Strategies for Improving Agricultural Productivity and Food Security in Africa. Amsterdam, the Netherlands.

International Assessment of Agricultural Knowledge, Science and Technology for Development (IAASTD). 2009. *Agriculture at a Crossroads*. Executive Summary of the Synthesis Report. Island Press: Washington, DC.

IPMS Ethiopia Project. www.ipms-ethiopia.org/

Klaasen, M. 2005. *Impact of Changing Norms and Standards for Livestock (product) Trade on Technological and Institutional Innovations along the Value Chain in Ethiopia*. MSc thesis, Wageningen University.

Lenne, J. and D. Thomas. 2006. Integrating Crop-Livestock Research and Development in Sub-Saharan Africa: Option, Imperative or Impossible? *Outlook on Agriculture* 35: 3, 167–175.

Marshall, F. and E. Hildebrand. 2002. Cattle Before Crops: The Beginnings of Food Production in Africa. *Journal of World History* 16: 2, 99–143.

Mekoya, K. 2008. *Multipurpose Fodder Trees in Ethiopia.* Thesis, Wageningen University.

Meridian Institute. 2013. *Innovation Platforms and Smallholder Farmers: Gaps and Opportunities*. Report prepared for the Bill and Melinda Gates Foundation. Seattle, USA.

Moll, A., S. Staal, and M. Ibrahim. 2007. Smallholder Dairy Production and Markets: A Comparison of Production Systems in Zambia, Kenya, and Sri Lanka. *Agricultural Systems* 94: 2, 593–603.

Mollenhorst, E. 2005. *How to House a Hen: Assessing Sustainable Development of Egg Production Systems*. Thesis, Wageningen University.

Mugabi, A. and J. Gulikers. 2009. Capacity Building of Teachers Towards Better Higher Education in sub-Saharan Africa: A Study of Eight Teachers doing an MSc in Wageningen University. *ECS Bulletin* 7: 5–6.

Mutsaers, H. 2007. *Peasants, Farmers and Scientists.* Springer: Dordrecht, The Netherlands.

Naylor, R., H. Steinfeld, W. Falcon, J. Galloway, V. Smil, E. Bradford, J. Alder, and H. Mooney. 2005. Losing the Links between Livestock and Land. *Science* 310: 5754, 1621–1622.

Nonhebel, S. 2007. Energy from Agricultural Residues and Consequences for Land Requirements for Food Production. *Agricultural Systems* 94: 586–592.

Phong, L. 2010. Dynamics of Sustainability in Integrated-Agriculture Aquaculture Systems in the Mekong Delta. PhD thesis, Wageningen University.

Ruben, R. 2007. *Farmers and Fishponds in South East Asia: Let the Fish Talk about the Water.* In Van der Zijpp, A., J. Verreth, T. Le Quang, M. van Mensvoort, R. Bosma and M. Beveridge, Eds. Fishponds in Farming Systems: Wageningen Academic Publishers: Wageningen.

Schiere, J. 1995. *Cattle, Straw and System Control.* Thesis, Landbouw Universiteit Wageningen.

Sere, C. and H. Steinfeld. 1996. *World Livestock Production Systems: Current Status, Issues and Trends.* FAO Animal Production and Health Paper no. 127. FAO, Rome.

Sere, C., A. van der Zijpp, G. Persley and E. Rege. 2007. Dynamics of Livestock Production Systems, Drivers of Change and Prospects for Animal Genetic Resources. *Animal Genetic Resources Information* 42: 3–27.

Staal, S. 2002. *Market Opportunities for Milk Producer Organizations*. ILRI News. www.cgiar.org/ilri/news/milkmarket.cfm.

Steinfeld, H., H. Mooney, F. Schneider and L. Neville. 2010. *Livestock in a Changing Landscape: Drivers, Consequences and Responses*. Island Press: Washington D.C.

Stoorvogel, J. and E. Smaling. 1990. *Assessment of Soil Nutrient Depletion in sub-Saharan Africa.* Winand Staring Centre, Report 28, Wageningen.

Stroebel, A., Swanepoel F., N. Nthakheni, A. Nesamvuni and G. Taylor. 2008. Benefits obtained from Cattle by Smallholder Farmers: A Case Study of Limpopo Province, South Africa. *Australian Journal of Experimental Agriculture,* 48: 825–828.

Stroebel, A., Swanepoel, F., and Pell, A. 2011. Sustainable Smallholder Livestock Systems: A Case Study of Limpopo Province, South Africa. *Livestock Science* 139, 186–190.

Study Handbook. 2009/2010. Wageningen University, Wageningen, the Netherlands.

Swanepoel, F. and A. Stroebel. 2012. *Capacity Building for African Agricultural Development: the Need for Transformation in Tertiary Agricultural Education.* Keynote address at the Third TAE-Partnership Workshop on Tertiary Agricultural Education in Africa. November 13–15, Stellenbosch, South Africa.

Thomassen, M. 2008. Environmental Impact of Dairy Cattle Production Systems: An Assessment. PhD Thesis, Wageningen University. Wageningen, the Netherlands.

Thompson, P. and A. Nardone. 1999. Sustainable Livestock Production: Methodological and Ethical Challenges. *Livestock Production Science* 61, 111–119.

Thorne, P., P. Thornton, R. Kruska, L. Reynolds, S. Waddington, A. Rutherford and A. Odero. 2003. *Maize as Food, Feed and Fertilizer in Intensifying Crop-Livestock Systems in East and Southern Africa: An Ex Ante Impact Assessment of Technology Interventions to Improve Smallholder Welfare.* ILRI Impact Assessment Series 11. ILRI, Nairobi, Kenya.

Van Huis, A., J. Jiggins, D. Kossou, C. Leeuwis, N. Roling, O. Sakyi-Dawson, P. Struik and R. Tossou. 2007. Can Convergence of Agricultural Sciences Support Innovation by Resource-Poor Farmers in Africa? The Cases of Benin and Ghana. *International Journal of Agricultural Sustainability*; 5 (2–3), 91–108.

Van der Zijpp, A. J. 1991. *Sustainable Animal Production: A Response to Environmental Problems*. In The Livestock Production Sector in Eastern Europe as affected by Current Changes. Proceedings of the Round-Table organised by EAAP/FAO. Budapest, April 14 – 17. EAAP Publication 57. Pudoc, Wageningen, the Netherlands.

Van der Zijpp, A. 2011. Assessment for Sustainable Development of Animal Production Systems (Guest Editor). *Livestock Science*, 139 (1–2), 1–190.

World Bank, 2009. *Minding the Stock: Bringing Public Policy to Bear on Livestock Sector Development.* Agriculture and Rural Development Department. Report no. 44010-GLB. World Bank, Washington D.C.

# CHAPTER THIRTEEN

# QUALITY ASSURANCE IN HIGHER EDUCATION IN LIFE SCIENCES

LUC D'HAESE

Defining and measuring quality in higher education, particularly in the life sciences, is a complex undertaking, especially given the different educational environments in which faculties of life sciences operate throughout the world. Quality assurance is often confused with evaluation or review, yet evaluation and review are only part of a much more complex and comprehensive system that is needed for quality assurance. Such a system has to establish stakeholder confidence in higher education inputs, processes, and outcomes to fulfil expectations or measure up to minimum requirements.

Quality assurance has many dimensions that can be applied to all components of a higher education institution—academic programmes, teaching, research, scholarships, staffing, students, buildings, facilities and equipment, services to the community and the academic environment, and more. There are also different levels of quality, based on stakeholder expectations, specific visions, and diverse interests within the institution. This makes it imperative to design quality assurance systems that build on the mission, vision, goals, and policies of individual institutions. Moreover, appropriate academic standards should reflect the attitudes, objectives, actions, and procedures the higher education institution has chosen (African Union, 2008).

In addition to the inherent difficulty of comparing standards across diverse higher education institutions in a system, in the absence of accreditation by peers the definition of higher education itself needs to be considered; thus it is difficult even to determine whether an institution with life science programmes is actually working at the university level.

This chapter looks at how diversity of institutions affects quality assurance systems in higher education, specifically in the life sciences. It explores the complexity of the concept of quality in this context, illustrates the differences among quality assurance systems by comparing those developed at the University of Antwerp (Belgium) with those at South African institutions, and then provides an overview of such systems in Europe, the U.S., Canada, and Africa.

***Key Messages***

- An *effective quality assurance system* is essential for efforts to establish and manage effective AET interventions. Yet such a system is often not implemented.
- In *defining quality* it is necessary to start by distinguishing absolute levels of quality. This allows ranking of outcomes on a scale based on relative quality, which involves judgment against a set standard.
- Quality assurance systems should provide *appropriate information* for use by different stakeholders in order to ensure that quality is assessed in a timely, effective and efficient manner.
- *Depending on the purpose,* internal or external systems may be put in place to assure quality:
  - *Internal* quality assurance systems are monitoring and evaluation efforts implemented by the programme or institution itself.
  - *External* quality assurance systems involve specialists, professionals and stakeholders who are not directly involved in the programme or institution, but are asked to make a quality assessment.
- The following structures are regarded as essential in streamlining the process of *internal quality assurance:*
  - structures of councils and commissions that accommodate the different groups of stakeholders interested in quality assurance;
  - structure for data collection of quality assurance parameters; and
  - structure for control of the study load.
- *External quality assurance* varies from system to system and can include different types of
  - institutional evaluations;
  - subject or programme evaluations;
  - accreditation at subject, programme, and institutional levels; or
  - any combination of these.
- The *effectiveness of such external evaluations* depends on
  - an explicit internal quality assurance strategy and
  - the use of mechanisms and methods aimed at achieving its objectives.
- The specific societal, economic, and educational imperatives in a country are *key drivers* that determine the relevance and usefulness of particular quality assurance approaches, mechanisms, and procedures.
- In future, greater mobility and communication among students, as well as flexibility in programming and curricula, will be needed. Agricultural education and training are no exception, and quality assurance systems can help *facilitate mobility and flexibility in the AET system.*
- No country, including those in sub-Saharan Africa, can afford to compromise the quality of its academic programmes. All efforts should therefore be made to *assure their quality in a credible and useful manner.*

## Quality in Higher Education: A Complex and Multidimensional Concept

In defining quality, it is necessary to start by distinguishing absolute levels of quality and then ranking outcomes on a relative scale. This process involves judgment against a set standard. The European Standards and Guidelines (ESGs) define quality as "principles to adhere to rather than something to be measured against" (ENQA, 2006).

The term quality assurance is common in business sectors such as manufacturing, where it refers to quality control. It also incorporates quality management, which refers to project managers, or academics in the case of higher education institutions, taking responsibility for ensuring quality levels. Opre and Opre (2006) defined quality assurance as "all the policies and processes directed to ensure the maintenance and enhancing of quality." Opre and Opre (2006) identified some of the lessons to be learned from the business sector, such as: 1) make the desire for quality an overarching principle in every operation; 2) be knowledgeable about the needs of students and academics; and 3) ensure the attractiveness of Higher Education Institutions (HEIs) through a focus on social and economic issues while maintaining a high level of academic integrity and quality.

The concept of quality assurance in academics is not new, but its implementation and certainly the methodologies it applies have evolved over time. The concept of quality assurance emerged from the business sector in the 1950s and early 1960s as a key factor of commercial success. Business theory predicts that profits will be maximised when a company is able to sell quality products or services at a reasonable price while controlling costs. Being an essential public good and financed by the government, higher education was not considered a profit-maximising business. As long as higher education institutions were seen as quasi-governmental institutions, quality assessment was not a concern, and universities were assumed to be defining and meeting their own social and economic responsibilities. Yet, with internationalisation of higher education and globalisation in general, the establishment of quality assurance systems has become a necessary tool to monitor the quality of the higher education delivered and to engage in international networks (UNESCO, 2005). Moreover, since the late 1980s, students and society have increasingly demanded good quality higher education.

The systems that are put into place to assure quality may be internal or external.

- Internal quality assurance systems are evaluations put into place by the study programme or the institution itself.
- External quality assurance systems involve specialists, professionals, and other stakeholders who are not directly involved in the study programme or institution being asked to make a quality assessment.

Additionally, the two types of systems are interlinked and their implementation depends on the goal of the evaluation (ENQA, 2006). Quality assurance systems should provide the necessary information to all stakeholders. According to the Flemish Inter-university Council and the Council of Flemish Institutions of Higher Education (VLIR/VLHORA, 2005), these functions of quality assurance are as follows.

- Improvement and guarantee function: using results of quality assurance to improve and anchor the quality of the study programme.
- Accountability function: using results of quality assurance as information for governments and consumers. This means accountability to the broader society on how public funds are efficiently and effectively used; institutions are not only encouraged to be responsible and economical, they are also compelled to communicate the results of the social advantages clearly. At the same time, this information is made available to students, their parents, and employers by means of public reports on the degree to which study programmes fulfil quality standards.
- Regulation of the higher education system: using results of quality assurance to attach concrete consequences in terms of steering and regulation. Quality indicators are applied, for example, as parameters for decision-making on the teaching qualifications of institutions, the acceptance of new study programmes, and the distribution of resources between education and research.

Higher education institutions around the world have started to develop quality assurance systems. Most often, they are specific to the local situation of the institution and to the stakeholders in the institution's region, country, or international environment. Two case studies of quality assurances systems—at the University of Antwerp and in the South African higher education institutions—illustrate the tools applied for internal and external quality assessment.

# Standards and Guidelines for Quality Assurance in Higher Education: Case studies in Belgium and South Africa

## Quality Assurance at the University of Antwerp

The system of internal quality assurance at the University of Antwerp (UA) is based on curriculum revisions (degree programme review) and on individual course evaluations. The UA uses a diagram for systematic and periodic curriculum revisions crafted on the framework of the Flemish Inter-University Council plan for external quality assurance. Its internal and external quality assurance systems are therefore inextricably linked.

As shown in Figure 13.1, the UA quality assurance cycle has a duration of eight years. A general curriculum review is supported by a thorough programme evaluation. The cycle of curriculum revisions is illustrated through a concrete time diagram. This process starts two years prior to the external evaluation. A curriculum revision is based on the findings derived from critical self-reflections and from the report of the external visiting commission. In principle, in-depth curriculum modifications should be made one or two years after the external evaluation.

Restricted interim curriculum modifications based on experience in specific parts of the education programme are also possible. Such small modifications are not bound to a certain year in the revision cycle.

To streamline the process the following structures are essential:

- Structures of councils and commissions that will accommodate the different groups of stakeholders interested in quality assurance;
- Structure for data collection of the quality assurance parameters;
- Structure for control of the study load, as it is a verifiable and objective indicator determined by legal decree.

## Education Council at university level

The Education Council is responsible for the overall education vision of the university. It is a discussion and reflection group that stimulates educational innovation and development. When an external quality assessment takes place, the Education Council gives the necessary intellectual and material support to the quality experts operating at faculty level. The Council also organises training sessions and is responsible for managing information regarding the outcomes of all internal and external quality assessments of degree programmes at the university.

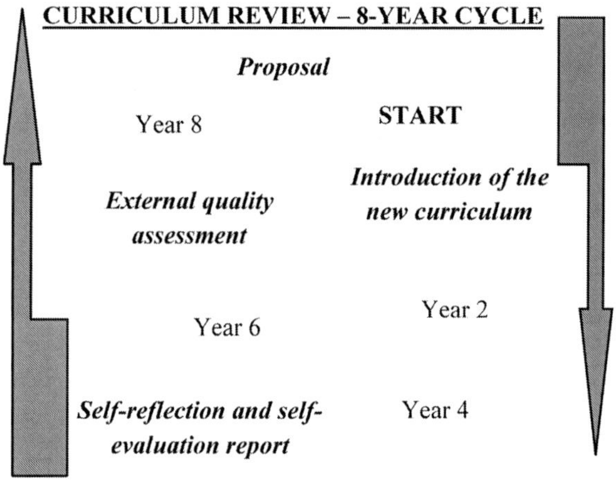


Figure 13.1. Quality assurance cycle UA

## Structure for Data Collection of Quality Assurance Parameters

Integrated quality assurance in education requires the following components:

- programme evaluation of the training;
- curriculum review and implementation;
- evaluation of the individual course units; and
- focus group meetings.

As shown in the curriculum review cycle (Figure 13.1), a programme evaluation starts two years prior to the external visit. The process starts with a self-evaluation report that provides a critical analysis of the study programme and an opportunity to identify strengths and weaknesses. It also highlights ideas for consolidating the strengths and tackling areas for improvement. It is important that all parties—students as well as academic, administrative, and technical personnel—are involved and active in drawing up the self-evaluation report. Such a report reflects the vision of the study programme, and any discussion points should be included in the text.

A self-evaluation report has to include information on the following six subjects:

- Objectives of the study programme;
- Curriculum;

- Deployment of personnel;
- Facilities;
- Internal quality assurance;
- Results.

A curriculum review also assesses the level and orientation of the study programme by checking the degree to which the overall study programme objectives correspond to the final qualifications of the academically-oriented bachelor's and master's study programmes, as articulated in Article 58 of the Decree on Flemish Higher Education. The outcome of this exercise is an overall picture of the relationships among the objectives and the content of the curriculum, the coherence of the study programme and the alignment (in terms of content and educational method) of the study programme components. The organisation of such an evaluation falls undr the responsibility of the departmental education commissions. Results are discussed in this commission with the support of the Faculty Quality Assurance Centre (Figure 13.2).

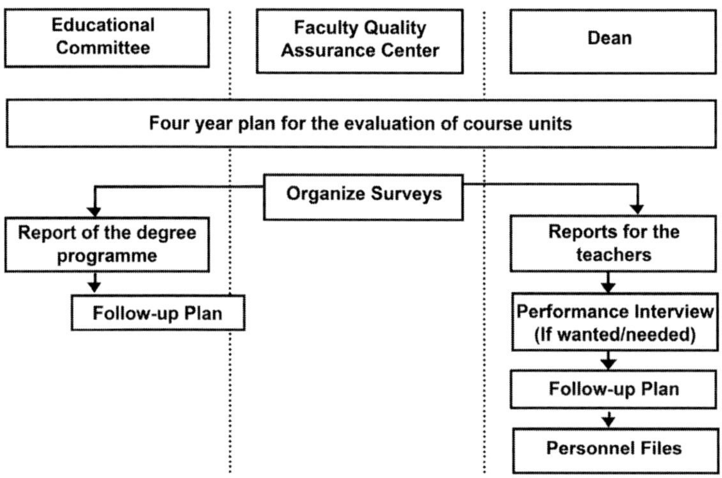


Figure 13.2. Internal quality assurance at the UA (Vos, 2008)

The evaluation of an individual course unit takes place at least every four years. Given that lecturers each have several courses, at least one course for each lecturer is evaluated every two years. At the end of each semester a focus group will discuss and evaluate the programme for that semester. Ten to twelve students are invited to take part in the group discussion, which covers the workload and the ease or difficulty of study and examination. A report is given to each lecturer and to the Education Committee chairman. If necessary, action is taken.

If requested by the Education Committee, an ad hoc evaluation can be organised to investigate specific educational problems. By official decree, each professor is evaluated at least once every five years, using a course unit questionnaire that covers twelve didactic dimensions and more than forty items. The students have to reply to closed and open questions. The reliability of the response is confirmed through Cronbach's *alpha* statistics.

For example: "Style of teaching," one of the didactic dimensions, has three items:

- The teacher explained the subject matter well.
- The teacher gave clear explanations during classes.
- The instruction of the teacher was bad.

The lecturer receives a graphical representation of the outcome of the survey (using EvaSys data-capturing software). An example is shown in Figure 13.3. The mean, standard deviation per dimension is then compared with the average of the similar surveys of all lecturers in that group.

The system is supported by the Faculty Quality Assurance Centre. The feedback of the evaluation report is split into two parts: one aimed at improving the lectures at degree-programme level, and one aimed at improving the educational performances of the individual lecturer. The evaluation feedback is as specific as possible, thus enabling the lecturers to take measures to improve their impacts. The lecturers also have the opportunity to comment on their weaker score(s).

As the evaluation report is based on student consultation, it means there will likely be a number of subjective aspects to positive or negative evaluations. But it is the task of the Faculty Quality Assurance Centre to put the results in the right context.

This is done, for example, in focus group meetings which consist of a range of structured conversations with a randomly chosen representative group of students for each study year concerning a number of subjects (communicated in advance). The main objective of the Faculty Quality Assurance Centre is to facilitate the discussion with the students. Systematically, all education aspects are taken in review including: the organisation of the lectures, the examinations and the study load. Individual lecturers receive a printout of the subjects' discussions of their contributions to the education programme.

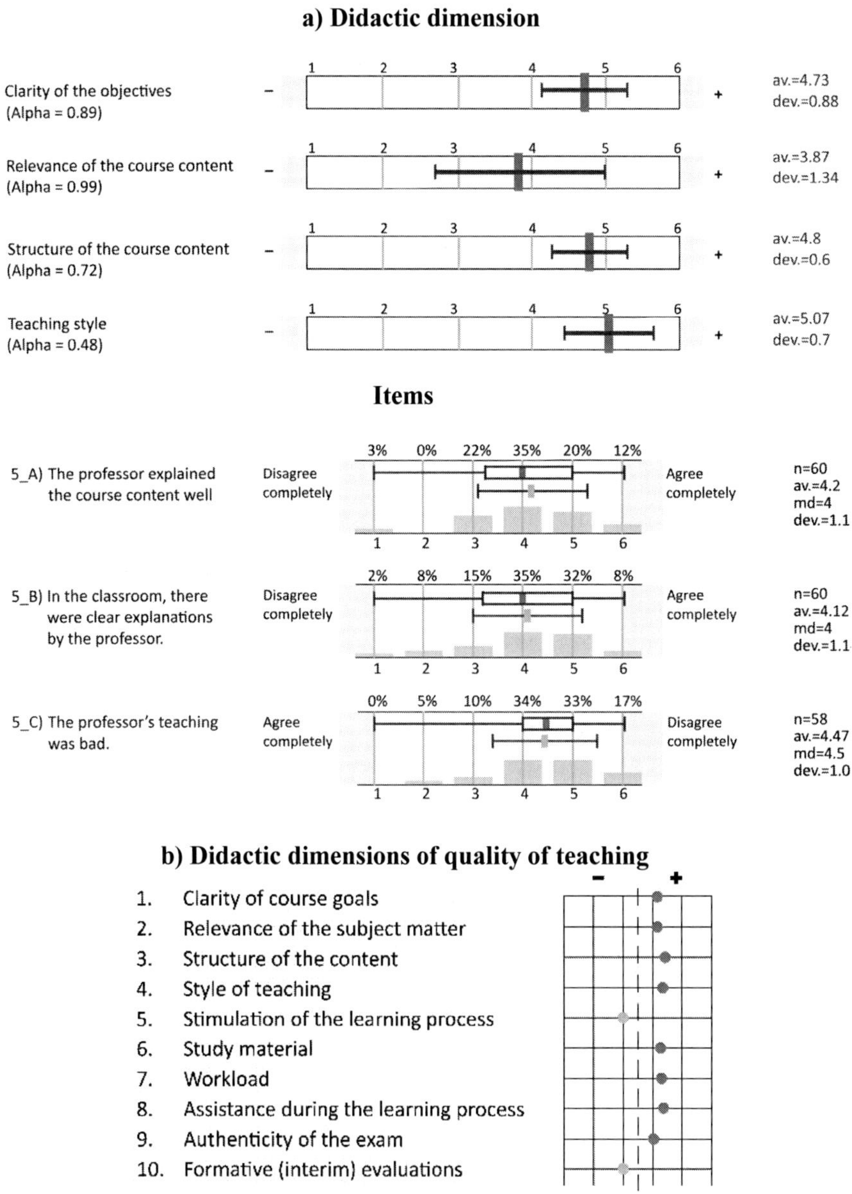


Figure 13.3. Individual professor evaluation

The focus group conversations are designed to complement the formal evaluation of the training. The combination of the two guarantees systematic and frequent controls and ensures that problems or student dissatisfaction on certain educational matters will be addressed. The students also receive feedback concerning the most important findings of the focus group conversations and the corresponding actions undertaken by the education commission. The commission uses the results of the focus group conversations as complementary information to optimise the training.

## Study Load Measurement

Study load measurement is a combination of theoretical analysis, focus group discussion, and so-called time writing. The total effective study time as measured by the European Credit Transfer System (ECTS) of points is one credit point for a study time of twenty-five to thirty hours. Study time comprises the time spent attending classes and in learning and evaluation activities, and ranges from 1,500 to 1,800 hours per year, per programme.

The number of ECTS points in a curriculum should be a good reflection of the student's actual study load. The theoretical analysis of the study load is calculated on the basis the so-called Vyt formula, which uses a weighing factor that considers the type of college, the number of contact hours, and total study load (including contact hours) to calculate how much time a modal student needs to succeed. For example, the applied weight factors for the training at the Faculty of Science are: 3 for *ex cathedra* lectures, 2 for seminars and exercises, 1.5 for practical exercises, 1.25 for training periods and 1 for self-study projects.

Currently, time spent writing and monitoring performance/results is applied in the first year of the bachelor's degree. In this programme, checking the study load is vital in order to give the best chance to new students to qualify. The writing component of the study in the second and third year of the bachelor's programme and also at the master's level is difficult since there are frequently more programme options and the students follow more individual routes. For these programmes, focus group conversations are organised to have a more quantitative measurement of study load.

The focus group discussion method is qualitative and complementary to the theoretical analysis of curricula and the time writing. The Education Commission takes into account observations of whether the study load the students experience is too high or too low, among other things, and organises specific improvements to the programme by, for example, increasing or lowering the number of credits of individual components.

## External Quality Assurance

The standards and guidelines proposed in this chapter envisage an important role for external quality assurance. The type of evaluation varies from system to system and can include different types of: 1) institutional evaluations; 2) subject or programme evaluations; 3) accreditation at subject, programme, and institutional levels; or 4) a combination of these. How effective such external evaluations are depends on an explicit internal quality assurance strategy and on the use of mechanisms and methods aimed at achieving its objectives (ENQA, 2006).

There are important reasons for involving external agencies in quality assurance. These include

- protection of national academic standards for higher education;
- accreditation of programmes or institutions;
- employer protection;
- public provision of independently verified quantitative and qualitative information about programmes or institutions; and
- improvement and enhancement of quality (ENQA, 2006).

European quality assurance agencies follow the legal, social, and cultural requirements of the jurisdictions and environments in which they operate and develop standards that are relevant for the local situations. Detailed guidelines usually differ by context, but some common elements emerge from collaboration and exchange of information across agencies. General principles of good practice in such external quality assurance processes require that

- institutional autonomy be respected;
- interests of students and other stakeholders, such as the labour market, are at the forefront of external quality assurance processes; and
- use should be made, wherever possible, of the results of the internal quality assessments of the institution itself (ENQA, 2006).

## Quality assurance in the South African higher education system

The South African Council on Higher Education (CHE) is an independent statutory body responsible for advising the Minister of Higher Education and Training on all higher education policy issues and for quality assurance in higher education and training. The CHE, established in May, 1998 under terms of Higher Education Act 101 of 1997, can be grouped into four main areas of work:

- advising the Minister of Education, on request or proactively, on all policy matters related to higher education;
- assuming executive responsibility for quality assurance and quality promotion within higher education and training, including programme accreditation, institutional audits, programme evaluation, quality promotion, and capacity building;
- monitoring and evaluating whether, how, to what extent, and with what consequences the vision, policy goals, and objectives for higher education are being realised, including reporting on the state of South African higher education; and
- contributing to the development of higher education by providing guidance on key national and systemic issues, producing publications, holding conferences, and conducting research to inform the government and stakeholders of the immediate and long-term challenges of higher education.

The CHE's one permanent committee, the Higher Education Quality Committee (HEQC), has executive responsibility for quality promotion and quality assurance in higher education. It is led by an executive director, with a deputy director responsible for coordinating HEQC work with that of higher education institutions and for relations with stakeholders in general. HEQC has four directorates, each with specific responsibilities:

- The Institutional Audits Directorate conducts audits of public and private higher education institutions' systems for ensuring good quality of provision in the three core functions of teaching and learning, research, and community engagement.
- The National Reviews Directorate re-accredits existing programmes in specific disciplines or qualification areas.
- The Programme Accreditation Directorate accredits the learning programmes of public and private higher education institutions.
- The Quality Promotion and Capacity Development Directorate implements the HEQC's quality promotion and capacity development programme, which disseminates information and knowledge about quality assurance and prepares individuals and institutions to participate in implementing the HEQC's quality assurance system.
- The work of the HEQC in programme accreditation aligns with that of the other key role players as follows.
- The Department of Education (DoE) approves the programme and qualification mixes (PQMs) of public higher education institutions, funds programmes of those institutions that are accredited by the HEQC, and

registers private higher education institutions for specific programmes before they are permitted to operate.

- The South African Qualifications Authority (SAQA) registers each qualification on the National Qualifications Framework (NQF).
- The HEQC accredits programmes leading to NQF-registered qualifications of higher education institutions, and where relevant, this is done in cooperation with the statutory professional councils and Sector Education and Training Authorities (SETAs).

The Institutional Audits Directorate of the HEQC audits the effectiveness of the arrangements for managing the quality of teaching and learning, research, community engagement with all public providers, and qualifying private providers of higher education.

HEQC also supports the principle that primary responsibility for quality provision rests with higher education institutions themselves. Thus, HEQC provides external validation of the claims made by a higher education institution in the self-evaluation report submitted as part of its audit portfolio. This portfolio is the basis of the institutional quality audit. An external panel of peers and experts carries out site visits, gathering information for comparison with the institution's self-evaluation report. Using the HEQC Audit Criteria, the panel then prepares an audit report based on its assessment of the effectiveness of the institutional quality management systems. Since 2004, all public and qualifying private higher education institutions have been audited once within every six-year audit cycle (CHE, 2004).

South African higher education institutions that offer programmes outside the country are subject to HEQC's programme accreditation requirements for all local as well as cross-border academic programmes. In addition, they must comply with the quality requirements of the host country. The HEQC cooperates with national quality assurance agencies in countries where South African higher education institutions have an operational presence in order to share relevant accreditation and audit information. Foreign institutions that offer higher education programmes in South Africa, including those subject to accreditation requirements of other national, regional, or international agencies, are also subject to the HEQC's programme accreditation requirements.

The HEQC requires an accredited programme to establish itself reasonably well before applying for the next higher level programme in the same field. However, if an institution's mid-term progress report for a new programme is satisfactory, application could be made for the next higher level programme in the same field. For example, application could be made for a master's programme once an institution has submitted a satisfactory mid-term progress report for the honours' programme.

To assist South African higher education institutions in designing academic programmes that will receive accreditation, the HEQC generated nineteen generic criteria that are used for programme reviews as well for the compilation of programme self-evaluation reports. These criteria for programme accreditation should be used as the basis for self-evaluation of the programme(s) submitted by the institution for accreditation. The HEQC uses the criteria, the self-evaluation report, and supporting evidence provided by the institution in the evaluation of applications for programme accreditation (new programmes) or re-accreditation (existing programmes). The HEQC criteria for programme accreditation are categorised using a model of input, process, output, impact, and review (CHE, 2004).

*Criterion 1:* The programme is consonant with the institution's mission, forms part of institutional planning and resource allocation, meets national requirements, the needs of students and other stakeholders, and is intellectually credible. It is designed coherently and articulates well with other relevant programmes, where possible.

*Criterion 2:* Recruitment documentation informs potential students of the programme accurately and sufficiently, and admission adheres to current legislation. Admission and selection of students are commensurate with the programme's academic requirements, within a framework of widened access and equity. The number of students selected takes into account the programme's intended learning outcomes, its capacity to offer good quality education, and the needs of the particular profession (in the case of professional and vocational programmes).

*Criterion 3:* Academic staff members responsible for the programme are suitably qualified and have sufficient relevant experience and teaching competence, and their assessment competence and research profiles are adequate for the nature and level of the programme. The institution and/or other recognised agencies contracted by the institution provide opportunities for academic staff members to enhance their competencies and to support their professional growth and development.

*Criterion 4:* The academic and support staff complement is of sufficient size and seniority for the nature and field of the programme and the size of the student body to ensure that all activities related to the programme can be carried out effectively. The ratio of full-time to part-time staff is appropriate. The recruitment and employment of staff follows relevant legislation and appropriate administrative procedures, including redress and equity considerations. Support staff are adequately qualified and their knowledge and skills are regularly updated.

*Criterion 5:* The institution gives recognition to the importance of promoting student learning. The teaching and learning strategy is appropriate for the institutional type (as reflected in its mission), mode(s) of delivery, and student composition and contains mechanisms to ensure the appropriateness of teaching and learning methods, making provision for staff members to upgrade their teaching methods. The strategy sets targets, plans for implementation, and sets mechanisms to monitor progress, evaluate impact, and effect improvement.

*Criterion 6:* The different delivery modes of the programme have appropriate policies and procedures for internal assessment; internal and external moderation; monitoring of student progress; explicitness, validity, and reliability of assessment practices; recording of assessment results; settling of disputes; rigor and security of the assessment system; recognition of prior learning; and development of staff competence in assessment.

*Criterion 7:* Suitable and sufficient venues, IT infrastructure, and library resources are available for students and staff in the programme. Policies ensure the proper management and maintenance of library resources, including support and access for students and staff. Staff development for library personnel takes place on a regular basis.

*Criterion 8:* The programme has effective administrative services for providing information, managing the information system, dealing with a diverse student population, and ensuring the integrity of processes leading to certification of qualifications obtained through the programme.

*Criterion 9:* Postgraduate programmes have appropriate policies, procedures and regulations for the admission and selection of students, the selection and appointment of supervisors, and the definition of the roles and responsibilities of supervisors and students, etc.

*Criterion 10:* The programme is effectively coordinated in order to facilitate the attainment of its intended purposes and outcomes.

*Criterion 11:* Academic development initiatives promote student, staff, and curriculum development and offer academic support for students where necessary.

*Criterion 12:* Effective teaching and learning methods and suitable learning materials and learning opportunities facilitate the achievement of the purposes and outcomes of the programme.

*Criterion 13:* The programme has effective assessment practices, which include internal (or external) assessment as well as internal and external moderation.

*Criterion 14:* The programme has taken measures to ensure the reliability, rigor, and security of the assessment system.

*Criterion 15:* The coordination of work-based learning is done effectively in all components of applicable programmes. This includes an adequate infrastructure, effective communication, recording of progress made, monitoring, and mentoring.

*Criterion 16:* The postgraduate programme is managed properly, offers opportunities for students to develop research competence, and ensures that research is properly assessed. Policies for student admission and selection, criteria for the selection and appointment of supervisors, and guidelines on the roles and responsibilities of supervisors and students are effectively implemented.

*Criterion 17:* Student retention and through-put rates in the programme are monitored, especially in terms of race and gender equity, and remedial measures are taken where necessary.

*Criterion 18:* The programme has taken steps to enhance the employability of students and to alleviate shortages of expertise in relevant fields in cases where these are the desired outcomes of the programme.

*Criterion 19:* User surveys, reviews, and impact studies on the effectiveness of the programme are undertaken at regular intervals. Results are used to improve the programme's design, delivery, and resourcing, and to provide for staff development and student support where necessary (CHE, 2010).

From the above it is clear that South Africa's higher education system has made substantial efforts to establish adequate policies and procedures to assure the quality of the system's academic programme offerings.

## The Diverse Landscape of Quality Assurance around the World

### Europe

Although each higher education institution in Europe is responsible for the implementation of its own quality assurance, over the years a European vision has emerged for quality assurance systems in higher education.

It is generally accepted that each institution is required to put systems into place that, in consultation with all relevant stakeholders including students, aim at developing "quality." These systems include internal and external quality assurance reviews that are interlinked, with internal quality assurance informing external procedures. To help institutions develop such systems, a set of standards and guidelines for quality assurance have been developed with support from the European Commission (ENQA, 2006).

In addition, a number of networks and platforms were formed for the implementation of the European system of higher education (ENQA, 2009):

- The *European Association for Quality Assurance in Higher Education (ENQA)* has two functions. It acts as a platform for exchanging information, sharing experience, and developing good practice; and it functions as a political actor representing the interests of agencies within different political processes and authorities.
- The *European Consortium for Accreditation (ECA)* explores possibilities for the implementation of joint accreditation procedures and the accreditation of joint degrees.
- The *European Quality Assurance Register for Higher Education (EQAR)* is a platform for establishing contacts among the different agencies active in the European Higher Education Area. It also facilitates contacts with agencies from outside.
- *Communication platforms* have been established by several networks that often serve specific regional interests, such as the Nordic Quality Assurance Network in Higher Education (NOQA) and the Central and Eastern European Network.
- *Quality labels* have been developed by several different initiatives as marketing tools. They indicate that a particular institution complies with a particular set of quality standards.

These different initiatives should be complementary and conducive to more transparency in the different quality assurance systems. Furthermore, the growing emphasis on external quality assurance increases the demand for international experts. As a central entity, the European Association for Quality Assurance in Higher Education (ENQA) brings together different views in order to increase trust among institutions. Furthermore, ENQA recently began developing specific European training schemes (ENQA, 2009).

## United States of America

In the U.S., quality assurance in higher education institutions has for many years been based on an accreditation system. Established accreditation organisations are responsible for developing quality standards and procedures for measuring the degree of compliance among higher education institutions. The accreditation organisations are "gatekeepers" in the complex public-private system. Their evaluations are critical in establishing eligibility for federal and state grants and loans, which annually total US$80 billion or more. They also form the regulatory bodies that inform and protect the "users" of higher education institutions from fraud and abuse.

Over time, the accreditation system in the U.S. has become a large and complex public-private system. However, the process of accreditation is not compulsory and is largely controlled by private organisations that have strong connections to the higher education community. Yet because higher education institutions often depend heavily on grants and loans provided by state and federal governments, they have little choice but to follow the accreditation organisations. The federal government is responsible for closely monitoring the process in order to protect the public interest in ensuring quality in higher education.

The call for accreditation systems to adapt to new public-private demands is the subject of a large debate. The debate centres on how to change accreditation, including improving performance, through open standards and processes securing consistency and transparency. Up to now, no consensus has been reached on how the accreditation system can respond to new demands.

Restructuring would face two major challenges: the extent of the current accreditation system and its diverse and decentralised structure. The accreditation system in the U.S. is large, with an estimated one hundred public and private accrediting organisations that are responsible for accrediting more than 6,400 institutions and 18,700 programmes. It is also diverse, with the various accreditation organisations using different standards and processes. The accreditation organisations are national, regional, or specialised in nature. While national and regional accreditation organisations review entire institutions, specialised accrediting agencies evaluate programmes, departments, or schools that are part of an institution. A number of specialised accreditation organisations are state government agencies responsible for certain subsectors, such as the agencies responsible for regulating healthcare professions.

Formal recognition of the accreditation organisations is based on quality standards that are included in the U.S. Higher Education Act. Recognition of the organisations by the state is necessary to ensure that federal funding is made available to quality institutions only. The U.S. Department of Education develops the criteria used in the recognition process. Responsibility within the department lies with the Secretary's National Advisory Committee on Institutional Quality and Integrity and the Accrediting Agency Evaluation Unit. These criteria guarantee that the organisation has put in place an accreditation process that "effectively addresses the quality of different institutions or programmes (curricula, faculty, facilities, equipment and supplies, fiscal and administrative capacity, student support services, etc.)." Accreditation organisations can also be recognised by the Council of Higher Education Accreditation (CHEA), which acts as a non-governmental body and publishes listings of higher education institutions that are recognised by the Department of Education, by CHEA, or by both (Schray, 2005).

## Canada

In Canada, quality assurance in education is well established. The Association of Universities and Colleges of Canada (AUCC) clusters Canadian higher education institutions that commit to offering high quality education that is competitive with the rest of the world. These institutions also commit to supplying complete, reliable information on the quality of the programmes in order to secure public confidence, enhance mobility of graduates, and guarantee credibility of Canadian universities. The AUCC quality assurance principles require that:

- the institution has in place a formal, approved, transparent policy committing it to ensuring the quality and continuous improvement of its academic programmes;
- where applicable, the institution is in compliance with the requirements of a provincial or regional authority having responsibility for quality assurance—either through that agency's verification of the institution's quality review processes or through the agency's own assessment of the programmes offered;
- the authority responsible for implementing the institutional policy, and for responding to recommendations resulting from the quality assurance process, is clearly identified; and
- there is a procedure in place for the regular review of the quality assurance policy and related processes.

The quality reviews proposed by AUCC are based on self-evaluation and peer review. In line with most European quality assurance systems, the evaluation procedures include internal and external stakeholders—students, faculty, administration, alumni, and community representatives—as well as external disciplinary experts. Accreditation and certification standards are often included in quality reviews. Not only are the procedures and assessment guidelines made public, but the results of the quality assessments are also published. For transparency and accountability purposes, the policy, relevant processes, and assessment calendar are made available in the institution and among external stakeholders such as the government and the public at large.

## Africa

All developing countries have increasingly recognised the importance of higher education for national development. In a globalised world, economic development thrives on knowledge. This breeds competition between countries, which means that governments should continuously invest in higher education if they want to make progress. This was confirmed by the UNESCO task force on higher education in developing countries, which reported that "the quality of knowledge generated within higher education institutions and its availability to the wider economy is becoming increasingly critical to national competitiveness" (UNESCO, 2005). Bloom et al. (2006) provided empirical evidence on the positive contribution of tertiary education to economic growth and poverty reduction.

In several African countries, higher education institutions have had difficulties due to harsh economic, social, political, and even conflict situations. Higher education passed through a long period of neglect and stagnation, which resulted in an overall decline in the quality of that education. For example, during a period when higher education institutions in Africa experienced increasing enrolments, they also had a decrease in available resources, as well as academic brain-drain. In many countries this resulted in overcrowding, deteriorating infrastructure, insufficient resources for non-salary expenditures such as textbooks and laboratory equipment, and a decline in the quality of teaching and research activities.

For many higher education institutions in Africa, setting up quality assurance systems will be an important step in their development trajectory. Institutionalising quality assurance mechanisms—including critical self-evaluation of programmes and institutional capacity—can be an important part of capacity-building efforts.

The Association of African Universities (AAU) has committed itself to facilitating this process. With the support of the World Bank, the AAU took the initiative to coordinate quality assurance processes in higher education institutions across sub-Saharan Africa. This programme, called the Quality Assurance Support Programme for African Higher Education, aims to institutionalise quality assurance systems, develop national quality assurance and accreditation agencies, support regional networks for coordination of cross-border protocols, and provide specialised capacity building in quality assurance. As part of this effort, the AAU initiated both the African Quality Assurance Networks (AfriQan) and the African Quality Assurance Agencies Network (Materu, 2007).

The current efforts to revitalise higher education in Africa are remarkable. Having quality assurance systems in place is a major step towards increasing quality levels in the system. Acceptable standards need to be set for educational programmes, scholarships, didactic approaches, administration, and infrastructure—and these standards should be maintained and enhanced.

## Concluding Comments

As illustrated in this chapter, quality assurance is a complex, multidimensional phenomenon which is often loaded and emotional. It raises questions around who should lead the debate, who defines it, and who decides and endorses the need for it. The specific societal, economic, and educational imperatives of a country are key drivers for determining the relevance of particular quality assurance approaches, mechanisms, and procedures. At present not all higher education systems and arrangements—such as programme accreditation and reviews—are equally developed, advanced, or rigorous. Yet there seems to be general acceptance that all academic programmes, regardless of the discipline or profession, should provide evidence of at least some generic quality criteria that can confirm their quality. These criteria include clear expected outcomes; curriculum alignment; reliable, applicable, and innovative assessment practices; and the proven ability to contribute to the development of effective graduates who will be responsible citizens.

Considering that future students will be even more mobile and connected, flexible programmes and curricula will be needed. Life science education and training are no exception. Increasingly, more will be expected from life sciences graduates in terms of their capacities to solve problems, to enhance current practices, and to plan for the future with all its unknown challenges. In developing countries, life science graduates at both undergraduate and postgraduate levels are expected to enhance the life quality of communities and individuals. Therefore, no country can afford to compromise the quality of its academic programmes. Instead, all efforts should be made to assure their quality.

## References

African Union. 2008. Developing an African Higher Education Quality Rating Mechanism and Harmonization Strategy. Presented at the Meeting of the African Union, March, Accra, Ghana.

Bloom, D., D. Canning, and K. Chan. 2006. Higher Education and Economic Development in Africa. Harvard University. Human Development Sector, Africa region. http://www.arp.harvard.edu/AfricaHigherEducation/Reports/BloomAndCanning.pdf

Council on Higher Education (CHE). 2004. Criteria for Programme Accreditation. Department of Education: Pretoria.

— 2010. Report on the National Review of Academic Professional Programmes in Education: Pretoria.

CHEA. 1998. CHEA Recognition: Recognition of Accrediting Organizations Policy and Procedures. Council for Higher Education Accreditation: Washington, DC.

European Association for Quality Assurance in Higher Education (ENQA). 2006. European Commission's Report on Progress in Quality Assurance. www.enqua.eu

—. 2009. Standards and Guidelines for Quality Assurance in the European Higher Education Area: Helsinki.

Opre, A. and D. Opre. 2006. Quality Assurance in Higher Education: Professional Development. Babes-Bolyai University, Cluj-Napoca, Romania.

Materu, P. 2007. Higher Education Quality Assurance in sub-Saharan Africa: Status, Challenges, Opportunities, and Promising Practices. World Bank working paper, n° 124, World Bank, Washington, DC.

Schray, V. 2005. Assuring Quality in Higher Education: Key Issues and Questions for Changing Accreditation in the United States. Part of a Series of Issues Papers in a National Dialogue: The Secretary of Education's Commission on the Future of Higher Education.

UNESCO. 2005. Guidelines for Quality Provision in Cross-border Higher Education. Paris, 2005.

Vlaamse Interuniversitaire Raad/Vlaamse Hogescholen Raad (VLIR/VLHORA). 2005. Handleiding Onderwijsvisitaties. Brussels, Belgium.

Vos, J. 2008. Internal Quality Assurance. Faculty of Science, University of Antwerp, Antwerp, Belgium.

# CHAPTER FOURTEEN

# FINANCING AGRICULTURAL EDUCATION AND TRAINING

LARRY ZUIDEMA

Agricultural education and training (AET) activities are essential to the overall economic development of African countries, most of which remain essentially agrarian in character. The economic benefits of AET are now better understood and justify increased financial support for this crucial component of the agricultural innovation system. The most salient rationale for increased support is the greater realised economic and social value produced by AET institutions, now perceived as vital and indispensable components of a wider system contributing to efficient food production, reduced rural poverty, increased economic activity, and improved social stability.

While funding for higher education and training is limited in most African countries, there are opportunities to mobilise additional financial resources and increase the efficiency of AET in many ways. However, even with improvements in funding and efficiency, government and donor support remain essential to operations and capacity development of AET institutions in sub-Saharan Africa.

This chapter addresses the challenge of securing continued and dependable financial support from multiple sources so that AET can fulfil its role in the wider agricultural innovation system to support agriculture's contribution to the overall health, welfare, and economic development of sub-Saharan African nations. Particular attention is given to higher education institutions and—to a lesser extent—to extension training. The first two sections of the chapter explore the institutional context for AET's financial support and some of the factors that have contributed to change within many sub-Saharan African countries. This is followed by a review of some perspectives and frameworks that can be used to explain and analyse agricultural education and training for development (AET4D) activities and to influence financial support in the future. The focus then turns to opportunities for, as well as challenges to, increasing funding and efficiency. The chapter concludes by identifying some assumptions, obstacles, and implications related to the financial status of AET institutions.

***Key Messages***

- Building and strengthening agricultural education and training require *sustained government support.* Such support can be justified by the economic value of AET to society, its vital role in an agricultural innovation system, and its key contributions to the value chain of agricultural activities.
- *Investment in AET yields clear benefits:* increased agricultural production, improved efficiencies, and enhanced value-added activities. These in turn generate revenues for governments that can be reinvested in AET. As institutions increasingly deliver the desired human capital for agriculture, their perceived value in the agricultural innovation system will grow and, as a consequence, their flows of funding.
- Over the past decades the low priority of agriculture among governments and donors in sub-Saharan Africa has meant that higher education institutions lacked *continuous, dependable, and adequate financial support for AET.* As a result, their performance has suffered.
- A *complex set of internal and external actors* are involved in AET financing—ministries of education, ministries of agriculture, international development organizations (donors), NGOs, the private sector, religious groups, and regional and sub-regional organisations.
- It is vital to review and assess *the potential for new sources of financing*—driven by the increased demand for higher education and training, improved market conditions, new private universities and colleges, and an increase in private enterprise activity in the agricultural sector—for AET institutions.
- *Two conceptual frameworks* help explain new financing possibilities: 1) alternative funding systems, and 2) demand-side financing.
- There are significant opportunities *to increase and streamline AET funding,* but also challenges in doing so. These encompass:
  - rationalising and harmonising donor support;
  - using new formulae to determine appropriate levels of government support and delivery mechanisms;
  - increasing contributions by AET beneficiaries;
  - innovative income-generating activities; and voluntary contributions.
- There are also significant opportunities to *increase the efficiency of AET institutions,* which include: improving internal and external management; reducing the cost of operations and programmes; improving inter-institutional cooperation; targeting and responding to client needs; and using SSA regional programs for staff development.
- The growing availability and use of *electronic modes of education and training* also show high potential to increase the quality and financial efficiency of AET activities.
- Ministries of agriculture and education and AET institutional leaders have to *collaborate closely* to review the dimensions of their AIS, to determine the focus and expected outputs of AET institutions, and to ensure that funding is available to meet these expectations.

When dealing with all of sub-Saharan Africa, some generalisations are necessary. It is fully recognised that each country has a unique set of policies, resources, and institutions, and that the legacy of colonialism continues to influence agricultural education and training and to affect the possibilities for change in the future. It is also recognised that many governments have severe financial constraints and multiple claims on their limited resources. Finally, while the focus is on public institutions, privately funded AET institutions are also considered.

## Institutional Context for AET Financing

A complex set of institutions provides, utilises, influences, finances, and otherwise supports agricultural education and training activities and services. Some already have defined relationships to AET, while others are becoming more important as partners with AET institutions. The following brief review of these institutions provides historical and current perspectives on their roles, particularly with respect to financing agricultural education and training activities.

### Agricultural education and training institutions

Before about 1960, several sub-Saharan Africa countries had agricultural education, research, and extension institutions that were financed by the colonial powers. However, many of these institutions were established to serve the needs of the colonial powers for an educated African citizenry and emerging ruling class within these countries, in part to support the production of export crops such as cotton, grown in Uganda for England. In short, these institutions were not specifically designed to support local farmers' needs in the context of national development.

Upon independence, most of which took place in the 1960s, many educational institutions continued to receive financial support from their countries' former colonisers. These funds decreased over time and, in most cases, were reduced to insignificance by the mid-1970s. Consequently, national budgets and donor support were required to sustain education and training institutions. For most countries, this was a difficult period for agricultural education and training, since there were many demands on national budgets, and budgetary efficiency of local institutions was weak. Since about the 1980s, national support for AET institutions has varied widely in most countries, leaving those without donor support facing severe budgetary problems and limited opportunities to improve their contributions to development.

The number of publicly funded African universities has grown from about one hundred in 1990 to about two hundred in 2009 (World Bank, 2009). More dramatically, the number of private universities and colleges has risen from about two dozen in 1990 to nearly five hundred in 2010, with a parallel increase in students enrolled in higher education. This rapid increase in enrolment and the sharp drop in recurrent budget support are reflected in declines in educational quality and concerns about relevance, discipline balance, equity of access, management, government-university relations, and sustainability (Saint, 1992). With public expenditures per student for higher education in Africa already significantly higher than in non-African countries that have similar levels of economic development, the rapid increase in student numbers has placed a huge burden on national budgets (Brossard and Foko, 2008). Most universities continue to rely on government funding and donor contributions while gradually increasing tuition fees and exploring income-generating activities. To complement dwindling support from governments, some universities are also seeking private sector support and investment income. In addition, competition between institutions for high quality students is beginning to sharpen the focus on quality of education, and tight budgets have stimulated efforts to improve efficiency of operations in universities.

Building nationally relevant agricultural education and training institutional capacity has been a long-term activity requiring significant financial resources, mostly from donors. Restructuring and reforming these institutions to meet current and future needs is equally dependent on long-term financial investments. Drawing lessons from an eight-country study of agricultural education and training, Eicher says that "building a system of agricultural institutions is a multi-generation process." To highlight this point, he recommends that "the World Bank prepare an African Agricultural Education and Training Plan with a thirty-year time frame for strengthening agricultural education and training in Africa" (Eicher, 2006).

## International development organisations

International development organisations (hereafter called donors) became active in the 1970s with grants and loans to support agricultural education and training activities, particularly for the development and strengthening of higher education institutions. This continued up to the early 1980s, when funding for higher education lost favour with many donors who considered this to be an elitist activity and shifted support to activities that "placed their hands on the shoulders of the poor." Donors began to focus on providing support to primary and secondary schools and to extension activities. In agriculture, non-governmental organisations (NGOs) began to receive funds formerly allocated to national governments for extension and training programmes, with the expectation that NGOs would be more effective in serving the poor.

From the perspective of an AIS, the World Bank found that from 1987 to 1997, 2 percent of a global investment of US$4.8 billion went to agricultural education, while the remaining 98 percent went to agricultural research and extension (World Bank, 2007). A study by Rygnestad et al. (2006) found that "World Bank lending to agricultural education components in projects in sub-Saharan Africa was very low between financial years 1998 and 2006, totalling just US$9.5 million. World Bank lending for agricultural training activities was found to be much higher during the same time period, totalling US$95.3 million."

With respect to overall support to higher education, it was not until the early 2000s that donors began to realise that higher education, which produces both public and private benefits, had been inappropriately left out of their priorities (Bloom et al., 2006). In fact, the Blair Commission for Africa recommended in March, 2005 that donors increase investments in Africa's capacity, starting with its system of higher education, particularly in science and technology (Blair et al., 2005). The Commission went on to urge the international community to provide US$500 million a year to strengthen sub-Saharan Africa's higher education institutions and up to US$3 billion over the following ten years to develop centres of excellence in science and technology (reported in Bloom et al., 2006). Most donors now agree that support for higher education is necessary to build national capacity for development, e.g. for training teachers and extension workers to address, among other priorities, the needs of poor rural areas. Funding by donors for all higher education in sub-Saharan Africa has increased dramatically since the mid-2000s, and national statistics show rapid overall growth in demand for higher education in many countries. Nevertheless, the vagaries and sporadic nature of donor support suggest that agricultural education and training institutions must retain financial support from governments and other sources to maintain quality programmes.

While not donors per se, several international groups finance agricultural training throughout sub-Saharan Africa, often in conjunction with their projects and programmes. Most notable are the Consultative Group on International Agricultural Research (CGIAR) Centres, which collaborate with agricultural education and training and research organisations on a number of projects and integrate training for selected audiences into their project activities. United Nations agencies such as the Food and Agriculture Organisation (FAO) and the UN Development Programme (UNDP) offer and fund a significant amount of training in sub-Saharan Africa nationally, sub-regionally, regionally, and abroad. Funding from these international groups is most useful if it supports ongoing activities or introduces new scientific methods and approaches at agricultural education and training institutions. However, the activities of these organizations are usually intermittent and may or may not be related to national priorities.

## Ministries of education

Financial support for AET is primarily governed by national policies on education administered by ministries of education, often with little influence from ministries of agriculture. Governments develop and enforce education policies, allocate funding (86 percent of AET funding came from various government programmes in 1990 according to a World Bank survey of sixteen universities in Africa), and negotiate with donors. They control policy for higher education in three ways: 1) setting access policy and tuition fee schedules; 2) controlling sources and mechanisms of finance; and 3) determining flexibility of universities to reallocate funds and use savings (Saint, 1992). "Most universities have little direction over the structure of their finances; the levels of student subsidies, fees and faculty salaries are regulated by an outside government agency. Budgetary flexibility is limited and, normally, income earned by the university must be returned to the Government treasury" (Maguire, 2000). In 2004, sub-Saharan African countries spent on average 4.5 percent of gross domestic product (GDP) on education, which is close to the 4.3 percent world average, making it difficult for African governments to consider increased spending for higher education. Higher education's share of national education budgets has been near the world average of 20 percent since the early 1980s, with some decline in East Africa and an increase in West Africa in the subsequent three decades (World Bank, 2009). University enrolments have increased dramatically over the past ten to fifteen years, despite many ministries of education following a policy of increased cost-sharing by students along with a decrease in welfare contributions.

## Ministries of agriculture

Following independence, ministries of agriculture were usually responsible for technical schools and training institutes for extension workers and functionaries in non-scientific positions. In many countries, ministries of agriculture were the "main clients" of the ministries of education since they usually determined the numbers of students to be admitted and supported. For example, in 1979 the Cameroon Ministry of Agriculture was active in selecting students for higher education, since these students would become civil servant employees in that ministry upon entry into the *Ecole Nationale Superieure Agronomique* (ENSA). All graduates were absorbed by the Ministry of Agriculture, whether or not positions within the Ministry or its parastatals were open at the time of graduation. At that time, it was not possible for a private individual to be enrolled for a degree in agriculture (Cornell University and ENSA, 1979). Today, the Faculty of Agronomy and Agricultural Sciences at Cameroon's University of Dschang offers degree programmes, including MSc and PhD, to anyone who can meet entry requirements. However, in many sub-Saharan African countries, privately funded students who pay the full cost of their education remain a distinct minority.

Since it is the responsibility of the ministries of agriculture in many countries to execute national agricultural policies, the lack of control over advanced agricultural education programmes often makes it difficult to reconcile human resource needs with degree programme content. In some countries, the only way for ministries of agriculture to influence on programmes of interest is to provide student and trainee scholarships. This disconnect between university agricultural programmes and national agricultural goals and programmes also occurs in the area of agricultural research, which is normally funded by ministries of agriculture only at their research stations (Michelsen and Hartwich, 2004). Eicher highlights this as a high transaction cost issue and recommends that "the skills and experience of both the education and agricultural ministries should be brought to bear in conceptualising, planning, and implementing agricultural education and training projects" (Eicher, 2006).

## Non-governmental organisations

National and international NGOs working in Africa vary greatly in terms of their origins, goals, capacities, funding, scopes for impact, and modes of operation. For the most part, NGOs work at local levels and focus on social and economic development among poverty groups. With a trend towards governments decentralising programme responsibility and financial support for agricultural extension and training activities, some NGOs are now working with local governments to support agricultural training activities. NGOs usually have at least modest financial resources for training, but may not have the depth of technical capacity to meet the challenges they encounter. While they usually have a focus for their activities, one common problem that many encounter is the difficulty of "scaling up" their development-oriented interventions and providing the training to support them.

NGOs have recently had the attention and support of international and national donor agencies and are now viewed as being generally more responsive to local needs than some government agencies in the rural areas where they are present. In many cases, they would make a much more significant contribution if they could develop ways to collaborate and cooperate with other AET institutions. Universities, colleges, and extension agencies could benefit from their local knowledge and experience with specific interventions, while NGOs could benefit from more technically oriented contributions, particularly in training. With more interaction, financial resources of all agencies would likely be utilised more efficiently and with greater overall impact.

## The private sector

One effect of past structural adjustment policies imposed by the international community was the break-up of parastatals, many of which were formed from the vestiges of colonial enterprises. This took most governments out of the business of agricultural production, but not out of input supply, credit, and/or marketing, especially for export crops. Slowly, this is changing to the point where, in most countries, the private sector is active in all aspects of agricultural production. With greater commercialisation, the private sector is increasingly involved in financing training activities in sub-Saharan African countries, and these training activities are usually well integrated with the inputs and services they provide and the marketing of commodities. Commodity producer and marketing associations also integrate training into their farmer-oriented activities. Still, for subsistence agricultural areas where commercial activities are limited, public education and extension systems continue to be the dependable source for financing agricultural education and training activities.

## Regional and sub-regional organisations

There are several AET-focused regional and sub-regional organisations that have grown in importance in supporting agricultural education and training since the late 2000s. Among them are the Association for Strengthening Agricultural Research in Eastern and Central Africa (ASARECA), West and Central African Council for Agricultural Research and Development (CORAF), Southern African Centre for Cooperation in Agricultural and Natural Resources Research and Training (SACCAR), and Forum for Agricultural Research in Africa (FARA).These organisations vary in purpose and operations, but many provide channels for donor funding of AET activities that serve several countries more efficiently and effectively than single-country endeavours could do. These organizations offer opportunities for national governments to supplement their financial resources for agricultural education and training through participation in the activities of these regional organisations.

## The Factors Contributing to Change

Current and projected national and global changes define the factors that contribute to the need for reform within AET institutions and in the financing that will enable them to fulfil their mandates effectively, efficiently, and sustainably in the future. While many of these factors are described in Chapter 1, some are highlighted here with reference to their financial implications.

### Demand for higher education and training

Universities in most sub-Saharan African countries have expanded both the size and scope of their programmes to meet rapidly rising demand. Reflecting this general increase in demand for higher education, the trend is toward supplementing government support with increased contributions by students to the costs of their education. Student contributions in sub-Saharan Africa were estimated to amount to about 25 percent of total education costs for the region in 2009 (World Bank, 2009). However, with rising enrolments, government support and subsidies also continue to grow in terms of budgetary totals.

### Globalisation and improving export markets

Concessions by importing countries and improving terms of trade are providing an economic stimulus which, in turn, increases demand for those with the higher education and training needed by the private sector. Increasingly, the private sector is willing to finance training and, in some cases, contribute to the cost of higher education for key personnel. However, this suggests the need for increased financial support for start-up costs related to the introduction of new curricula.

### Rural stagnation

Persistent rural poverty could be partly remedied by financial assistance that would enable AET institutions to provide manpower capable of dealing with rural development in a changing economic, ecological, and social environment. Of particular importance is funding for farmer training activities to enable farmers to produce a market surplus, thereby increasing rural income and welfare.

## Declining public support for agricultural education and training:

It appears that AET institutions have been "removed from the market place and from the rest of education and training curricula and, as such, have witnessed the growing irrelevance of education and training curricula, decreases in educational quality, the unemployment of agricultural graduates and, consequently, reduced investment support" (Gasperini, 2000, in Atchoarena and Gasperini, 2003).Today, governments remain challenged to find ways of producing human resources relevant to an emerging agenda for the future. To this end, some governments are promoting the co-financing of training for skill development with employers.

## Increase in private universities

The dramatic increase in private universities in sub-Saharan Africa means that local educational institutions can be more responsive to local needs in terms of curriculum and manpower requirements. Also, more competition with public educational institutions will likely make education more affordable to a wider audience. However, the maintenance and operation of expensive multiple agricultural education facilities may spread scarce financial resources (from all sources) over too many institutions, thereby reducing sustainability. A current issue is how to bring these private institutions into a national education policy framework; what level of public financial support, such as student scholarships, is appropriate to ensure their continued financial viability?

## Increase in private extension services and training suppliers

Non-public actors providing training for farmers and others in the agricultural value chain in sub-Saharan Africa include national and international NGOs, producer associations, marketing firms, input suppliers, and cooperatives. Since many of these private firms are linked to production, they are often able to recoup the costs of training from those getting the benefits and/or from market returns. For example, Uganda is gradually reducing the share of public financing in farm advisory costs and is using public finance to contract privately delivered advisory services (Nahdy, 2004). Public sector extension can also introduce cost sharing by farmers, but often the extension service's contribution—such as training related to protecting the environment—is more of a public good, where all citizens of a country benefit. In the end, public sector training costs are reduced, while costs for private sector training are at least partly paid for by increased economic activity.

### Electronic modes of education and training

Since the turn of the century there has been a phenomenal increase in the availability of electronic resources for agricultural education and training, mostly supported by international agencies and donors. This not only saves financial resources that would otherwise be needed for books and expensive scientific journals, it also increases the value of AET activities, making them more attractive for increased funding. Satellite cellular phones and distance education, including simultaneous lecture delivery involving advanced educational institutions abroad collaborating with African universities, may also be employed to provide quality education at reduced cost.

## Perspectives and Frameworks for Change

This section presents five useful perspectives and conceptual frameworks on which to base and justify recommendations for increasing financial support for AET in sub-Saharan Africa. The first explains the relationship between higher education and economic growth—education's worth to society. This is followed by a brief exploration of two important perspectives that can lead to an understanding of how to enhance the economic value of agricultural education and training: the agricultural innovation system and the value chain perspectives. Finally, two schematic representations of frameworks for financing both higher education and extension training are presented to illustrate the principles on which recommendations are developed for increasing funding for agricultural education and training.

### Economic Value Perspective

Any discussion of financing AET should include an assessment of the economic value of higher education and training to the overall economic and social development of a country. Since much of the economic activity of many sub-Saharan African countries is from the agriculture sector and since most rural poor are engaged in agriculture, at least for survival, the extent to which AET activities contribute to development needs to be explored. This then becomes an important criterion that governments can use to decide the magnitude of their financial support for these activities.

Studies commissioned by the World Bank and other donors in the mid-1980s and early 1990s found that economic returns from higher education, while significant, were less than those from primary and secondary education (Psacharopoulos and Woodhall, 1985; Psacharopoulos, 1994). This caused a shift in donor funding away from higher education in sub-Saharan Africa, partly justified on the basis that higher education was more of a private good (benefiting individuals) and less of a public good (benefiting society).

In the early 2000s, several studies took issue with these conclusions. Studies by Robert Evenson found that higher education programmes in agriculture that create skills related to science and technology "have a public externality value much higher than the private value of these skills in labour markets" (Evenson, 2004). On this basis, he found strong justification for public investment in the creation of graduate programmes in innovation-related fields of study in sub-Saharan Africa. A more recent comprehensive study found that 32.1 percent of the benefits from a bachelor's degree can be categorised as private market, 39.3 percent as private non-market, and 28.6 percent as non-market social externalities (McMahon, 2009). According to a World Bank-commissioned study by Bloom et al. (2006), "[h]igher education may create greater tax revenue, increase savings and investment, and lead to a more entrepreneurial society."

Based on the analysis of their data, the authors indicated that "a one-year increase in the tertiary education stock would raise the long-run steady-state level of African GDP per capita due to factor inputs by 12.2 percent," concluding that "expanding tertiary education may promote faster technological catch-up and improve a country's ability to maximise its economic output." The conceptual framework they used for linking higher education to economic growth is shown in Figure 14.1.

These and other studies have convinced many donors and some African governments of the need to increase support for higher education, which includes agricultural education and training institutions and their activities. In fact, a recent World Bank report stated that "[s]ince most African countries remain agrarian societies where agriculture contributes significantly to both economic and social development, the need for renewed support to agricultural education and training is both obvious and necessary" (World Bank, 2007).

## The agricultural innovation systems perspective

Placing agricultural education in the context of its relationship to other institutions and processes is important for exploring how AET is to be financed in the future. The focus here is mainly on cooperation, partnerships, and ensuring that all components of a system are functioning well to meet defined needs and that, together, they are financially able meet their responsibilities within the system.

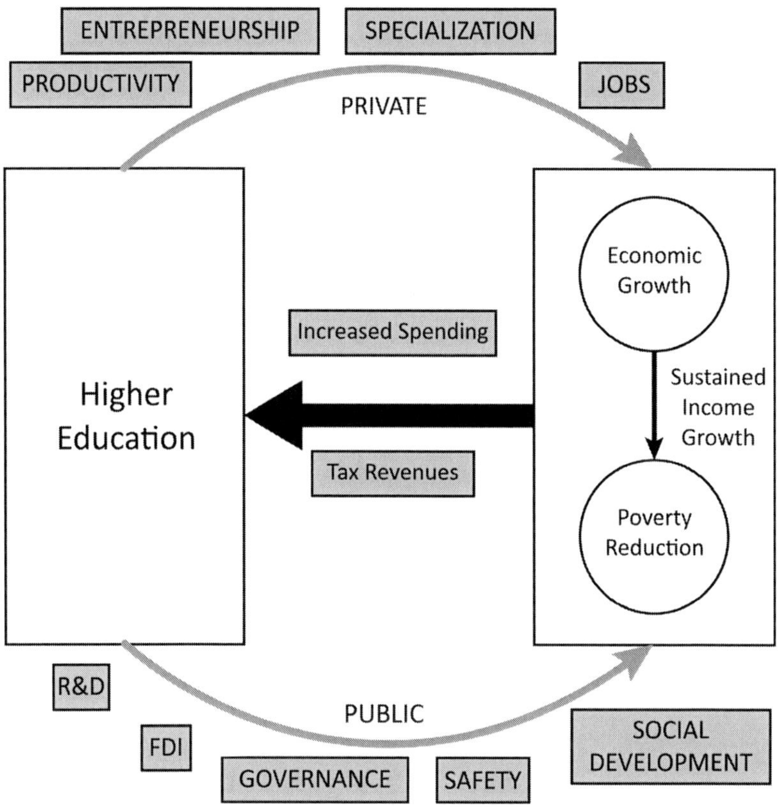


Figure 14.1. Conceptual framework for higher education and economic growth
Source: Bloom et al., 2006

From about the 1990s, the agricultural knowledge and information system (AKIS) was the prevailing framework. Röling (1989) defined AKIS as "a set of agricultural organisations or persons, and the links and interactions between them, engaged in processes such as the generation, transformation, transmission, storage, retrieval, integration, diffusion, and utilisation of knowledge and information, with the purpose of working synergistically to support decision-making, problem-solving and innovation in a given country's agriculture or a domain thereof." He also indicated that the AKIS perspective is a way of conceptualising the task relationships and information flows between various actors involved in agricultural technology systems.

The former International Service for National Agricultural Research (ISNAR), in conceptualising how universities can partner with National Agricultural Research Systems (NARS), developed a diagram that demonstrates how agricultural education relates to the overall agricultural knowledge and information system (Michelsen and Hartwich, 2004).

In the 2000s, the AKIS framework received less attention than a newer framework called the agricultural innovation system (AIS). A 2008 World Bank publication defined an AIS as "the network of organizations, enterprises, and individuals focused on bringing new products, new processes, and new forms of organization into economic use, together with the institutions and policies that affect the system's behaviour and performance" (Rajalahti et al., 2008).

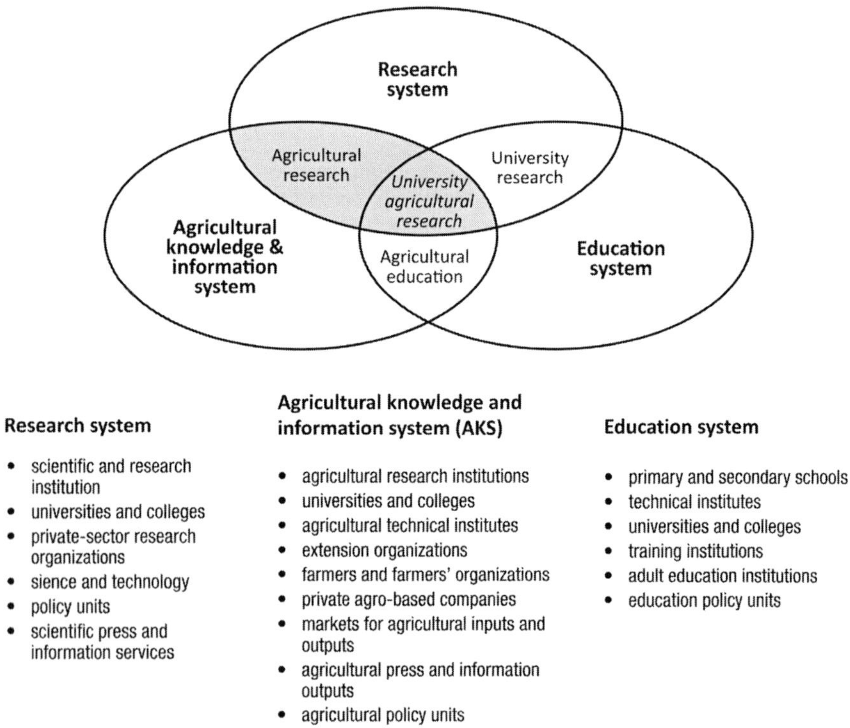


Figure 14.2. Systems context for university participation in agricultural research
Source: Michelsen and Hartwich, 2004

Here the focus is on the dynamic process of technological change. An International Food Policy Research Institute (IFPRI) paper linked agricultural education and training in Africa to the AIS framework and showed how "an innovation system embeds technological change within a larger, more complex system of actions and interactions among diverse actors, social and economic institutions, and organizational cultures and practices" (Davis et al., 2007). The following diagram illustrates how agricultural education and training contribute to an AIS.

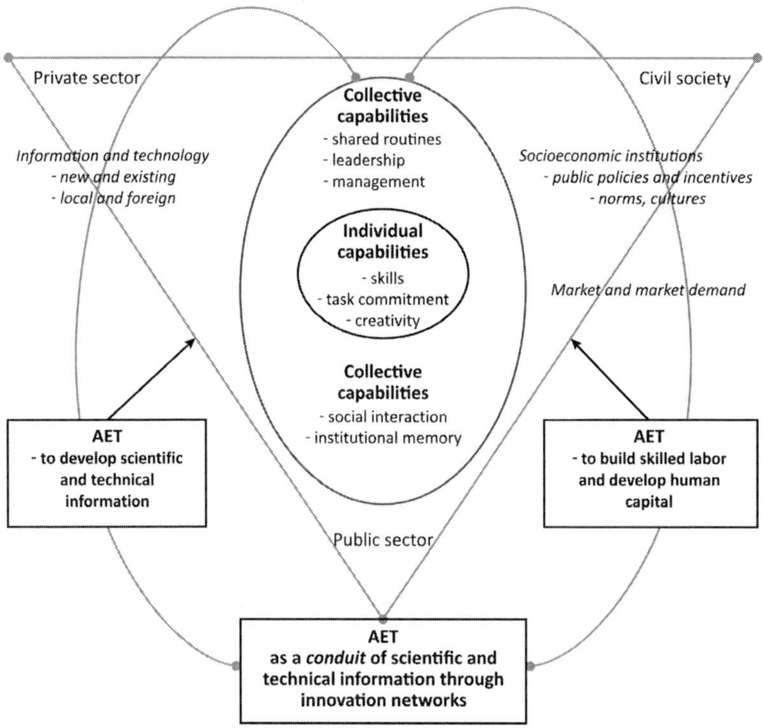


Figure 14.3. Agricultural education and training in a system of innovation
Source: Davis et al., 2007

Spielman et al. (2008) use the AIS perspective to "show how agricultural education and training in sub-Saharan Africa can contribute to agricultural development by strengthening innovative capacity or the ability of individuals and organizations to introduce new products and processes that are socially or economically relevant, particularly with respect to smallholder farmers who represent the largest group of agricultural producers in the region." They identify three key elements of an innovation system: "strengthening individual and collective capabilities to innovate; improving organisational cultures and behaviours in support of such capabilities; and fostering networks and linkages with other innovation agents."

This AIS perspective suggests several reforms for agricultural education and training. Among them is one that deals with economic constraints (the scarcity of financial resources) by suggesting that actors (agents) within the AIS "integrate into innovation networks to achieve economies of scale and scope, reallocate labour and human capital more efficiently, reduce transactions costs, exploit complementarities and realise synergies in the innovation process" (Spielman et al., 2008). It is this final point that will be further explored in this chapter as one way for AET to become more effective without utilising significantly greater financial resources.

## Value chain perspective

Financing for agricultural education and training activities becomes more feasible and attractive with a country's increased involvement in the value chain of selected commodities. A value chain in agriculture is a way of assessing, understanding, and describing the entire range of activities and actors necessary to produce, process, and market a commodity for consumption. These activities usually involve processes for product transformation and various inputs and services from different sectors of an economy that add value to a product until it reaches a consumer. To increase their benefits from world markets (about 13 percent of African agricultural production is exported), African countries will need to move away from raw product exports towards increased processing in an efficient manner. Where it is not possible to do this within a country, regional approaches may be possible. The *2009 Economic Report on Africa*, published by the UN Economic Commission for Africa (UNECA) and the African Union Commission (AU), contains a full discussion of value chains along with the value chain diagram shown below.

This UNECA/AU report indicated that "there is great potential for maximizing values from horizontal and vertical integration at every level of the value chain strategy. In the long run, expected gains from the strategy would undoubtedly justify intervention costs in terms of economic diversification, increased productivity, food security, job creation, and poverty alleviation." Among the policy recommendations of the UNECA/AU report, which focused on the possibility of regionally based value chains for selected commodities, was the following recommendation relating to skills development: "enhancing technical, managerial and financial skills is important, and governments could, for instance, promote public funded programmes or provide tax incentives to encourage private sector involvement, entrepreneurship development, and training" (UNECA/AU, 2009).

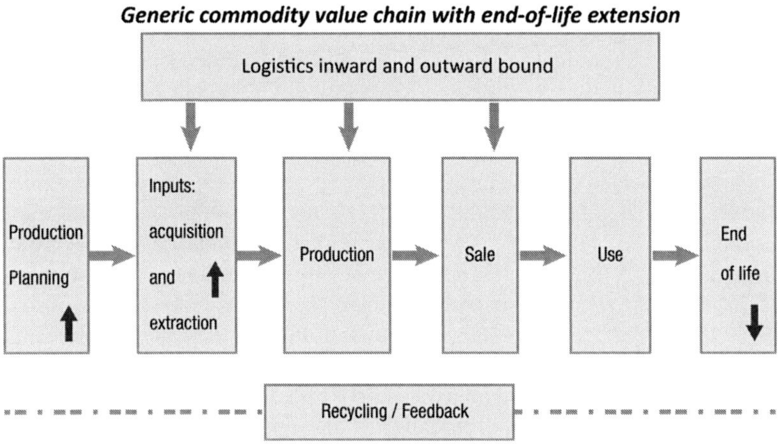


Figure 14.4. Value Chain Diagram. Source: UNECA/AU, 2009

Using the value chain approach, governments can assess the feasibility of increased participation in the value chain of commodities for which they have a comparative advantage for production. In promoting higher value chain activities, the need for skilled human resources along the chain will increase demand for AET activities at all levels. Furthermore, governments will have more tax revenues from this increased economic activity, some of which would properly be reinvested in financing agricultural education and training institutions. The private sector would also be stimulated to make direct and indirect payments for AET services to assure the necessary human resources would be available to private enterprise.

## Framework for financing higher education

Given the fact that most African governments already spend appropriate proportions for higher education relative to all education expenses and to GDP, government resources need to be used more efficiently and effectively, and other sources for agricultural education and training need to be explored. Below is a framework for considering different options for funding higher education, not all of which are now in use (Jongbloed, 2004). The diagram shown is an adapted version of this framework (Davis et al., 2007).

The major issues are 1) what items governments fund for higher education, and 2) how they fund them (mechanisms involving governments, students, and private entities). In Figure 14.5, Q1 represents a centralised traditional system of funding based on negotiated budgets; Q2 is a performance-based funding system based on a formula that takes account of credits accumulated by students; Q3 represents a market-oriented system where institutions are invited to submit price-competitive tenders for a supply of graduates or of research activities in specific fields of study; and Q4 is the ultimate demand-driven system based on government-issued vouchers that students can use to obtain educational services at an institution of their choice (under discussion for the Netherlands). It is noted that the Organisation for Economic Cooperation and Development (OECD) trends are to move away from the budget-oriented (Q1) and towards the programme-oriented approach (Q2) and even towards the competitive-based system (Q3). In some countries, such as the U.S., financial support comes from a variety of sources using mechanisms represented by Q1, Q2, and, to a lesser extent, by Q3 (Jongbloed, 2004). All mechanisms have their positive and negative outcomes, and most would require some modification to ensure that critical areas of social importance continue to be supported in the interest of equitable national development.

**Centralised** (regulated) approaches

| **Budget oriented**<br>(input)<br>Q1 | **Programme oriented**<br>(output)<br>Q2 |
|---|---|
| Q4<br>**Student centred**<br>(input) | Q3<br>**Supply driven**<br>(output) |

**Decentralised** (market) approaches

Figure 14.5. Alternative education funding systems. Source: Davis et al., 2007

This framework is particularly useful for exploring funding mechanisms for AET in higher education institutions, particularly for financial support from governments. However, it is also important to review the possible mix of funding sources (government, student tuition fees, private sector, NGOs, etc.) and the mechanisms for delivering each of them to allow agricultural education and training institutions to fulfil their mandates and contribute effectively in the context of an AIS. This will be further explored in the next section of this chapter.

## Framework for financing extension training

Training activities in agriculture are among the services offered by a host of organisations including higher education institutions, extension agents, NGOs, commercial companies, cooperative associations, religious groups, and consulting firms. Who does the training, what they offer, and who pays for it mostly depend upon the activity and the target audience involved. A very comprehensive and widely accepted framework on financing agricultural and rural extension activities was developed by the Neuchatel Group (2007), an informal group of development agencies involved in agricultural development in sub-Saharan Africa. This group published several charts that illustrate the framework, one of which shows various approaches for demand-side financing of training to increase accountability and serve both public and private interests (see Figure 14.6).

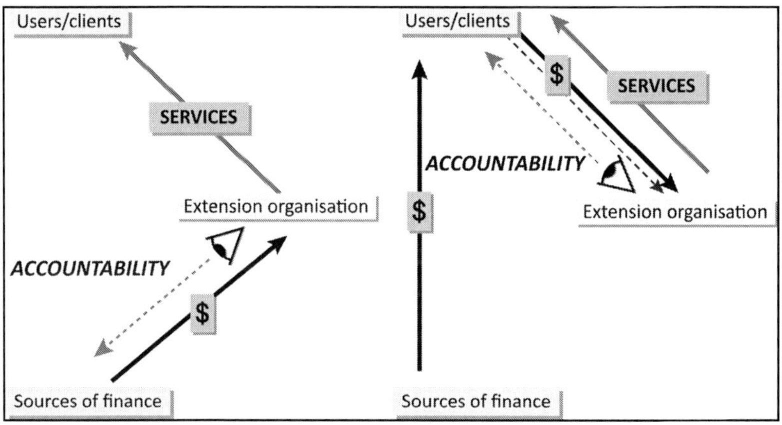


Figure 14.6. Empowerment and accountability through demand-side financing
Source: Neuchatel Group, 2007

There are several methods by which extension and training organisations can be financed, and each has implications for the quality of training provided to farmers and other intended beneficiaries. Among these methods are tax revenues, levies on products, combined commercial services, farmer association membership fees or government subsidies, NGO external donations or government subsidies, and consulting firms (Van den Ban, 2000). Central to the issue of financing training is the concept of public versus private good, i.e. determining when training is of general benefit to society and when it is exclusively of benefit to an individual. One interesting approach explained in the

Neuchatel framework is a voucher system that allows farmers and other user clients to obtain training and other services from public or private training organisations of their choice.

## Opportunities and Challenges for Increasing Funding

This section explores some of the opportunities and challenges for funding agricultural education and training activities, primarily of higher education institutions, to serve the needs of agricultural and rural development in countries within sub-Saharan Africa. It looks not only at sources of support, but also at several means of ensuring the efficient use of funds. While funding sources are reviewed individually, the long-term viability of AET institutions depends on their access to multiple resources and their ability to increase efficiency by having the freedom to employ cost reduction measures.

In focusing on higher education institutions, their changing roles within an agricultural innovation system, and their relationships with other organisations within the system, it is important to note that education and training in agriculture and related fields of study are often more expensive than in other disciplines (with the exception of related social science disciplines). Field work and the training of farmers and others often involve the cost of travel to rural areas. Costs for educating students in these disciplines include scientific and technical equipment and associated specialised laboratories, farm equipment and operations, and specialised supplies and teaching aides. Often forgotten or deferred is the maintenance, up-grading and eventual replacement of donor- and government-provided buildings, laboratories, and other facilities. Many agricultural education and training institutions in sub-Saharan Africa are currently experiencing financial crises requiring them to put most of their financial support into salaries to retain qualified staff. This reduces the funds available for critical infrastructure, equipment, and supplies needed for a quality education.

One important factor related to funding is the level of autonomy public agricultural higher education institutions have to receive and manage financial resources beyond those provided by the government. Increased fiscal autonomy gives AET institutions room to manoeuvre to source new funds and employ appropriate expenditure methods, thereby allowing them to improve their programme flexibility and responsiveness within their national AIS.

Another critical factor related to financing is the extent to which an institution provides a public or private good, or both. Generally, this would be the primary criterion in determining who pays the cost of agricultural education and training activities and the amount of empowerment given to students and trainees to determine the agenda—"he who pays the piper, calls the tune." Public and private goods can be explained by the following examples.

- Private good: If a farmer hires a consultant, thereby excluding others from benefiting from the training, it would be a private good. The farmer would likely pay the entire cost directly. If a company or cooperative provides training as part of a producer contract, the farmer pays indirectly through fees or an assessment on products sold.
- Public good: A training programme for farmers on controlling soil erosion would be a public good, since this would benefit not only the farmer, but also society as a whole. Here, it would be expected that the government would pay the cost through extension or another agency.

For agricultural education and training activities, governments need to focus their financial resources on public goods and shift activities that constitute purely private goods to the emerging private sector. However, governments need to recognise that higher education for students in agricultural disciplines has both public and private good dimensions, meaning that cost sharing is appropriate.

The following section identifies the opportunities and challenges of securing financial support from donors, the government, beneficiaries, income-generating activities, and voluntary contributions.

## Donor support

It is clear that donor support for agricultural education and training continues to be necessary and justified in most countries of sub-Saharan Africa. However, this donor support could be more effective and efficient if donors from various agencies could work together and with governments to define complementary rather than disjointed or competitive responses to agricultural education and training needs. The following are some examples of areas where donor support can be effective and responsive to national needs.

- *Foreign exchange.* Expenditures that utilise scarce foreign exchange are an aspect of donor support that represents the comparative advantage of national donors such as the U.S. Agency for International Development (USAID) and international donors and lenders such as the World Bank. This includes expenditures for capacity development of current and potential AET staff attending universities abroad for postgraduate degrees, and for critical up-to-date laboratory and computer-related equipment. The latter is necessary to take advantage of increasingly available e-resources on the Internet. These are traditional areas for external financial support and, given the current state of finances for agricultural education and training and the vital roles of AET within an AIS, this continues to be an urgent need.

- *PhD programmes.* The establishment of PhD programmes in selected disciplines is another important area for donor support at a few large and academically reputable sub-Saharan African AET institutions. The initial focus would be on capacity development for these institutions, but this could readily be extended to building technical capacity for private sector activities. From the donors' perspective, this would constitute a shift of funding away from individual and risky scholarships for training abroad, toward support for a long-term investment in building and strengthening African capacity for advanced education within SSA, nationally or regionally. This has previously been the focus of donor support in other regions of the world and to a limited extent in Africa, but many more African AET institutions have progressed to the stage of their development where such donor investment would be viable and fruitful.
- *Facility improvements.* Maintenance and upgrading of buildings and facilities used for agricultural education and training may be a less palatable area for the donors and lenders who built these facilities, since most of them required governments to cover these costs over time (World Bank, 2007). However, as a result of the continued inability of governments to meet their obligations, there remains a need for donors to protect their previous investments and to upgrade existing facilities. One possibility is for donors to focus on the facilities that are needed for the new agenda, such as for food science laboratories, in order to encourage more value-added enterprises within the country (moving up the value chain). It is anticipated that increased commercial activity in agriculture and the resulting tax revenues will eventually encourage governments to underwrite these maintenance expenses.
- *Research funds.* The establishment of a competitive research fund for joint, farmer-centred research activities involving staff of agricultural colleges and national agricultural research units presents an important as well as novel opportunity for donor support of AET activities. This would bring significant additional funding to agricultural education and training institutions and increase their exposure to farmer problems and possibilities, thereby improving the relevance of student education. If postgraduate programmes were offered, they would allow African graduate students to conduct applied and adaptive research on priority national issues. Programmes such as this would also give ministries of agriculture access to highly qualified university professionals who would be guided to focus on and assist in meeting the nationally determined agricultural research

agenda. How could this be accomplished? ISNAR has produced a study guide for improving agricultural research at universities in sub-Saharan Africa that includes a range of strategies and actions to improve funding for this research in the context of national issues (Michelsen et al., 2003). This recommended donor-supported competitive research fund would not only increase funding for AET institutions, but would also strengthen the relationship between two key institutions, often based in different ministries, within a country's AIS.

## Government support

Most African governments know that effective support for the agriculture sector will stimulate economic and social development and produce increased tax revenues. Since agriculture is a knowledge-based activity, these governments also know that they need to maintain support for AET activities. However, because of their scarce financial resources, many governments now employ several different delivery modes. Some are moving towards more output-based support, such as basing the level of government support on the number of students graduating. Incentives and accountability are key concepts for future funding by governments.

The total recurrent cost of higher education is highly variable for each sub-Saharan African country and is based in part on each country's stage of development and current economy. The World Bank estimates for 2006 show a range of all recurrent public expenditures from US$455 in Togo to US$7,742 in Lesotho, with half of the thirty-two countries surveyed being below US$1,000 and seven above US$2,000 (World Bank, 2009). In reviewing government support for agricultural education and training activities, two elements need to be considered: sources and delivery of funds.

- *Sources of funds.* To pay their share of expenses for higher education, government funding sources include various general and earmarked taxes, commodity check-offs for export commodities, specific and general donor contributions, and so forth. An important question for governments with respect to agricultural education and training institutions is how to determine how much to invest. One policy option for governments is to support AET institutions on the basis of a set percentage of Agricultural Gross Domestic Product (AgGDP). This would assure that governments were paying an appropriate amount and, at the same time, would stimulate these institutions to make their contributions to future increases in AgGDP to assure additional support as demand for their services increases.

- *Delivery of funds.* A number of traditional methods can be employed for direct delivery of funds to AET institutions, including line item budgets, lump sum budgets, grants, and formula funding. However, there are some newer approaches to funding higher education. The Jongbloed (2004) framework for financing higher education (introduced previously) focuses on "the extent funding would have to be supply-driven versus demand-driven and whether it should be input-oriented or performance-based." In this thoughtful conceptual paper, Jongbloed describes three options for funding that go beyond the traditional budget negotiation based on cost projections: student centred, supply driven, and programme oriented.

Public agricultural education and training institutions in Africa are usually funded through budget negotiations. In an effort to increase results from educational investments, governments may soon shift to more output-oriented systems and introduce an element of competition. Formula funding, based on number of student completions in selected programme areas, is becoming more common elsewhere. Mozambique has been moving away from sole dependence on negotiated budgets, with a reform model that is more dependent on diversified funding. Among the elements of Mozambique's new model for financing higher education are:

- greater cost recovery through the introduction of higher fees at all levels;
- the earmarking of institutional funds for special projects; and
- scholarship funds for students that are administered through the student instead of through the university (Davis et al., 2007).

The World Bank (2009) notes that "South Africa adopted a formula-based funding system in 2001, combining block grants based on teaching inputs, graduate outputs, and research productivity with earmarked funds for redress and development." This source then identifies Nigeria, Ethiopia, Ghana, and Kenya as examples of other countries working towards formula-based approaches.

Needs-based and competitive scholarships and loans for students provide indirect support for agricultural education and training institutions by covering all or a portion of direct costs of students, including welfare costs and tuition fees. To broaden access to higher education, governments may provide scholarships for those, for example, who are from low-income families in rural areas. Here, the assessment of who is in need is a challenge, particularly in cultures where the extended family supports those selected for higher education. Scholarships can also be used selectively as incentives to encourage students to study in areas of primary importance for future economic and social development, such as small agri-business establishments in rural areas.

Competitive scholarships usually ensure that the best qualified students in a category are able to continue studying, thereby allowing for higher standards of education and training. Loans are not common and experience shows that repayment rates are relatively low, often due to poor employment opportunities following graduation. However, properly managed government or government-subsidized student loan programs for AET in promising new growth areas will broaden educational opportunity while ensuring that repayment will be based on potential future employment with high income.

With respect to extension-level training in agriculture, there are now many institutions other than governments providing financial support for services, which brings the above discussion of the public versus private good criterion into focus. For example, in related policy discussions in Kenya, it has been suggested that the public sector should not duplicate training in areas that are served efficiently by the private sector. “This would leave more public resources for concentrating extension services on farmers who are in areas that are remote or poorly served by the commercial system” (Muyanga and Jayne, 2008). Undoubtedly, significant savings of government resources can be achieved by exploring and evaluating such opportunities within the more pluralistic institutional landscape now emerging to provide AET services. This is best accomplished through dialogue, agreement, and collaboration rather than through unnecessary competition. In addition to providing support for extension-level training to public sector units, governments can also contract such services with NGOs and the private sector to meet strategic objectives and gain access to other financial resources through cost-sharing. To target specific skill areas, governments can offer needs-based and competitive scholarships for training, similar to scholarships offered to students in degree programmes.

## Beneficiary contributions (tuition fees)

Most sub-Saharan Africa countries are shifting toward increased tuition-based support for agricultural education and training, although direct support from ministries of education remains paramount. The shift to increased tuition fees is due to the poor state of government revenues, the perceived increase in students’ abilities to pay, increased demand, and the recognition that substantial private benefit can be gained from publicly supported agricultural education and training institutions. Francophone countries still have low tuition rates, so there is a greater opportunity to shift financial support from public to private sources in these countries than in Anglophone countries, where tuition fee charges have already been increasing.

As curriculum changes respond to the human resource needs of commercial agriculture, demand for higher education will continue to increase, as will the ability to pay. Those without the ability to pay can be offered partial to full needs-based scholarships, loans, or even vouchers. Although private benefits justify partial payment by students for higher education, governments need to continue to support them proportionately, given the substantial public benefit of AET institutions.

## Income-generating activities

Since agricultural production can be part of the agricultural education process, income-generating possibilities for AET institutions are greater than for those in other areas of higher education. However, experience shows that most university-run agricultural production enterprises provide only modest income and more often produce losses. In some cases, they may even deprive local farmers of an opportunity for income. Accordingly, if substantial land is available, it is often better to rent it out or form an independent for-profit entity that channels funds back to AET activities. Other possible income-generating activities, such as the printing of publications, are also better left to the private sector.

The more attractive income-generating activities are those that focus on utilising the intellectual capacity of AET institutions to assist partner institutions with activities that improve the agriculture sector's contribution to overall economic and social development. One such activity is contract research—which utilises AET's comparative institutional advantage with respect to highly trained human resources, specialised equipment, and laboratories—for both the public and private sectors. Another interesting activity is consulting by staff, with fees being split on some basis with the AET institution. Shared royalties and patent rewards require a strong policy and legal framework for protecting intellectual property rights, but can produce modest amounts of income and an incentive for innovation on the part of both the staff and the institution. And finally, AET institutions can charge fees for the organisation of or participation in private good-type training, for example, training for commercial firms and cooperative associations. One of the important benefits of these activities is that they allow staff to become more familiar with real agriculture-related problems in the country and bring these experiences to students and trainees. A major constraint to university AET staff participation is the heavy teaching load usually assigned; university or government policy-level intervention is required to permit and encourage the recommended income-generating activities.

## Voluntary contributions

As agricultural activities move higher up the value chain and involve more private sector entities, the potential for voluntary contributions to AET institutions increases. Currently an insignificant source of funding in sub-Saharan Africa, contributions will likely become increasingly important as mutual benefits for contributors and recipients alike can be realized. Contributors may use funding to direct manpower development and to provide the private sector with the necessary access to practical application of science and technology. Financial contributions may come in the form of general donations, restricted or "tagged" donations, scholarships, in-kind donations, and named buildings and professorships.

NGOs are another source of contributions, as they need trained manpower for the increasingly technical agricultural support they provide their targeted clientele, particularly to those in poor rural areas. Given the mission of many NGOs to serve rural areas, AET institutions can initially focus on obtaining scholarships for students from such areas. These students would otherwise have relatively few opportunities for agricultural education and training, and trained human resources from these areas may increase income-generating opportunities there in the future. Wealthy alumni can also be cultivated to provide financial support in a number of ways. This is an important source for U.S. universities, but less so for those from Europe, where government support is still significant and expected.

One of the challenges to obtaining voluntary contributions is the need for both government and universities to adopt and administer a viable policy framework. Commercial entities making voluntary contributions are usually encouraged to do so if they receive tax benefits from the government. Rigorous university policies are needed: first, to allow for contributions; and second, to ensure that there are fair and equitable ground rules for everyone wishing to donate. Without such a framework, compromise of the university's mission and the creation of unfair advantages within the private sector are distinct possibilities.

## Opportunities and Challenges for Increasing Efficiency

The other side of the financial coin (so to speak) is the ability of AET institutions to increase their efficiency. There are several ways this can be achieved, both directly and indirectly: through improvements in fiscal and overall management, reduced operational and programme costs, improvements in institutional cooperation within the AIS system, targeting clients to respond to new potential areas of demand, and utilising regional programmes for staff development. These are discussed below.

## Improving management

Agricultural education and training institutions need to manage both their internal and external environments (an old axiom of management) to increase public confidence that they are using currently available resources efficiently and that they deserve access to new resources.

- *Internal management.* As previously noted, since the 2000s the number of private AET institutions, both for-profit and not-for-profit, has grown. This suggests that there is a market for trained human resources that can respond to increasingly technical and management-oriented agricultural human resource needs. Those institutions that use good business management practices will likely survive, along with those, such as religious organisations, whose sources of income are externally generated. In this competitive market, public AET institutions will need to adopt financial practices that allow them to make a sustainable contribution to the education and training needs of the agricultural sector. These practices include the development of long-term financial plans (including fund sourcing and cost management), realistic annual budgets, decentralised expenditure authority and accountability, and improved expenditure reporting and analysis.
- *External management.* Leaders of agricultural education and training institutions need to understand and address the institutional environments in which they function—both that of the AIS broadly and of other AET institutions specifically. This will allow institutions to position themselves in the marketplace and to establish the kinds of relationships among themselves that can be depended upon to help each meet its mandates effectively. Competitors, allies, clients, and funders are among those in the external environment whose activities usually impact AET institutions. A better understanding and management of the external environment will likely improve institutional relationships, which will surely result in more efficient operations for all involved.

## Reducing operational and programme costs

While agricultural education and training institutions are often aware of where they might individually reduce costs, there are often institutional and legal barriers to accomplishing this task. Nonetheless, there are areas in which AET institutions could reduce costs, as follows:

- *Operations*. Cost savings related to operations are an important method of reallocating financial resources to preserve an AET organisation's core mandate and objectives. While operating within a legal and administrative framework usually defined by government, AET institutions may have some flexibility for improving their financial positions internally through changes in operational procedures. Examples include: higher student-teacher ratios—which are already often occurring by default; outsourcing of student services to the private sector—which is practiced in a few countries; and shifting of welfare costs (food and lodging) to students who are able to pay. These and other operational cost savings may serve to free scarce resources for improving and adding educational programmes.
- *Programmes*. One opportunity for cost saving is that of agreements with other institutions (public and private) to handle peripheral programme areas. This would allow AET institutions to concentrate on core AET objectives and programmes. Such inter-institutional agreements could serve to rationalize programme areas, reducing program duplication and increasing overall financial efficiency. Another opportunity, since globalisation and the Internet have rapidly increased access to information about agricultural innovations from all over the world, is the strategic use of electronic training resources. Most AET institutions have a unique capability to gather, organise and share technical information electronically; information from all sources (including from the farmers themselves) about innovations for increasing farmer productivity can contribute to economic development. The increased capability of AET institutions to utilise e-resources has two financial implications: 1) cost savings resulting from the use of e-resources as opposed to traditional methods of curricula and materials development; and 2) the ability to attract financial resources as AET institutions appear to be ahead of the curve, demonstrating their relevance and value to other key institutions within an AIS.

## Improving institutional cooperation

The AIS perspective used in this book, as well as the ISNAR institutional context shown above, is a systems construct. These constructs only move from theory to practice when the institutions within them define and form cooperative, efficient, and effective relationships of various types and intensity. While there are many impediments to institutional cooperation, the ability of any one institution in the system to reduce its costs and increase its value to society is a strong motivation for that institution to pursue cooperative measures. Agricultural education and training institutions that operate actively and functionally within an organisational system will prove that "the whole is greater than the sum of its parts."

Fred Wiersema (1996), a prominent business management consultant, has suggested three evolutionary models of partnership based on successful business practices to reduce costs and improve effectiveness. These models may be applied to agricultural education and training within an AIS as follows:

- collaborate on design—to develop responsive educational materials and activities;
- synchronise operations—to evaluate and refine these materials and activities;
- integrate businesses or institutions—to improve the processes and operations. Each partner redesigns its organisation and then works with other institutions in the system to provide AET services more efficiently and effectively.

Ultimately, this process will lead to shared responsibility and durable linkages of AET institutions with other strategic organisations in the AIS. Establishing and maintaining strategic partnerships is usually a sign of organisational maturity and has high potential for generating increased funding to AET organisations. Some challenges to partnerships include: the possibility of high transaction costs, mandate creep beyond institutional competency, poor communications due primarily to status differences, and the difficulty of sharing financial and other resources across organisational boundaries. Some of these challenges can be mitigated with the formation of a separate linkage entity, such as a formal network, that facilitates institutional relationships.

A specific case can be made for improved public-private institutional cooperation and coordination "to diffuse innovations for agricultural productivity and to advance people-centred, production-oriented, environmentally sensitive agricultural education, knowledge transfer, and training and development programmes" (Rivera and Alex, 2008). Others stress the need for innovation networks, partnerships, and linkages among AIS institutions and focus on how AET institutions can address the innovation needs of different groups of farmers, from subsistence to commercial. In order to address the financial constraint, these researchers propose that AET institutions "integrate into innovation networks (of AIS organisations) to achieve economies of scale and scope, reallocate labour and human capital more efficiently, reduce transactions costs, exploit complementarities, and realise synergies in the innovation process" (Davis et al., 2007).

## Targeting clients

It takes a major institutional culture shift to move from supplying human resources for government positions to meeting broad-based demands from both the private and public sectors for highly trained and educated personnel to work in rural and agricultural areas. The move towards more demand-driven approaches, as defined by applicants and national goals, will require constant interaction with other institutions involved in the agricultural innovation system as well as with producers themselves. Adopting a process for demand assessment and response is recommended for all AET institutions. The result will likely mean the introduction of new curricula and training programmes, such as those related to entrepreneurship or marketing, to meet the needs of new clients in cooperation with new partners.

When public AET institutions, both universities and extension units, better define and sharpen their focus on target audiences and develop a client orientation, opportunities for both income generation and cost management will increase. For example, the steady shift towards more commercialised agriculture suggests that students may be willing to pay for the type of agricultural education that gives them opportunities in agribusiness, where future income potential is relatively high. At the same time, governments, donor groups, and NGOs are increasing their focus on poor rural areas by defining and paying for the type of agricultural education (e.g. local agricultural teachers) and training (e.g. marketing) needed to help improve incomes of the poor in rural areas. Extension units, for example, now operate among a plethora of organisations that provide agricultural training and information services in most countries. Many are associated with the private sector, such as production associations and agricultural supply companies, which provide training as a private good on the basis of direct or indirect fees. There is no need for public extension organisations to duplicate or unnecessarily compete with these private groups—it is better to save scarce resources for other training endeavours. There may, however, be opportunities for extension services to complement this private training by providing, for example, training-of-trainer programmes (for a fee) to increase technical competence beyond the level a private organisation could offer.

On the other hand, NGOs operating at local, regional, and national levels often focus on poverty groups, usually in a large rural development context. In essence, these organisations are generally providing public goods, such as soil and water management for an improved environment. Since public agricultural extensions usually also provide public goods to farmers in rural areas, there may be opportunities for extension services to provide strategic training support for NGO programmes. One area that is ripe for exploration is the technical training to backstop and scale up successful NGO-tested innovations that they and others may be willing to finance.

Another aspect of client targeting is the opportunity for educators to make joint assessments with clients of AET needs at the local level. Several authors stress the need for participatory and collaborative learning approaches to assure the relevance of education and training curricula, materials and programmes. Here, AET institutions can learn some lessons from the business world. The previously mentioned business management consultant, Wiersema, suggests that "customer-intimate companies bring an entirely fresh perspective. They discover unsuspected problems, detect unrealised potential, and create a dynamic synergy with customers. In the integration of their operations, suppliers become more than merely useful: they become indispensable" (Wiersema, 1996).

By targeting agricultural education and training to specific clients, both efficiency and effectiveness are likely to improve. When this happens, it is much easier to demonstrate the value of AET organisations to society. This is a primary basis for seeking and securing financial support from all sources since it ensures that agricultural education and training are functioning at a level commensurate with goals for the agriculture sector in the context of overall economic and social development.

## Using regional programmes for staff development

While donors have, in the past, provided substantial funding to staff of agricultural higher education institutions in sub-Saharan Africa for postgraduate (MSc and PhD) education abroad (usually in the U.S. and Europe, but also in other non-SSA countries), such scholarships have more recently been substantially reduced in most countries. The primary reasons are a general reduction in donor support for university development, the high cost of overseas degree programmes, programme relevance issues, and brain drain problems. The result has been an interest in supporting postgraduate education at in-country institutions that can train their own staff at a much lower cost. Since most of these institutions have limited disciplinary scope for postgraduate degree education and very few fledgling advanced degree programmes, issues of quality and in-breeding immediately arise. This has led to the creation and increased use of a growing range of regional and sub-regional advanced degree programmes—supported by donor contributions to reduce the cost of capacity development—within Africa.

The World Bank notes that "regional programmes make sense where fixed costs are high and demand is clear but insufficient to reach a critical mass in each country, e.g., narrowly specialised technical fields, emerging new cross-disciplinary areas, small nations. Regional programmes promise economies in assembling a critical mass of specialists." Further, the Bank identifies two conditions for the establishment of regional centres: demonstrable economies of scale and the availability of financing to allow participation by non-host countries (World Bank, 2007). In 2009, the World Bank listed twenty-three regional master's degree programmes and ten PhD programmes in agriculture in Africa, indicating that this is a growing phenomenon. "Whereas all of these regional programmes are based on cooperation among existing tertiary institutions, plans are advancing to construct regional campuses for several African Institutes of Science and Technology, initially in Nigeria, subsequently in Tanzania, and then in Southern Africa" (World Bank, 2009).

Many current, agriculturally-focused postgraduate programmes in Africa are being facilitated and coordinated by sub-regional organisations. Among these are the Regional Universities Forum for Capacity Building in Agriculture (RUFORUM), a consortium of twenty-five universities in Eastern and Southern Africa established in 2004, and the *Conseil Africain et Malgache pour l'Enseignement Supereur* (CAMES), a coordination group for seventeen francophone countries in West Africa. A paper presenting the rationale for and experience of RUFORM highlights the benefits of regional programmes in developing capacity for agricultural transformation (Ekwamu et al., 2009). These efforts are designed to reduce the cost of staff development for AET institutions in Africa substantially, for both donors and governments. The long-term experience of the Southeast Asia Regional Center for Graduate Study and Research in Agriculture (SEARCA), located at the University of the Philippines College of Agriculture at Los Baños (UPLB), could provide some useful lessons for regional programmes located at existing national institutions in Africa. SEARCA was created in 1966 by the Southeast Asian Ministers of Education Organisation (SEAMEO) to "provide to the participating countries high quality graduate study in agriculture; promote, undertake, and coordinate research programmes related to the needs and problems of the Southeast Asian region; and disseminate the findings of agricultural research and experimentation." The Philippines provides the operating funds for administration, and SEAMEO provides scholarships for graduate programmes and training.

While very promising, there are some finance-related cautions to the development of regional centres. The first is the risk of civil unrest, which can make these centres inaccessible for long periods of time, with little financial relief for those paying the costs. The second, according to Eicher (2006), concerns long-term commitment, because "experience has shown that even though donors may underwrite a regional institution of higher education for a decade or longer, it has been difficult for national political leaders to make a financial commitment to continue the programme after foreign aid is finished." According to a study of regional training strategies in African agriculture, problems include financial sustainability, accreditation, language barriers, quality, and programme sustainability (Tongoona and Mudhara, 2007).

Despite these cautions, most studies (including Tongoona and Eicher) and other assessments show some very positive benefits from regional agricultural MSc and PhD programmes in sub-Saharan Africa. The most successful are likely to be those programmes promoted and supported at existing universities in larger countries, where the range of academic areas is greatest. Programmes of sufficient quality and reputation will be utilised by other countries for staff development in the areas of agricultural teaching and research; these programmes will be able to attract external financial resources not only from students but also from donors, governments, and sub-regional organisations. Long-term investments and sustained financial support, including competitive scholarships, will be required for these programmes to succeed. The key measure of success will be the number of qualified graduates who return to fill high-level positions in AET institutions in their home countries.

## Conclusions

It is clear that no single action will increase financial resources for AET organisations in proportion to their value in terms of economic and social development in sub-Saharan Africa. Assessing this value and recognising that agriculture is an engine of growth and development as well as a livelihood is an important first step for governments. Focusing on financing AET, this chapter has provided a justification for investment in agricultural education and training, a discussion of factors contributing to change, some conceptual frameworks for considering AET in perspective, and several opportunities for and challenges to increasing financial resources through fund sourcing, cost reductions, better management, and institutional cooperation. The concluding section of the chapter focuses on some of the assumptions about conditions and obstacles that will influence, for better or worse, the financing of AET activities in the future.

1. Agriculture is perceived as a low-status occupation for those entering institutions of higher learning; this perception adversely influences demand and the ability of AET institutions to attract further financial resources. It is assumed that increased value-added activities for both domestic and export consumption will stimulate more private sector commercial activity in

agriculture, thereby increasing demand for agricultural education and training. It is further assumed that AET institutions will recognise this opportunity and make the necessary changes in their curricula to attract more students and better financial support to serve their needs.

2. Agricultural education and training institutions are often poorly integrated with other institutions in an agricultural innovation system; this is partly due to the way AET institutions are financed. Many publicly funded agricultural higher education institutions are financed through ministries of education (except in some francophone countries). However, much of the need for agriculturally trained human resources is defined by ministries of agriculture, which are directly or indirectly responsible for activities in the agriculture sector. In other words, the policy framework and financing of many AET institutions is provided by the ministry of education, but the needs, scope and focus of AET activities are (or should be) of prime interest to the ministry of agriculture. This creates an organisational obstacle that can inhibit the match of financial resources with needs and opportunity. Since many ministries of agriculture do not control the financial resources and policies of AET institutions, it is difficult for them to integrate education, research, extension, and other services to meet the agriculture and rural development goals of the country. Another dimension of the poor integration problem is the difficulty AET institutions have developing and financing relevant postgraduate research programmes. This was one of the issues highlighted in a major study on improving agricultural research at universities through better integration with National Agricultural Research Organisations administered by ministries of agriculture (Michelsen et al., 2004). The assumption here is that there will be better communication and a process for reconciling financial resources with the needs of AET institutions in the national context. Accomplishing this will require the development and effective functioning of inter-ministerial mechanisms for making public financing decisions regarding agricultural education and research in institutions of higher learning.
3. Over the past two decades donors have reduced their support for strengthening and developing agricultural higher education institutions in Africa. There is a need to increase human resource capacity (academic staff) for all levels of higher education in agriculture, and governments are hard pressed to generate the funds needed for renewal. This is particularly important for addressing an expanded agenda in postgraduate education, as well as for the expansion, updating, and maintenance of facilities necessary for AET programmes. The assumption is that the donor community will continue the recent upward trend, making strategic contributions to AET in areas such as capacity development and supporting activities de-

signed to increase agricultural productivity and income in a wider development context.
4. More institutional cooperation and partnering can help rationalise and streamline the programmes and operations within the emerging landscape of public and private AET institutions; this process will likely reduce duplication and budgetary requirements. However, these same activities can also increase transaction costs by consuming the valuable time of highly trained manpower. It is assumed that ministries, along with the private sector, can minimise these costs by agreeing on clear roles and responsibilities for those institutions involved, and that efforts will be made to control mandate drift and other behaviours that undermine cooperation.
5. The value of agricultural education and training will only be realised when agricultural productivity improves and the sector makes a contribution to the overall development of a country. This improvement alone will facilitate increased funding for AET activities. But there are many factors that contribute to agricultural and rural development as well as to overall economic and social development. Should the agriculture sector remain weak, tax and other revenues may decline and the ability of governments to finance AET institutions will also decline. This is a chicken-and-egg situation. The assumption is that governments will realise that they need to increase the level and scope of education in agriculture in order to capitalise on new science and technology and that this is a long-term activity worth the sustained investment required.

In the future, key government and university officials will deal with many issues, obstacles, and opportunities in the sourcing, delivery, and utilisation of financial support for agricultural education and training. But what should they do in the present?

Ministers of agriculture could analyse the AIS in which they operate to assess to what extent and how the organisations within the interlinked system are functioning (called network analysis). Knowing that AET institutions are a vital component of an innovation system for stimulating productivity and improving the welfare of farmers and others in rural areas, ministers need to make every effort to ensure the institutions' effectiveness, efficiency, and long-term financial viability. They could also provide incentives for other units (e.g. ministry research units) to establish effective working relationships with AET institutions. Since many of these institutions are regulated and financed by ministries of education, ministers of agriculture need to work closely with their counterparts in education to ensure the institutions will have the financial support needed to remain viable as they fulfil their role in society of producing the human resources required by both government and the emerging private sector. Ministers of education could also be asked to approve a broader curriculum for agricultural education and training that goes beyond production agriculture to include a wider spectrum of concerns including agribusiness, rural development, environment, and ecology. To that end, ministers of agriculture need to work closely with AET institutions to support these new curricula offerings, which would eventually help demonstrate the value of AET institutions to the economy and society.

Ministers of education, in turn, need to work closely with ministers of agriculture to determine the amount of funding required to meet the projected education and training needs of AET organisations and to decide how these needs can be met through multiple funding sources and policy interventions. They also need to analyse the new institutional landscape of AET providers and develop policies that will help ministers of agriculture and education leaders rationalise the AIS in order to reduce overall costs and improve both effectiveness and efficiency. One such policy would be to develop standards for accreditation of diverse organisations involved in AET activities. Another would be to encourage the reduction of redundancy in the system.

University leaders involved with agricultural education and training would be well advised to move toward more client-oriented policies and embrace new agricultural curricula that respond to emerging needs within the economy and society as a whole. They should also improve organisational standards so that the quality of AET output demonstrates their organisation's value in terms of contributions to improved health, welfare, and economic outlook for those involved in the agricultural sector, many of whom remain trapped in poverty. Finally, since knowledge, science, and technology are at the core of any effort to improve agricultural productivity, they need to take leadership in establishing a more formalised network of institutions involved in AIS. Such a network would have high potential not only for reducing overall costs and increasing the efficiency of all involved but might even offer opportunities for increased overall funding.

## References

Atchoarena, D. and L. Gasperini, Eds. 2003. *Education for Rural Development: Towards New Policy Responses.* Chapter IV: Strategies and Institutions for Promoting Skills for Rural Development. FAO and UNESCO, International Institute for Educational Planning: Rome and Paris.

Blair, T., F. Adeola, K.Y. Amoako, N. Kassebaum-Baker, H. Benn, G. Brown, M. Camdessus, B. Geldof, R. Goodale, J. Peiding, W.S. Kalema, T. Manuel, B. Mkapa, L. Mohohlo, T. Thiam, A. Tibaijuka, M. Zenawi. 2005. Commission for Africa, *Our Common Interest*, 11 March 2005. Executive Summary, 14, 136–137.

Bloom, D., D. Channing, and K. Chan. 2006. *Higher Education and Economic Development in Africa.* Human Development Sector, Africa Region. World Bank, Washington, DC.

Brossard, M. and B. Foko. 2008. *Costs and Financing of Higher Education in Francophone Africa.* African Human Development Series 40100. World Bank, Washington DC.

Cornell University and ENSA. 1979. *A Study of the Development of the University Center for Agriculture at Dschang, Cameroon: Agricultural Manpower Needs Assessment and Implications for Participatory Development.* Cornell University, Ithaca, NY.

Davis, K., J. Ekboir, W. Mekasha, C. Ocieng, D. Spielman, and E. Zerfu. 2007. *Strengthening Agricultural Education and Training in Sub-Saharan Africa from an Innovation Systems Perspective: Case Studies of Ethiopia and Mozambique.* IFPRI Discussion Paper 00736. ISNAR Division, IFPRI, Washington, DC.

Eicher, C. 2006. *The Evolution of Agricultural Education and Training: Global Insights of Relevance for Africa.* Paper commissioned by the World Bank, Washington, DC.

Ekwamu, A., W. Ochola, W. Ekaya, M. Osiru, and N. Dhlamini. 2009. *Capacity Development for Agricultural Transformation: Making Postgraduate Level Training Relevant to Africa's Agricultural and Rural Sector Development.* Regional Universities Forum for Capacity Building in Agriculture (RURORUM), Kampala, Uganda.

Evenson, R. 2004. Private and Public Values of Higher Education in Developing Countries: Guidelines for Investment. *Journal of Higher Education in Africa* 2:1, 151–176.

Gasperini, L. 2000. *From Agricultural Education to Education for Rural Development and Food Security: All for Education and Food for All.* Proceedings of the Fifth European Conference on Higher Agricultural Education, September 10–13, 2000, Seale-Hayne Campus, University of Plymouth, UK. As reported in Atchoarena and Gasperini, 2003.

Jongbloed, B. 2004. *Funding Higher Education: Options, Trade-offs and Dilemmas.* Unpublished paper for Fulbright Brainstorms 2004, New Trends in Higher Education, Maputo, Mozambique.

Maguire, C. 2000. *Agricultural Education in Africa: Managing Change.* Workshop Paper: Sasakawa Africa Association, Accra and Cape Coast, Ghana.

McMahon, W. 2009. *The Private and Social Benefits of Higher Education: Higher Learning, Greater Good.* The John Hopkins Press: Baltimore, MD.

Michelsen, H. and F. Hartwich. 2004. *University-based Agricultural Research: A Comparative Study in Sub-Saharan Africa*. ISNAR Research Report 26, The Hague, Netherlands.

Michelsen, H., L. Zuidema, C. Hoste, and D. Shapiro. 2003. *Improving Agricultural Research at Universities in Sub-Saharan Africa: A Study Guide.* ISNAR Research Management Guide 6, The Hague, Netherlands.

Muyanga, M. and T. Jayne. 2008. Private Agricultural Extension Systems in Kenya: Practices and Policy Lessons. *Journal of Agricultural Education and Extension* 14: 2, 111–124.

Nahdy, S. 2004. Uganda: the Ugandan National Agricultural Advisory Services (NAADS). A case study in W. Rivera and G. Alex, Eds. *Volume I: Decentralized Systems, Agriculture and Rural Development.* Discussion Paper 8: Extension Reform for Rural Development. World Bank, Washington DC.

Neuchatel Group. 2007. *Common Framework on Financing Agricultural and Rural Extension.* Second Edition, Switzerland.

Psacharopoulos, G. and M. Woodhall. 1985. *Education for Development: An Analysis of Investment Choices.* Published for the World Bank, Oxford University Press.

Psacharopoulos, G. 1994. Returns to Investment in Education: A Global Update. *World Development* 22: 9, 1325–1343.

Rajalahti, R., W. Janssen, and E. Pehu. 2008. *Agricultural Innovation Systems: from Diagnostics toward Operational Practices*. Agriculture and Rural Development Discussion Paper 38. World Bank, Washington, DC.

Rivera, W. and G. Alex. 2008. Human Resource Development for Modernizing the Agricultural Workforce. *Human Resource Development Review* 7: 4, 374–386.

Röling, N. 1989. *Why Farmers Matter: the Role of User Participation in Technology Development and Delivery.* ISNAR Discussion Paper, The Hague, Netherlands.

Rygnestad, H., W. Saint, and E. Pehu. 2006. *Agricultural Education and Training in Africa.* FY 98–06.World Bank, Washington, DC.

Saint, W. 1992. *Universities in Africa: Strategies for Stabilization and Revitalization.* World Bank Technical Paper 194, Africa Technical Department Series. World Bank, Washington, DC.

Spielman, D., J. Ekboir, K. Davis, and C. Ochieng. 2008. An Innovation Systems Perspective on Strengthening Agricultural Education and Training in Sub-Saharan Africa. *Agricultural Systems* 98: 1, 1–9.

Tongoona, P. and M. Mudhara. 2007. *Regional Training Strategies for Post-secondary Agricultural Education in Sub-Saharan Africa: When do they Work and How Should they be Done?* University of KwaZulu-Natal, South Africa.

UNECA/AU. 2009. *An Integrated Regional Value Chains Approach to Agricultural Development in Africa.* Chapter 5 in the Economic Report on Africa, "Developing African Agriculture through Regional Value Chains." Addis Ababa, Ethiopia.

Van den Ban, A. 2000. *Different Ways of Financing Agricultural Extension.* ODI Agricultural Research and Extension Network (AgREN) Paper 106b, DFID, England.

Wiersema, F. 1996. *Customer Intimacy: Pick Your Partners, Shape Your Culture, Win Together*. Knowledge Exchange: Santa Monica, CA.

World Bank, 2007. *Cultivating Knowledge and Skills to Grow African Agriculture: A Synthesis of an Institutional, Regional, and International Review.* World Bank Report 40997-AFR, Washington, DC.

—, 2009. *Accelerating Catch-up: Tertiary Education for Growth in Sub-Saharan Africa.* World Bank, Washington, DC.

# CHAPTER FIFTEEN

# INNOVATION AND ENTREPRENEURSHIP IN AGRICULTURAL EDUCATION AND TRAINING

JOHAN JORDAAN AND GLEN TAYLOR

Agriculture will remain important in the economic development of African countries since it forms the basis of most African economies and provides the largest source of livelihoods and employment. Evidence shows that Gross Domestic Product (GDP) growth originating in agriculture has been, on average, two to four times as effective in raising incomes among the poor as growth from non-agricultural sectors (Ivanic and Martin, 2008). Despite this, many African economies have either grown slowly or have stagnated. Many African countries are historically associated with economic sectors relying on natural resources and raw materials. The promotion of multi-sectoral linkages between agriculture and other sectors of the economy is vital to supporting sustainable economic development. The Technology and Innovation Report recently published by UNCTAD (2010) highlights the importance of identifying and supporting linkages and processes that promote innovation and productivity increases. The main challenge is seen to be the strengthening of innovation capabilities in African agricultural systems, including agricultural education and training (AET) institutions, in order to address poverty successfully, improve food security, and achieve broad economic growth and development. Agricultural development depends to a great extent on how successfully knowledge is generated and applied. Investments in knowledge—especially science and technology—have featured prominently in most strategies to promote sustainable agricultural development. Although many of these investments have been successful, the context for agriculture is changing rapidly, forcing the process of knowledge generation and use to change as well. It is increasingly recognised that the value of traditional agricultural science and technology investments in research, extension, and education were not sufficient to enable widespread agricultural innovation

(World Bank, 2006). There is a rising awareness of the role of human capital formation in entrepreneurship to enable innovation and productivity growth.

***Key Messages***

- Agriculture sectors in developing countries must be responsive and competitive in order to benefit from the changing global system. Agricultural development depends to a great extent on how successfully knowledge is applied. There is a renewed worldwide interest in *entrepreneurship and innovation as drivers of economic growth* and regional development for poverty alleviation. However, based on existing innovation measures, Africa compares unfavourably with other developing regions.
- The World Economic Forum (2009) notes that *"entrepreneurship education plays an essential role* in shaping attitudes, skills and culture—from the primary level up... We believe entrepreneurial skills, attitudes and behaviours can be learned."
- There is rising awareness of the role of human capital development in enabling innovation and productivity growth. Lessons from recent agricultural reform experiences indicate that AET should and can play an enormously important role by producing *agricultural graduates who can successfully engage in entrepreneurship and innovation.*
- Recent views on innovation emphasize a *systems approach,* where innovation does not solely rely on the research and science supply-side, but includes the demand-side and all factors affecting the use of new and existing knowledge to create novel applications.
- Informative in this context is the *Triple Helix concept,* which is widely recognized for its significance in sustainable development. The concept shows that innovation occurs through a continuous exchange of ideas among the three main innovation actors, i.e. academia, industry (including the community), and government. These actors exist in all developing economies. The missing link, however, is a strategy to integrate the different elements to enhance innovation.
- Innovation education in AET must be addressed at two main levels *in academic institutions:* at the organisational level (where structures, processes, and strategies need to allow for entrepreneurial and innovative activities) and at the operational level (where entrepreneurship and innovation need to be integrated within curricula). Key requisites for success are individual and collective capabilities in innovation; innovation-supportive organisational cultures (i.e. a culture that incorporates entrepreneurship and innovation as a core institutional strategy); external networks; and pedagogical approaches (to curriculum development).
- Knowledge and skills imperative for *innovation curricula* include a systems orientation, the ability and willingness to work across disciplines, experiential training, and work-integrated learning.

Across the world actions are taken to support entrepreneurship and innovation. Many Organisation for Economic Cooperation and Development (OECD) countries have made entrepreneurship an explicit policy priority by stating that entrepreneurship is important for economic growth, productivity, innovation, and employment (OECD, 2009b). Entrepreneurship has for decades been an important part of EU agricultural policy to increase value creation in agricultural production (Alsos et al., 2003). Entrepreneurship is believed to offer ways to meet economic, social, and environmental challenges within a globalised and uncertain economy. In the current economic crisis, entrepreneurship is gaining renewed attention, and new evidence shows that it is a significant driver of productivity growth (Oosterwijk, 2008). In the Comprehensive African Agriculture Development Programme (CAADP), African leaders state their vision for improved agricultural productivity through accelerated innovation (FARA, 2006). It appears therefore that entrepreneurship is increasingly being acknowledged as important for motivating individuals and businesses to perceive and develop new opportunities that can promote positive change and create economic growth. This entrepreneurial spirit is regarded as the main source of innovation (Martinez et al., 2010).

In the context of education, a recent report emphasises the importance of entrepreneurship education and training as follows:

> While education is one of the most important foundations for economic development, entrepreneurship is a major driver of innovation and economic growth. Entrepreneurship education plays an essential role in shaping attitudes, skills and culture—from the primary level up. ...We believe entrepreneurial skills, attitudes and behaviours can be learned, and that exposure to entrepreneurship education throughout an individual's lifelong learning path, starting from youth and continuing through adulthood into higher education—as well as reaching out to those economically or socially excluded—is imperative" (World Economic Forum, 2009).

Indeed, across the world, many universities are assuming new roles and responsibilities in development initiatives and are engaging actively in educating students about entrepreneurship and in providing innovation support infrastructure (Etzkowitz, 2008; OECD, 2009a). In the knowledge economy, research specifically directed to innovation is increasingly shifting from company research and development (R&D) laboratories to universities, leading to entrepreneurial cultures within universities and also new organisational structures like technology parks, incubators, offices for the transfer of technology, and research commercialisation offices (Etzkowitz et al., 2000).

In their Framework for Agricultural Productivity, the Forum for Agricultural Research in Africa (FARA, 2006) highlights the critical importance of the quality of agricultural higher education in establishing the necessary competencies of agricultural scientists, researchers, extension officers, and farmers in raising their capacity to generate new knowledge or to access existing knowledge and to adapt, apply, and transfer knowledge to others. Lessons from recent agricultural reform experiences point to AET and the role it needs to play in entrepreneurship education, innovation, productivity, and growth in the agriculture sector (World Bank, 2007; OECD, 2009a). In a new, knowledge-intensive agriculture sector, AET increasingly needs to produce agricultural graduates capable of engaging actively in entrepreneurship and innovation.

The purpose of this chapter is to explore the importance of innovation and entrepreneurship in AET. After briefly outlining the evolving global context of agriculture and the need for responding to global changes, the concepts of entrepreneurship and innovation are discussed from a theoretical perspective. This is followed by a brief assessment of the state of entrepreneurship research and innovation in Africa. Practical strategies and actions for the transformation of organisational systems and cultures, the creation of innovation-supportive networks, and the introduction of approaches and methods for integrating entrepreneurship into agricultural education are discussed next. The chapter concludes with a case study on networks linking a university, a government ministry, and a peri-urban agricultural community to help develop entrepreneurship capacity by using business incubation methods.

## The Evolving Global Context of Agriculture as a Driver for Entrepreneurship and Innovation

The evolving global landscape of agriculture presents many challenges for the agricultural sectors of developing countries. Producers need to re-orient themselves towards meeting demands of a changing market instead of continuing with traditional methods and products. They need to engage in value-added activities (vertical diversification) or transform existing production systems and processes. Individual farmers in particular have to contend with issues such as open and increasingly competitive domestic and world markets, with integrated food chains that have changed the way food is produced, distributed, and consumed. The traditional food sectors in developing countries are not isolated from these phenomena. They need to connect to markets, which implies the management of market information and intelligence.

Advances in technologies have led to opportunities for improved up- and down-stream linkages and supply chain integration, which in turn create opportunities for value-added activities beyond production. The challenge is for the farmer to make a transition from being mainly an agricultural producer to one of being an agricultural entrepreneur doing business in a market-led environment. The major challenge for commercial farmers is to develop and sustain competitiveness in the global arena. For the subsistence and smallholder farmer, in addition to the above, the challenge is to move beyond the subsistence stage of the venture to capitalise on new opportunities provided by changes in the market landscape. Smallholders need to learn to adjust to emerging opportunities, implying a need for knowledge and a capacity to improve products, processes, and systems.

Similarly, other players in the agricultural value chain need to respond to changes in the agricultural landscape. Firms increasingly compete less on the basis of price and more on the basis of their ability to design novel products or improve the quality of management in their production. It is widely accepted that business enterprises subscribing to an entrepreneurial approach achieve competitiveness through creative and innovative business practices (Porter, 1985; Hamel and Prahalad, 1994; Drucker, 1998). Firms that anticipate or quickly adapt to changing consumer demand or changing production conditions are in a better position to navigate among dynamic markets.

For developing countries to benefit from the changing global system, they must have a responsive and competitive agricultural sector. Different types of knowledge, skills, and mindsets among farmers, educators, extension officers, researchers, and public servants are required to operate in a rapidly changing environment. Technical expertise needs to be complemented with functional expertise in areas such as business, finance, markets, and law, which will place strong demands on AET for human capital development (World Bank, 2006). Developing countries will need innovative policies, programs, and investments just to keep abreast. In this context, agricultural entrepreneurship and innovation seem imperative (Spielman and Kelemework, 2009).

## A Theoretical basis of entrepreneurship

Despite an increased focus on entrepreneurship and innovation worldwide, the OECD-Eurostat Entrepreneurship Indicators Programme (OECD, 2009b) reports that a common understanding of entrepreneurship is constrained by a lack of internationally comparable empirical evidence. Interpretations of entrepreneurship are based on different policy needs and diverse views and thus differ considerably among countries. Early research took different approaches and offered various definitions of entrepreneurship. The most common definitions include reference to behaviours such as initiative, organising or reorganising of social and economic mechanisms, creativity, innovation, new-product development, new venture creation, management of change, small business management, and risk-taking. The rewards for assuming such behaviour may come in the form of money, personal satisfaction, social change, or independence gained through creation and ownership of a business (Schumpeter, 1942; Thompson et al., 2000). Although many early research studies of entrepreneurship focused on identifying personality characteristics and behaviours that were believed to distinguish entrepreneurial from non-entrepreneurial individuals, this model has increasingly been regarded an over-simplification of the matter (Drucker, 1998; Timmons, 1999). Nonetheless, testimony given by successful entrepreneurs confirms that there are certain traits, attitudes, and behaviours that entrepreneurs do have in common (Schölhamer and Kuriloff, 1979; Kuratko and Hodgetts, 1995; Timmons, 1999). Some of the most cited entrepreneurial characteristics and behaviours are listed in Table 15.1.

The common understanding of entrepreneurship gradually shifted from identifying and describing personality characteristics of the individual (who the entrepreneur is) to actions and outcomes of entrepreneurship (the process of entrepreneurship) and the factors affecting it (Gartner, 1989; Sexton and Bowman-Upton, 1991). Definitions emerged that narrowed the concept of entrepreneurship to be associated with the creation of new businesses. Even today, many still associate entrepreneurship with the creation of new businesses. In some countries, entrepreneurship is linked to regional development programmes that create new ventures to increase employment and output. In others, entrepreneurship means an increase in opportunities for certain designated groups, such as women or minorities, to participate in the economy. Some countries simply seek to increase venture creation as such, be it high-growth firms, technology-based firms, or any form of venture that provides self-employment (Naude and Havenga, 2005; OECD, 2009b). The dynamic process of new venture creation remains an important part of entrepreneurship and innovation; these ventures ensure that new products, processes, and organisational structures are introduced throughout the economy (OECD, 2009b).

Entrepreneurship is, however, a multi-faceted phenomenon and involves more than creating new ventures. The essence of entrepreneurship seems to be the recognition and exploitation of opportunity, without regard to the control of resources. This is emphasised by the fact that entrepreneurs are also to be found in businesses they do not own or control (Sexton and Bowman-Upton, 1991). Awareness of this reality gave rise to the concept of corporate entrepreneurship, or intrapreneurship, referring to entrepreneurship that occurs within organisations, as opposed to an entrepreneur establishing his or her own venture (Guth and Ginsberg, 1990; Zahra, 1991; Pinchot and Pellman, 1999).

| Characteristics | Attitudes and behaviours |
|---|---|
| Creativity, innovation, ability to adapt | • Open-minded, lateral thinking, creative<br>• Ability to adapt and change<br>• Ability to conceptualise and integrate |
| Opportunity orientation | • Intimate knowledge of customer needs<br>• Market-driven<br>• Obsessed with value creation |
| Tolerance of risk, ambiguity and uncertainty | • Calculated risk taker / sharer<br>• Ability to integrate and conceptualise<br>• Tolerant of conflict, stress and uncertainty |
| Motivation to excel | • Achievement and growth-driven<br>• Results-driven with high but realistic goals<br>• Supportive and stimulates interaction<br>• Internal locus of control |
| Commitment and determination | • Total immersion<br>• Persists in solving problems<br>• Decisiveness |
| Leadership | • Integrity, reliability, builder of trust<br>• Inspirational<br>• Superior learner, mentor and team builder |

**Table 15.1. Dominant attitudes and behaviours of entrepreneurs . Source: Adapted from Schölhamer and Kuriloff (1979); Kuratko and Hodgetts (1995); and Timmons (1999)**

Corporate entrepreneurship is seen as the process of introducing and producing new products, processes, and services within an existing organisation to grow and develop the organisation as a whole. When an entrepreneurial philosophy underpins an organisation's outlook and operations, this is defined as corporate entrepreneurship. There is a larger focus on the entrepreneurial behaviour of the organisation, as compared to the individual, with emphasis on recognising opportunities within an existing organisation and utilising such opportunities through innovation for the benefit of the organisation. This broader view of entrepreneurship is confirmed by Sharma (1999), who describes the process of entrepreneurship as follows:

> Entrepreneurship encompasses acts of organizational creation, strategic renewal, or innovation that occur within or outside an existing organization. Entrepreneurs are individuals or groups of individuals, acting independently or as part of a corporate system, who create new organizations or instigate renewal or innovation within an existing organization.

There has been a growing awareness in recent years of the successful application of entrepreneurial principles in general society and community development settings. This has led to the concept of social entrepreneurship. A social entrepreneur is regarded as an enterprising person who applies entrepreneurial principles and skills in resolving social and community problems and improving society in general. Cherwitz (2002) confirms this by stating:

> Creating material wealth is only one expression of entrepreneurship. Entrepreneurship isn't a business. It is an attitude to engaging the world—a process of cultural innovation.

The emphasis is upon forming partnerships with the community to solve problems and identify opportunities. Social entrepreneurs also take risks and engage in innovative approaches, usually to advocate change or to improve the lives of people, but not for personal gain. In most aspects they are not different from other entrepreneurs.

The broader concept of organisational and social entrepreneurship increasingly emphasises entrepreneurship as a mind-set or an orientation—an orientation that favours proactiveness, risk-taking, competitiveness, autonomy, and innovation (Lumpkin and Dess, 1996; Wong et al., 2005). Entrepreneurs act on opportunity in creative ways; they combine (and manage) production factors in unique ways to create value through innovation.

## A Theoretical Basis of Innovation

Innovation in general refers to the successful implementation of creative ideas in the form of new products, new processes, or new services to the market (Amabile, 1998; Fagerberg, 2005). The creative process is the starting point of innovation. Creativity in general refers to the cognitive processes involved in creating new and novel outcomes, either through discovery or through a reformation of existing knowledge (Sternberg and Lubart, 1999; Rickards, 1999). These novel outcomes can be in the form of ideas, products, processes, or anything that is useful or functional in practice, including new technologies and institutional or organisational changes (Amabile et al., 1996). Zhao (2005) regards innovation as the specific tool by which entrepreneurs exploit change. Innovation has to address market needs and requires entrepreneurship if it is to achieve commercial success. A key concept in innovation is to understand that innovation can be derived from existing information and need not necessarily be derived from new information or new discoveries. Often innovations are not new to the world, but may be new to the agents who adopt them (Nelson and Rosenberg, 1993; OECD, 2013).

In the context of corporate entrepreneurship (intrapreneurship), Timmons (1999) suggests that innovation be expressed in terms of the creation of value. This involves being inventive; organising or applying resources in new ways; or launching or reintroducing products or technologies that were previously unavailable or non-existent in the market. Innovation also extends to the ways in which enterprise interacts with the market. These concepts seem to integrate creativity into the business and strategy domain. In the modern business environment, intrapreneurship calls for constant innovation in pursuit of sustainable competitive advantage (Peters and Waterman, 1982; Hamel and Prahalad, 1994; Drucker 1998). It is widely accepted that innovation is the engine of social and economic development, in both developed and developing countries.

There are, however, different views on the best way for countries and firms to foster innovation. The traditional linear model of innovation interprets innovation as a sequence of steps starting with research and ending with dissemination (World Bank, 2006). A more recent model of innovation accentuates a systems approach. Some key elements of the systems approach to innovation are contrasted with the more traditional linear view in Table 15.2.

The common view of the linear approach is that scientific research is the main driver of innovation: as more research is being done, new knowledge and technology is created which is then transferred and adopted and so leads to increased innovation. It is increasingly recognised in the agriculture sector that the linear model is not necessarily sufficient to enable sustained innovation. As the context of agriculture has changed, so have views of innovation. The more recent view of innovation emphasises a systems approach—where innovation does not solely rely on the research and science supply side, but includes the demand side and all factors affecting the use of new and existing knowledge to create novel applications (Davis et al., 2007).

The innovation systems approach includes the interactions amongst all actors involved in innovation. An innovation system can be defined as "a network of organizations, enterprises, and individuals focused on bringing new products, new processes, and new forms of organization into economic use, together with the institutions and policies that affect their behaviour and performance. It extends beyond the creation of knowledge to encompass the factors affecting demand for and use of knowledge in novel and useful ways" (World Bank, 2006).

| | Linear Approach | Innovation Systems Approach |
|---|---|---|
| **Objectives** | Emphasis on<br>• advanced technology / radical innovations<br>• technology that changes production mode | Emphasis on<br>• learning within organisations to innovate<br>• strengthening capabilities to innovate |
| **Strategies** | • Supply-driven science and technology<br>• Focus on conventional research path: basic, strategic, applied, adaptive<br>• Hierarchical knowledge dissemination: from research to extension to user<br>• R&D undertaken by public institutes<br>• Centralised management of innovation | • Demand- *and* supply-driven science and technology<br>• Focus on the roles and interactions of diverse agents in society<br>• Focus on complex and dynamic interactions among agents<br>• Network-based knowledge dissemination<br>• R&D publicly and privately undertaken<br>• Decentralised management of innovation |
| **Instruments** | • Direct public financing<br>• Indirect public financing: subsidy programs, incentive schemes<br>• Private investment | • Competitive grant programs<br>• Public-private-civil society partnerships<br>• Advisory and consultancy services<br>• Participatory research projects<br>• Changes to individual and organisational practices, behaviours, and cultures to promote mobility of educators and researchers and their integration into networks |

**Table 15.2. Innovation as a linear versus a systems approach: key elements. Source: Adapted from Davis et al. (2007)**

Figure 15.1 presents a conceptual framework of such an innovation system for the agriculture sector (Spielman and Birner, 2008). The agricultural innovation system includes three distinctive elements: an education and knowledge element, a business and enterprise element, and an element representing bridging institutions. The knowledge and education element consists of the agricultural research and education systems. The business and enterprise element comprises the agricultural value chain. Between these elements are the bridging institutions that facilitate the transfer of knowledge and technology between them. This typically includes extension services, NGOs, and other stakeholders. Within each of the three elements of the system, actors simultaneously serve as consumers and producers of knowledge and information, as producers and consumers of goods and services in the value chain, and as bridging institutions between the various components.

This innovation systems framework emphasises the systems approach rather than the conventional approach of one-way knowledge and information transfer. The framework also makes provision for aspects that may encourage or inhibit innovation, such as informal institutions (which establish the rules, norms, and culture of a society); behaviours, attitudes, and practices (which determine the ways in which individuals and organisations within each element act and interact); and government policies. In addition, innovation is influenced by other external factors such as forward and backward linkages to other sectors of the economy and the international market and business environment, the political system, and policies that impact innovation (e.g. science and technology policies). A successful innovation system therefore relies on the capacity of individuals and organisations to learn and to change the ways in which they are used to operate. It also relies on the capacity of innovation agents to interact and form networks among themselves and on the existence of formal and informal institutions that regulate the interactions among different agents (Davis et al., 2007).

Informative in this context is the Triple Helix concept (Ezkowitz, 2008), which promotes the interactions among three main actors in innovation: the university, industry, and government. The Triple Helix framework (Figure 15.2) shows that innovation occurs through continuous exchange of ideas among the three relatively equal intersecting spheres of government, academia, and industry, the latter including the community. At the intersection, each sphere potentially adopts the role of another sphere in addition to its traditional role, e.g. the university forming firms, the government acting as a venture capitalist, and industry involved in training and research.

The institutional spheres still perform their traditional functions, but increasingly assume the task of advancing innovation and development. The institutional overlapping creates a productive environment of interactions. The strong emphasis on interactions, linkages, and collaboration is conducive to innovation.

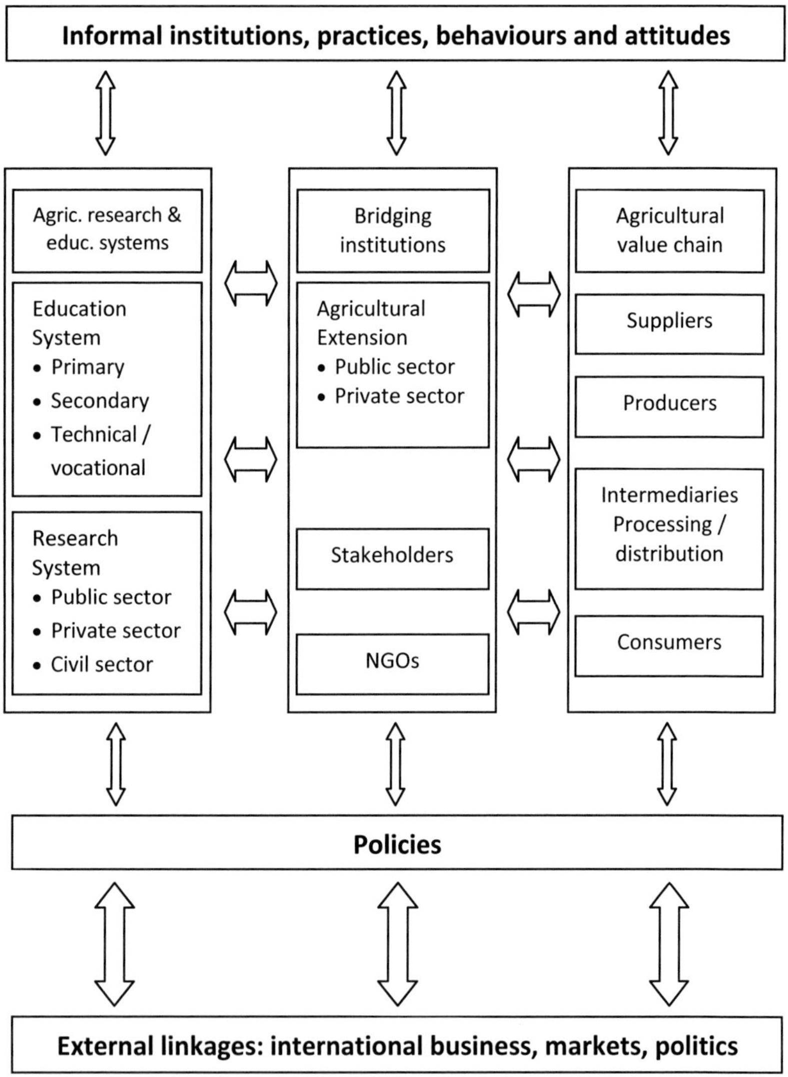


Figure 15.1. Conceptual framework of an agricultural innovation system. Source: Adapted from Spielman and Birner, 2008

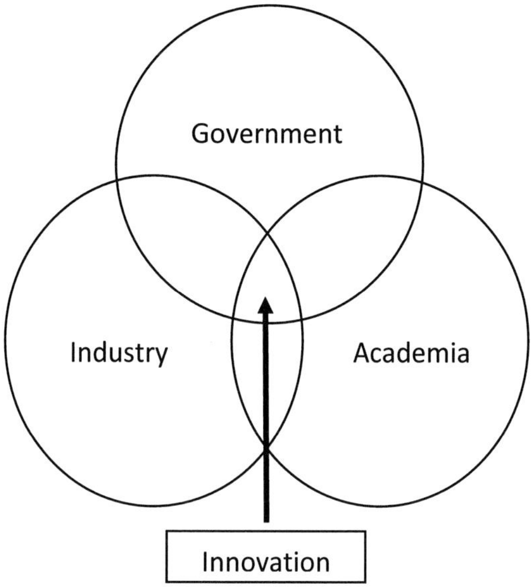


Figure 15.2. The Triple Helix framework. Source: Dzisah and Etzkowitz, 2009

The Triple Helix is widely recognised for its significance in sustainable development through the generation and effective application of knowledge. In all developing economies, the essential elements of the "triple helix" exist—the missing link is often a strategy to integrate the different elements to enhance innovation (Dzisah and Etzkowitz, 2009).

## The State of Entrepreneurship and Innovation in Africa

Elsewhere in the world research shows that entrepreneurship education and training play an important role in enhancing the entrepreneurial propensities of individuals (Rabbior, 1990; GEM, 2002). This is reflected in the many schools and institutions of further- and higher education that offer courses in entrepreneurship and small business management (Hisrich and O'Cinneide, 1996; Brockhaus et al., 2001). According to a study by Naude and Havenga (2005), little research on entrepreneurship in Africa is available compared to such research elsewhere in the world.

Table 15.3 shows a distribution of entrepreneurship research studies in ten-year intervals for the period 1963 to 2001 for selected African countries. More than 76 percent of the research originated from five countries: 61.2 percent from South Africa, 5.0 percent from Zimbabwe, 3.8 percent from Nigeria, 3.7 percent from Kenya, and 3.1 percent from Ghana. The majority of entrepreneurship research in Africa dealt with education, management, and skills issues (14.2 percent), followed by research on opportunities and constraints to entrepreneurship (14.0 percent), and the contribution of entrepreneurship to economic development (10.3 percent). Most research publications were produced between 1990 and 2001, probably due to an increased awareness of the contribution that entrepreneurship and small business can make to growth and development. The current world economic crisis has spurred a renewed interest in entrepreneurship and innovation as drivers of economic development, and AET in Africa needs to use this opportunity to engage more meaningfully in programmes and policies that will flow from this renewed interest.

| Country | Total 1963 - 2001 | 1960 - 1969 | 1970 - 1979 | 1980 - 1989 | 1990 - 2001 | Percentage |
|---|---|---|---|---|---|---|
| Ghana | 16 | 1 | 6 | 4 | 5 | 3.08 |
| Kenya | 19 | - | 2 | 7 | 10 | 3.65 |
| Nigeria | 20 | 2 | 4 | 8 | 6 | 3.85 |
| South Africa | 318 | 3 | 25 | 49 | 241 | 61.15 |
| Zimbabwe | 26 | - | - | 5 | 21 | 5.00 |
| Rest of Africa | 121 | 3 | 15 | 21 | 82 | 23.27 |
| Total | 520 | 9 | 52 | 94 | 365 | 100.00 |

**Table 15.3. Entrepreneurship research studies for selected countries in Africa. Source: Naude and Havenga (2005)**

Spielman and Birner (2008) attempt to give an account of the state of innovation in Africa using two indicators: the World Bank's Knowledge Economy Index (KEI) and CGIAR's Public sector Agricultural research and development (R&D) intensity indicators. Spielman and Kelemework (2009), however, contend that the true underlying processes of innovation in developing agriculture are not sufficiently captured by most of the existing innovation indicators due to their bias towards research and technology. They suggest a composite Agriculture, Development, and Innovation Index (ADII), with four sub-indices based on the four domains of the innovation systems framework, namely knowledge and education, business and enterprise, bridging institutions, and enabling environment. (For a full discussion of ADII methodology, see Spielman and Kelemework, 2009.)

Table 15.4 combines innovation indicators for selected developing-agriculture countries based on the World Bank's KEI, CGIAR's Public sector Agricultural R&D intensity indicators, and the ADII-index of Spielman and Kelemework (2009). In general, on the knowledge economy indicators, the overall innovation climate in Africa seems to be lagging behind both Asia and Latin America, but in terms of knowledge creation in the agriculture sector (R&D intensity), Africa and Asia are very similar. Data from Latin America is higher than both other regions, but needs to be interpreted in the context of 2006 updated figures. There doesn't seem to be much difference between regions on the composite ADII scores. However, some differences are apparent in the underlying domains; for example, in Africa the enabling environment domain is much stronger than the domains of knowledge and education, business and enterprise, and bridging institutions, while in Latin America the bridging institution domain is weak relative to the other domains.

A comparison of Africa, Latin America, and East Asia on the Innovation Index, calculated from the World Bank 2010 KAM database (KAM 2010), is presented in Figure 15.3. Davis et al. (2007) note some limitations of the KAM database: it tends to include measures that may not be immediately relevant to small-scale agriculture—for example patent applications and high-tech exports. These measures may overlook more local innovation processes that are not captured in formal measurements. UNESCO (2009) echoes this concern and states that recent literature indicates that innovation in developing countries tends to be incremental, informal, and mostly below the technology frontier. Based on existing measurements, Africa overall seems to compare unfavourably with other developing regions and only outperforms these regions in the category "Science and Technology articles with foreign co-authorship."

In light of the current state of entrepreneurship and innovation in Africa and a renewed worldwide interest in entrepreneurship and innovation as drivers of economic growth and regional development for poverty alleviation, policy makers turn to academia to contribute to development goals. The need to revise institutional curricula to teach and promote entrepreneurial skills and risk-taking is clear (UNESCO, 2009). Since agriculture plays a major role in most African economies, it is argued that AET can play an important role in developing and sustaining competitiveness in the global market through entrepreneurship education.

| Type | Overall innovation climate | | | | | | Investment in knowledge creation |
|---|---|---|---|---|---|---|---|
| Sector | General | Developing Agriculture | | | | | Agriculture |
| Source | World Bank | Spielman and Kelemework | | | | | ASTI / CGIAR |
| Measure | Knowledge Economy Index score | ADII and Innovation Framework domain indices | | | | | Public sector agricultural R&D intensity |
| Index [a] | KEI | ADII | BI | BE | EE | KE | R&D intensity |
| Year | 2010 | 2009 | | | | | 2000 |
| *Africa* | | | | | | | |
| Nigeria | 1.84 | 3.46 | 2.80 | 3.31 | 4.74 | 2.98 | 0.38 |
| Kenya | 2.77 | 5.27 | 4.88 | 4.23 | 7.02 | 4.95 | 2.68 |
| Ethiopia | 1.30 | 3.81 | 3.14 | 3.57 | 5.53 | 2.98 | 0.38 |
| Ghana | 2.46 | 4.50 | 4.54 | 3.93 | 7.41 | 2.14 | 0.47 |
| Senegal | 2.57 | 5.87 | 6.85 | 4.29 | 7.87 | 4.48 | - |
| South Africa | 5.38 | 5.14 | 4.50 | 4.50 | 6.47 | 5.08 | 3.04 |
| *Asia* | | | | | | | |
| India | 3.09 | 4.71 | 4.10 | 4.17 | 6.12 | 4.46 | 0.48 |
| China | 4.47 | 4.81 | 4.70 | 4.52 | 6.48 | 3.53 | 0.37 |
| Vietnam | 3.51 | 4.37 | 3.55 | 5.37 | 5.78 | 2.80 | 0.67 |
| Thailand | 5.52 | 6.15 | 7.00 | 4.79 | 7.19 | 5.60 | - |
| Malaysia | 6.07 | 5.88 | 5.80 | 5.61 | 6.71 | 5.40 | 3.64 |
| Indonesia | 3.29 | 4.29 | 4.00 | 3.56 | 5.58 | 4.02 | - |
| *Latin America* | | | | | | | 2006 |
| Brazil | 5.66 | 5.07 | 3.31 | 5.37 | 7.27 | 4.32 | 1.68 |
| Mexico | 5.33 | 4.78 | 2.96 | 5.06 | 6.92 | 4.16 | 1.21 |
| Colombia | 4.84 | 4.83 | 3.00 | 4.98 | 6.88 | 4.45 | 0.50 |

*Notes:* [a] KEI is the knowledge economy index; ADII is the agriculture development and innovation index; BI is the bridging institutions domain; BE is the business and enterprise domain; EE is the enabling environment domain; KE is the knowledge and education domain; R&D intensity is the gross domestic expenditure on R&D as percentage of agricultural GDP (latest available ASTI figures).

**Table 15.4. Innovation indicators for selected countries in Africa, Asia and Latin America. Source: Adapted from Spielman and Birner (2008); KAM (2010); ASTI/ CGIAR (2010); Spielman and Kelemework (2009)**

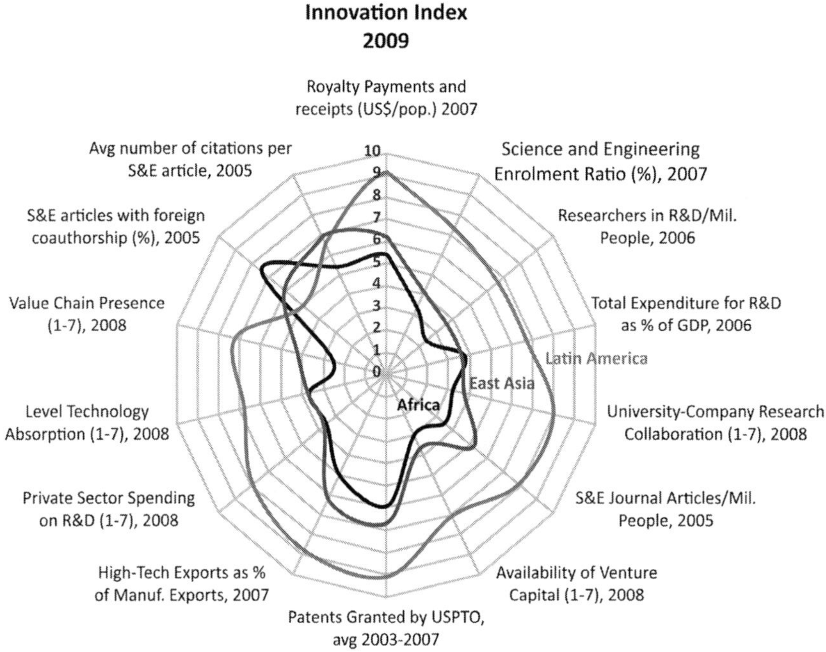


Figure 15.3. Africa innovation index compared to Latin America and East Asia
Source: http://go.worldbank.org/JGAO5XE940

## The Role of AET in Developing Entrepreneurship and Innovation in Agriculture

A study by Bloom et al. (2006) concluded that higher education in general will contribute significantly to Africa's potential to achieve economic growth. By investing in higher education, Africa can accelerate technological diffusion, which in turn will decrease knowledge gaps and assist in reducing poverty. Evenson (2004) suggests a need—based on studies of the return rates on agricultural research—for public investment in graduate programs in "innovation-related fields of study" in Africa. From the systems framework in Figure 15.1, with AET firmly integrated into a larger, more complex innovation system, we conclude that it is imperative for higher education to play a role in fostering innovation and entrepreneurship in Africa.

Developing innovation in AET needs to be addressed at two main levels: at the organisational level—where structures, processes, and strategies of

AET need to be adapted in order to allow for entrepreneurial and innovative activities; and at the operational level—where entrepreneurship and innovation need to be integrated within AET curricula to help develop innovative capabilities of students. The innovation systems approach can be helpful in determining what actions need to be taken within AET to contribute to the development of entrepreneurship and innovation in agriculture. Three key elements are suggested by Davis et al. (2007) and Spielman and Birner (2008) for creating successful innovation systems: individual and collective innovative capabilities, innovation-supportive organisational cultures, and the development of outside networks with other innovation agents. The following section builds on this idea, putting forward some actions through which AET might reform its organisational systems and cultures to allow for greater involvement in outside networks and partnerships and to introduce teaching approaches and practices that will contribute to an increased innovation capacity in Africa. A framework for guiding entrepreneurship and innovation development in African AET is presented in Figure 15.4.

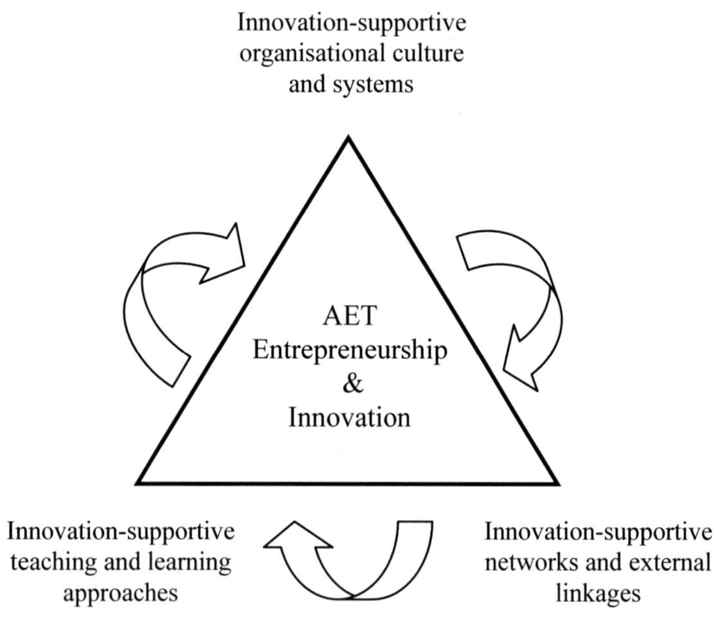


Figure 15.4. Framework for developing entrepreneurship and innovation in AET in Africa

## Developing Innovation-Supportive Organisational Cultures and Systems in AET

The first important step in developing an innovation-supportive culture is to create an enabling institutional environment where systems and processes of AET are conducive to entrepreneurial and innovative behaviour. This includes the presence of an organisational culture that fosters entrepreneurship and innovation. Such an "entrepreneurial university" is characterised by three distinctive elements embedded in its operations (Trajtenberg, 2008):

- the extent to which the university is innovative in terms of how it is structured and governed, and how it adapts to change;
- the extent to which the university can drive entrepreneurship in the broader economy by training entrepreneurs, developing ideas, and engaging with industry; and
- the extent to which the university has the ability as an entrepreneurial institution to effect broader change throughout society at large.

Organisational culture and hierarchies often inhibit innovative behaviour. Top-down control, public service regulations, administrative rules, and bureaucratic practices often lead to an inward focus on internal administrative issues and internal competition for resources, rather than on opportunities in the external environment. In general, AET in Africa needs to transform cultures, systems, processes, and policies across the entire institution to allow for entrepreneurial and innovative behaviour (Spielman and Birner, 2008). Visible, strong leadership is needed to transform institutions and ensure that strategies are implemented. It is important for the top management of the institution to demonstrate commitment to entrepreneurship and innovation. To this purpose, it is vital to introduce entrepreneurship and innovation as a core institutional strategy and to demonstrate it in the context of embeddedness in day-to-day operations.

A range of actions and practices to support and sustain innovation-supportive institutional strategies for AET in Africa are proposed:

- create an overarching institutional entrepreneurship and innovation framework to guide governance systems and processes: innovation and entrepreneurship are broadly understood as a strategic objective of the institution, with top-down support;
- develop links with local entrepreneurs, businesses, and the community;
- promote close relationships and networks with local entrepreneurs, businesses, and the community that support entrepreneurship learning and in-

novation opportunities—including facilities for incubation or links and partnerships to external incubators;
- appoint industry and community entrepreneurial leaders to serve on governing boards;
- decrease levels of management hierarchy and streamline bureaucratic administrative rules, policies, and rigidities;
- develop a whole-campus approach, targeting students and staff in all subjects and at all levels of learning;
- develop an entrepreneurial approach to both teaching and research by academics—giving development and training support for academics with narrow and strong scientific/technical backgrounds to engage in entrepreneurship and innovation practices;
- dedicate structures within AET to coordinate, integrate, and support cross-faculty collaboration and cross-disciplinary research and teaching groups;
- allow for employees to experiment—encourage multi-disciplinary teams in order to promote creativity and encourage innovative solutions to problems;
- redesign discipline-focused departments into cross-disciplinary units;
- take into account entrepreneurial attitudes, behaviours, and experience in recruitment of staff—both academic and administrative;
- design faculty-level recognition and reward systems in terms of performance, career development, and engagement in entrepreneurial activities;
- allow and promote the application of a wide range of pedagogic tools; and
- empower employees to participate in management and decision-making—employees need to take initiative and add value to the operations of the university.

When an institution commits to the above principles, entrepreneurship and innovation can become part of an AET core strategy and organisational culture. An environment can be created that supports the development of entrepreneurship and encourages innovative behaviour while reducing the risks involved for staff. This in turn can lead to a culture that is open to change and to opportunities for innovation. It has been argued (Gibb and Hannon, 2006) that entrepreneurial universities are capable of responding flexibly and strategically to opportunities in the environment. This includes active engagement and networking with stakeholder communities in a variety of forms such as consultancy, training, research, technology transfer, and engagement in science parks and incubators. It is essential for AET in Africa to develop an organisational culture of networking and partnering in order to link with a range of stakeholders in the agricultural innovation system.

## Developing Innovation-supportive Networking Systems in AET

One of the fundamental constraints for innovation in AET in Africa is the scarcity of resources. One way of overcoming this constraint is to integrate into innovation networks to utilise human and financial resources more efficiently. Network linkages can be sideways to AET, backward to secondary education, and forward to industry partners. An innovation-supportive networking system may include universities, colleges, technical/vocational centres, research centres, extension agents, agri-businesses, farmer organisations, non-governmental organisations (NGOs), and various other actors who create, exchange, or use knowledge and technology (Davis et al., 2007). Networks form part of the entrepreneurial university's core strategy and organisational culture—it has working alliances with individuals and groups concerned with innovation and forms part of a larger "ecosystem of networked connectivity" (Crow, 2008).

A range of actions and practices to develop and sustain innovation-supportive networking strategies for AET in Africa are proposed:

- form AET consortia that work together to leverage human capital and inter-disciplinary collaboration to strengthen research and innovation capacities of AET;
- develop inter-university forums on teaching in order to share best-practice knowledge and practices;
- promote research-driven collaboration between the university and external agricultural research organisations, extension services, and producers to facilitate useful and applied research;
- develop collaboration between the university and public sector agricultural extension services and producers to strengthen agricultural extension with regard to the integration of modern technologies in indigenous knowledge systems and practices;
- introduce consultative processes with stakeholders and user communities to assist in agricultural research and curriculum priority setting;
- use experiential training and work-integrated learning programmes to link farmers with students and academics to facilitate a multi-directional flow of knowledge and information;
- engage with diverse user communities outside of academia to exchange knowledge and information on development imperatives, including consulting by academics;
- provide learning opportunities geared to the specific needs of different actors in the innovation system through short learning programs (SLPs),

thereby facilitating a two-way exchange of knowledge and information to the benefit of both the university and the learners; and
- promote stronger linkages between AET and other innovation actors through internships, exchange programs, and sabbaticals for academics.

When organisations commit to the above, the formal or informal linkages among those with similar or complementary mandates can facilitate the access and flow of information among different actors, help communities adapt to existing knowledge and technology, and ultimately contribute to the accomplishment of innovation goals while sharing costs, risks, and benefits. Networks in all their different forms offer three specific benefits to their participants: they prevent isolation, they enable mobilisation of capacity and resources, and they play a role in quality assurance through information-sharing and peer review (World Bank, 2007).

## Developing Entrepreneurial and Innovation-supportive Teaching Systems in AET

Due to its origins in the economics and business disciplines, the concept of entrepreneurship and innovation is reasonably familiar to commerce educators and students. Yet there does not seem to be consensus on the best approach to teaching entrepreneurship. The most common approach in Europe and the U.S. seems to focus on the traditional business school model (Gibb and Hannon, 2006), which emphasises new venture creation, business growth, business planning, and the traditional functional areas of management. Although necessary in terms of scientific content, this approach can be regarded as insufficient in terms of pedagogies and practices that need to "stimulate entrepreneurial attributes and values, provide insights into the entrepreneurial-life world, allow for practice of entrepreneurial behaviours, develop emotional intelligence, and promote the value of acquiring of tacit (experiential) knowledge under pressure" (Gibb, 2006).

Gibb (1998) offers a set of entrepreneurial core capacities that should guide entrepreneurship teaching. He describes these as *"capacities that constitute the basic necessary and sufficient conditions for the pursuit of effective entrepreneurial behaviour individually, organisationally, and societally in an increasingly turbulent and global environment.*" Teaching entrepreneurship should move beyond teaching the science of entrepreneurship only, towards teaching the art of entrepreneurship as well. Where science refers to the theory of business based on the analysis of practical experience, the art of entrepreneurship involves the ability to apply theoretical knowledge creatively and with initiative. It deals with perception, creative thinking, learning-by-doing and creating useful outcomes. In line with the broader concept of organisational entrepreneurship defined earlier, where the emphasis is on an entrepreneurial orientation, a clear distinction needs to be made between the conventional approach to teaching and an approach designed to develop an entrepreneurial orientation (Gibb, 1998).

Widespread confusion exists about introducing entrepreneurship and innovation in scientific and non-business settings in general (Jordaan, 2001). Some propose that an entrepreneurship module be introduced as part of an existing subject offering; some propose a semester course as a full subject offering; some propose an extracurricular short course; some argue that such a course is not necessary in a particular field of study. When entrepreneurship is introduced in a particular non-business degree or diploma, it is often as a once-off intervention over a specified, limited time period, managed and taught exclusively by academics from the business administration field, with limited application to the disciplines of the particular non-business course (Dolabela, 1999). Overall, we still lack a well-developed theoretical and conceptual framework to direct entrepreneurship teaching (Barr et al., 2009). Since little evidence exists on entrepreneurship teaching in agricultural courses per se, the following section draws on practices and processes common to entrepreneurship development in higher education across the world (Rabbior, 1990; Laukkanen, 2000; Jordaan, 2001; Etzkowitz and Zhou, 2008; Harkema and Schout, 2008; Almeida, 2008; Mustar, 2009; OECD, 2009a).

A range of actions and practices to support entrepreneurial teaching strategies for AETs in Africa is proposed:

- Various teaching styles and approaches need to be used to help develop an entrepreneurial mindset tailored to the needs of different groups of students. Entrepreneurial pedagogies may include methods that are highly participatory and allow for inter-disciplinary teamwork, problem solving, and integration of concepts.
- Teaching methods and assessments should be constructed in ways that facilitate the development of an entrepreneurial orientation—a focus on attitudes, behaviours, and skills in terms of opportunity recognition, creativity, innovation, handling of ambiguity, dealing with uncertainty and pressure, and so forth.
- The lecturer needs to assume the role of facilitator and stimulate students through mentoring and coaching to be co-constructors of their own learning and career development. This calls for a learner-centred approach and action learning through debates, critical discussion, critical thinking, and continuous feedback.
- Education must be real-time and real-world to help produce graduates ready for innovation, be it in the workplace or as self-employed entrepre-

neurs. Science and technology need to be combined with business competencies and even practical life skills to develop analytical competence, decision-making abilities, and independent. creative thought.
- Experiential learning and attachments to private firms through work-integrated learning need to be emphasised; this allows for competence-based training provided by firms to students and also allows for establishing linkages between AETs and industry, which is important in terms of identifying priorities for research, extension, and education.
- Entrepreneurial education needs to be integrated across faculties, disciplines, and departmental boundaries and needs to make provision for electives and non-traditional agricultural sciences such as, for example, project management, agribusiness, human resource management, law, leadership, and strategic management, among others.
- The introduction of entrepreneurship into AET curricula needs to be formalised and supported by top management through endorsement at the strategic and operational levels; commitment of the necessary resources, departmental buy-in, and ownership by academics are all equally important.
- Curriculum development processes need to be linked to industry to ensure demand-side needs are captured through stakeholder workshops, consultations, and so forth.
- Developing an entrepreneurship research thrust across departmental boundaries may be valuable in the process of stimulating an entrepreneurial orientation amongst academics and also provide a valuable resource for teaching material.
- Students can be involved in industry-consulting projects or community outreach to further strengthen real-life educational experiences; and
- When preparing assignments, actual business environments can be simulated and supported with guest lecturers from industry.

Committing to the above creates a supportive environment for the development of an entrepreneurial orientation, which in turn may encourage innovative behaviour amongst both staff and students. There are implications for resources and for lecturer development. Lecturers may not be knowledgeable on entrepreneurial teaching or modern advances in education and learning, or they may not have sufficient knowledge of the modern work world and the skills required by students. An effective agricultural innovation system requires professionals with a new skills set and mind set. Lecturers, especially those locked into narrow disciplinary structures, will need to develop modern approaches to cross-disciplinary content, especially when teaching for entrepreneurship development.

## Concluding Comments

The evolving global landscape of agriculture presents many challenges for the agriculture sector in developing countries, where continuous, incremental innovation is required to keep pace. Different types of knowledge, skills, and mindsets are required from farmers, extension officers, researchers, public servants, academics, and other participants in the agricultural value chain. An entrepreneurial spirit is increasingly acknowledged as important for motivating individuals and organisations to perceive and develop new opportunities that can promote positive change and create economic growth. This entrepreneurial spirit is seen as the main source of innovation. With AET firmly integrated within a larger, more complex innovation system, fostering innovation and entrepreneurship is imperative. This calls for transforming organisational systems and cultures of AET, the creation of innovation-supportive networks, and introducing approaches and methods to assist with integrating entrepreneurship into agricultural education. A number of strategies and actions have been proposed to this end.

Preconditions for success include demonstrated support from top management through incorporation of entrepreneurship and innovation as a core strategy, backed by administrative systems and an organisational culture conducive to entrepreneurial behaviour. Crucial to this are staff training workshops and development support for academics to provide the necessary knowledge and skills to effect an entrepreneurial strategy. Knowledge and skills imperative for entrepreneurship education include a systems orientation and the ability and willingness to work across disciplines. Curriculum restructuring to allow for the integration of entrepreneurship principles, practices, and pedagogies in existing curricula is implicated, including appropriate assessment practices. Of particular importance will be the provision for experiential training and work-integrated learning. Close relationships and networks with local farmer entrepreneurs, businesses, government ministries, research institutions, and the community should be actively pursued.

AET can play an enormously important role in assisting the agriculture sector in developing and sustaining innovative capacities. Not only does AET have a captive audience of potential new entrants to agriculture, but it also has access to an important intellectual faction of the future agricultural fraternity. Introducing entrepreneurship and innovation principles and practices to students may provide the necessary input for greater innovation capacity in the future. Making entrepreneurship development part of its strategy may help AET institutions to become true entrepreneurial institutions playing their roles in a wider system of innovation and successfully contribute to addressing poverty, managing ecosystems, improving food security, and achieving broad economic growth and development. Entrepreneurship orientation is the start; innovation will follow.

## Case study: Networking between a university, a government ministry and a peri-urban agricultural community in South Africa—adapting business incubation methods for developing agricultural entrepreneurship capacity

This case study reports experiences of a community project jointly managed by an academic institution (Nelson Mandela Metropolitan University), a government ministry, and a peri-urban community of land reform beneficiaries—the Thembalethu agricultural community—near the city of George in South Africa. The main objective was to facilitate enterprise development and entrepreneurship education by adapting conventional business incubation methods to the local context. The chosen business was a small broiler-rearing operation of two hundred birds, beginning with day-old chicks and running through an eight-week cycle to finished broiler. Five people from the community were able to "buy" into the business as "trainee entrepreneurs," each paying R120 (US$17). This entitled trainees to own forty broiler chickens each and to participate in a training programme. The university provided seed money in the form of an interest-free loan to be paid back at the end of each rearing cycle. The trainee entrepreneurs managed the operation as a team. At the end of the rearing cycle, each entrepreneur was to sell her birds and pay back her portion of the loan (the actual costs incurred). Entrepreneurs had the option of retaining their profits or of re-investing in a new cycle. A maximum of five cycles (eight weeks each) were permitted before graduation. During each cycle, academics and students from the university offered hands-on demonstration and training, with back-up from an extension officer of the Ministry of Agriculture. The main focus was on business and entrepreneurship training, practically simulating a normal business process in each cycle. This process, from start-up when the day-old chicks were ordered to selling finished broilers, involved day-to-day management of the operation, marketing of the mature birds, assessing the profitability of the venture, and finally, at the end of each cycle, "liquidation" of the operation when the loan was to be repaid.

Performance of the operation improved progressively with each rearing cycle, indicating positive learning curve effects. From a range of critical incidents recorded, a number of entrepreneurial attitudes and innovative behaviours could be identified, *inter alia* opportunity recognition, initiative, creativity, ability to adapt and change, and value creation. The broiler-rearing venture was found to be particularly suited to the purpose of learning about entrepreneurial business ventures; the complete venture cycle and business process, from start-up to harvest, was simulated over each of the broiler-rearing cycles.

The venture also served as an example of successful technology transfer and innovation, where the possibilities of combining traditional rural systems with high-tech genetics and feed technologies were demonstrated practically—a case of introducing new-to-the-context technology (as opposed to new-to-the-world technological innovation). The "methodology" of combining demonstration, training, and hands-on participation in a supported learning environment can in itself be regarded as an example of innovation at the community level.

For the university, the experience gained was invaluable from an innovation point of view. The institution, staff, and students all increased their capacities to provide entrepreneurship education to non-traditional university target groups; and all gained an understanding of the centrality of education and training to the economic and social development of a region. In addition, partnering with public extension services and commercial financing institutions strengthened the innovation networks of the university.

## References


Almeida, M. 2008. Innovation and Entrepreneurship in Brazilian Universities. *International Journal of Technology Management and Sustainable Development* 7: 1, 39–58.

Alsos, G., E. Ljunggren, and L. Pettersen. 2003. Farm-based Entrepreneurs: What Triggers the Start-Up of New Business Activities. *Journal of Small Business and Enterprise Development* 10: 4, 435–443.

Amabile, T., R. Conti, H. Coon, J. Lazenby, and M. Herron. 1996. Assessing the Work Environment for Creativity. *Academy of Management Journal* 39: 4, 1154–1184.

Amabile, T. 1998. How to Kill Creativity. *Harvard Business Review* 76: 5, 77–88.

Agricultural Science and Technology Indicators (ASTI). 2010. *ASTI Database*. http://www.asti.cgiar.org/index.cfm.

Barr, S., T. Baker, S. Markham, and A. Kingon. 2009. Bridging the Valley of Death: Lessons Learned from Fourteen Years of Commercialization of Technology Education. *Academy of Management Learning and Education* 8: 3, 370–388.

Bloom, D., D. Canning, and K. Chan. 2006. *Higher Education and Economic Development in Africa.* Africa Region Human Development Series No. 102. World Bank: Washington, D.C.

Brockhaus, H., G. Hills, H. Klandt, and H. Welsch. 2001. *Entrepreneurship Education: A Global View.* Ashgate: Aldershot, UK.
Cherwitz, A. 2002. *Intellectual Entrepreneurship: A Vision for Graduate Education.* Change, November/December.
Crow, M. 2008. *Building an Entrepreneurial University.* In Proceedings of the Future of the Research University Meeting: Global Challenges of the 21st Century. 2008 Kauffman-Planck Summit on Entrepreneurship Research and Policy, June 8–11, Bavaria, Germany.
Davis, K., J. Ekboir, W. Mekasha, C. Ochieng, D. Spielman, and E. Zerfu. 2007. *Strengthening Agricultural Education and Training in Sub-Saharan Africa from an Innovation Systems Perspective: Case Studies of Ethiopia and Mozambique.* IFPRI Discussion paper 00736, IFPRI, Washington, DC.
Dolabela, F. 1999. *Entrepreneur Workshop Methodology: A New Way of Teaching.* Proceedings of the 44th ICSB World Conference, June 20–23, Naples, Italy.
Drucker, P. 1998. The Discipline of Innovation. *Harvard Business Review* 76: 6, 149–154.
Dzisah, J. and H. Etzkowitz. 2009. *Triple Helix Circulation: The Heart of Innovation and Development.* Conference Theme Paper, Triple Helix VII Conference, June 17–19, Glasgow.
Etzkowitz, H., A. Webster, C. Gebhart, and B. Terra. 2000. The Future of the University and the University of the Future: The Evolution of the Ivory Tower to Entrepreneurial Paradigm. *Research Policy* 29: 2, 313–330.
Etzkowitz, H. 2008. *Triple Helix Innovation: Industry, University and Government in Action.* Routledge: London.
Etzkowitz, H. and C. Zhou. 2008. Building the Entrepreneurial University: A Global Perspective. *Science and Public Policy* 35: 9, 627–635.
Evenson, R. 2004. Private and Public Values of Higher Education in Developing Countries: Guidelines for Investment. *Journal of Higher Education in Africa* 2: 1, 151–176.
Fagerberg, J. 2005. Innovation: A Guide to the Literature. In Fagerberg, J., D. Mowery, and R. Nelson, Eds. *The Oxford Handbook of Innovation.* Oxford University Press: Oxford.
FARA. 2006. *Framework for African Agricultural Productivity.* Accra, Ghana.
Gartner, W. 1989. Some Suggestions for Research on Entrepreneurial Traits and Characteristics. *Entrepreneurship Theory and Practice* 14: Fall, 27–37.
Global Entrepreneurship Monitor (GEM). 2002. *South African Executive Report.* Graduate School of Business, University of Cape Town, Cape Town.

Gibb, A. 1998. *Entrepreneurial Core Capacities, Competitiveness and Management Development in the 21st Century.* Proceedings of the Internationalizing Entrepreneurship Education and Training Conference, July 26–28, Oestrich-Winkel, Germany.

—. 2006. *Entrepreneurship: Unique Solutions for Unique Environments. Can this be Achieved with the Existing Paradigm?* Background Paper to Plenary Presentation, ICSB World Conference, June, Melbourne, Australia.

Gibb, A. and P. Hannon. 2006. Towards the Entrepreneurial University. *International Journal of Entrepreneurship Education* 4, 73–110.

Guth, W. and A. Ginsberg. 1990. Corporate Entrepreneurship. *Strategic Management Journal* 11: Summer, 5–15.

Hamel, G. and C. Prahalad. 1994. *Competing for the Future*. Harvard Business School Press: Boston.

Harkema, S. and H. Schout. 2008. Incorporating Student-Centred Learning in Innovation and Entrepreneurship Education. *European Journal of Education* 43: 4, 513–526.

Hisrich, R. and B. O'Cinneide. 1996. Entrepreneurial Activities in Europe-Oriented Institutions. *Journal of Managerial Psychology* 11: 2, 45–65.

Ivanic, M. and W. Martin. 2008. Implications of Higher Global Food Prices for Poverty Reduction in Low-Income Countries. *Agricultural Economics* 39: 405–416.

Jordaan, J. 2001. *The Challenge of Integrating Entrepreneurship in Tertiary Agricultural Education: A South African Experience.* In Miettinen, A. and H. Klandt, Eds. Entrepreneurship Research Monographien, Band 25, Edited Proceedings of the IntEnt 2000 Conference, July 9–12, Tampere, Finland. Josef Eul Verlag: Lohmar-Köln, Germany.

KAM. 2010. *The World Bank Knowledge for Development (K4D) website: Knowledge Assessment Methodology.* http://info.worldbank.org/—etools/kam2/KAM_page3.asp?default=1

Kuratko, D. and R. Hodgetts. 1995. *Entrepreneurship: A Contemporary Approach.* Third edition. Dryden Press: New York.

Laukkanen, M. 2000. Exploring Alternative Approaches in High-Level Entrepreneurship Education: Creating Micro-Mechanisms for Endogenous Regional Growth. *Entrepreneurship and Regional Development* 12: 25–47.

Lumpkin, G. and G. Dess. 1996. Clarifying the Entrepreneurial Orientation Construct and Linking it to Performance. *Academy of Management Review* 21: 1, 135–172.

Martinez, A., J. Levie, D. Kelly, R. Saemundsson, and T. Schott. 2010. *Global Entrepreneurship Monitor (GEM) Special Report: A Global Perspective on Entrepreneurship Education and Training.* GEM: Babson Park, MA.

Mustar, P. 2009. Technology Management Education: Innovation and Entrepreneurship at MINES Paris Tech. *Academy of Management Learning and Education* 8: 3, 418–425.

Naude, W. and J. Havenga. 2005. An Overview of African Entrepreneurship and Small Business Research. *Journal of Small Business and Entrepreneurship* 18:1, 101–120.

Nelson, R. and N. Rosenberg. 1993. Technical Innovation and National Systems. In Nelson, R., Ed. *National Innovation Systems: A Comparative Analysis.* Oxford University Press: New York.

OECD, 2009a. *Universities, Innovation and Entrepreneurship: Criteria and Examples of Good Practice.* OECD, Paris.

—. 2009b. *Measuring Entrepreneurship: A Collection of Indicators.* OECD-Eurostat entrepreneurship indicators programme. OECD, Paris.

—. 2013. *Innovation and Inclusive Development.* Conference Discussion Report, November 21, Cape Town, South Africa.

Oosterwijk, J. 2008. *Moving from a Managed Economy to an Entrepreneurial Economy: The Challenge for Europe and its Universities*. In Proceedings of the Future of the Research University Meeting the Global Challenges of the 21st Century, 2008 Kauffman-Planck Summit on Entrepreneurship Research and Policy. June 8-11, Bavaria, Germany.

Peters, T. and R. Waterman. 1982. *In Search of Excellence*. Harper and Row: New York.

Pinchot, G. and R. Pellman. 1999. *Intrapreneuring in Action*. Berret-Koehler: San Francisco.

Porter, M. 1985. *Competitive Advantage: Creating and Sustaining Superior Performance.* Free Press: New York.

Rabbior, G. 1990. Elements of a Successful Entrepreneurship/Economics Education Program. In Kent, C., Ed. *Entrepreneurship Education: Current Developments, Future Directions.* Quorum Books: London.

Rickards, T. 1999. *Creativity and the Management of Change.* Blackwell: Oxford.

Schölhamer, H. and A. Kuriloff. 1979. *Entrepreneurship and Small Business Management.* Wiley: New York.

Schumpeter, J. 1942. *Capitalism, Socialism and Democracy.* Harper and Brothers: New York.

Sexton, D. and N. Bowman-Upton. 1991. Entrepreneurship: Creativity and Growth. MacMillan: New York.

Sharma, P. 1999. Toward Reconciliation of the Definitional Issues in the Field of Corporate Entrepreneurship. *Entrepreneurship Theory and Practice* 23: 3, 11–28.

Spielman, D. and R. Birner. 2008. *How Innovative is your Agriculture? Using Innovation Indicators and Benchmarks to Strengthen National Agricultural Innovation Systems.* Agriculture and Rural Development Discussion Paper no. 41. World Bank, Washington, DC.

Spielman, D. and D. Kelemework. 2009. *Measuring Agricultural Innovation System Properties and Performance: Illustrations from Ethiopia and Vietnam.* IFPRI Discussion paper 00851.

Sternberg, R. and T. Lubart. 1999. The Concept of Creativity: Prospects and Paradigms. In Sternberg, R., Ed. *Handbook of Creativity.* Cambridge University Press, Cambridge.

Thompson, J., G. Alvy, and A. Lees. 2000. Social Entrepreneurship: A New Look at the People and the Potential. *Management Decision* 38: 5, 348–338.

Timmons, J. 1999. *New Venture Creation: Entrepreneurship for the Twenty-first Century.* Fifth Ed. McGraw-Hill: Boston.

Trajtenberg, M. 2008. *Entrepreneurial Universities: The View from Israel.* In Proceedings of the Future of the Research University: Meeting the Global Challenges of the 21st Century. 2008 Kauffman-Planck Summit on Entrepreneurship Research and Policy. June 8–11, Bavaria, Germany.

UNCTAD. 2010. *Technology and Innovation Report 2010: Enhancing Food Security in Africa through Science, Technology and Innovation.* United Nations Publication UNCTAD/TIR/2009, Geneva, Switzerland.

UNESCO. 2009. *Innovation for Development: Converting Knowledge to Value.* Workshop on Innovation for Development. January 28–30, Paris.

Wong, P., Y. Ho, and E. Autio. 2005. Entrepreneurship, Innovation and Economic Growth: Evidence from GEM Data. *Small Business Economics* 24, 335–50.

World Bank. 2006. *Enhancing Agricultural Innovation: How to go Beyond the Strengthening of Research Systems.* World Bank, Washington, DC.

—. 2007. *Cultivating Knowledge and Skills to Grow African Agriculture: A Synthesis of an Institutional, Regional and International Review.* World Bank, Washington, DC.

World Economic Forum. 2009. *Educating the Next Wave of Entrepreneurs: Unlocking Entrepreneurial Capabilities to Meet the Global Challenges of the 21st Century.* A Report of the Global Education Initiative, World Economic Forum, Switzerland.

Zahra, S. 1991. Predictors and Financial Outcomes of Corporate Entrepreneurship: An Exploratory Study. *Journal of Business Venturing* 6: 259–286.

Zhao, F. 2005. Exploring the Synergy between Entrepreneurship and Innovation. *International Journal of Entrepreneurial Behaviour and Research* 11: 1, 25–41.

# CHAPTER SIXTEEN

# GENDER IN AGRICULTURAL EDUCATION AND TRAINING IN SUB-SAHARAN AFRICA

MARGARET KROMA

Agricultural education and training (AET) in sub-Saharan Africa is once more moving centre stage in public discourse on development. Following decades of diminished support by major funding institutions, including the World Bank and USAID, AET is again recognised as a vital component of development (Eicher, 2006; World Bank, 2007; ACBF, 2012). At the threshold of the new millennium, these institutions are actively re-engaging efforts to improve the human capacity base for agriculture in the region. A number of factors drive the renewed focus, including a serious dearth of agricultural innovation capacity and of effective policies for bringing about much-needed transformation in agriculture and rural development. Over-arching global economic restructuring processes, manifested by expanding market power and trade liberalisation, are also important factors to which AET must respond.

As the revitalisation of AET gathers momentum, it becomes necessary to draw attention to the reality that agriculture and natural resource management systems in sub-Saharan Africa are neither gender- nor scale-neutral. Past efforts did not acknowledge or subsequently take into account scale and gender imbalances in the development of curricula and capacity- building. The results of these oversights are evident in the historically slow uptake of technologies among smallholder farmers in Africa (FAO, 2012; Dixon et al., 2001; Doss, 2001). The central positioning of gender dimensions in sub-Saharan Africa is imperative for allowing new AET initiatives to succeed.

This chapter examines the role of gender in AET in sub-Saharan Africa. It presents a case for approaching investment and policy-making through a critical gender lens. It analyses key gender trends and issues and proposes strategic options for positioning gender as a central research theme in developing sound agricultural education and development programmes.

***Key Messages***

- Transforming the gender dynamic at all levels in AET in Africa is a necessary condition for the *revitalisation of the continent's economies and social fabric.*
    - Rural women predominate in food and agricultural systems in Africa as producers and providers of labour. Yet neither the underlying knowledge system nor the female representation in AET reflects this reality. *Insufficient attention to gender in agriculture* has been identified as the most important factor explaining this paradox and the resulting low productivity and profitability of African agriculture.
- Revitalisation strategies must be directed *at both numerical expansion of women's participation and curriculum revision.* This will involve two mutually-reinforcing processes: integration of mechanisms that will ensure balanced engagement of women; and transformation in underlying knowledge systems.
    - Low representation of women in AET implies that *Africa is not taking advantage of the full range of human capital resources available.* Several interplaying factors limit the participation of women in agricultural education. These include poverty and socio-cultural beliefs, lack of female participation in science at secondary education levels, and factors at the higher education and professional levels.
    - Beyond representation lies a deeper concern: the putatively objective and *non-gendered nature of the conventional agricultural knowledge system* that undergirds AET in Africa. This system has been successful at generating technologies that were widely adopted in Asia and Latin America (regions that experienced the Green Revolution in the 1970s). In sub-Saharan Africa, however, these technologies largely bypassed female smallholder farmers because they were unaffordable or irrelevant. Thus, an integrated, gender-responsive agricultural knowledge system that more realistically mirrors the diversity of Africa's production is needed.
- At the *higher education level*, AET must recognise and respond to infrastructure needs for female students (such as female hostels). Appropriate actions include: designing flexible course schedules (i.e. evening, weekend and part-time study opportunities); ensuring a non-discriminatory institutional environment for women; mainstreaming gender in institutional management and recruitment policies; targeting and earmarking fellowships and scholarships for women; and revising curricula.
- *Curriculum revision* must go beyond traditional agricultural science knowledge and result in subject matter that integrates knowledge on women's production and natural resource management roles and needs. It must involve courses on the importance of women's roles and contributions to agriculture and on working with rural women; on interdisciplinary teaching and learning; and on teaching materials directed specifically to the needs of women farmers.

The chapter is divided into five sections. The first conceptually reviews the context of gender and AET in Africa. Next the trends and current status of gender in AET are assessed, while the third section examines the factors underlying poor integration of gender in AET and the challenges to building an AET system that can effectively serve the needs of Africa's diverse rural and agricultural clientele. Section four draws on three case studies of AET innovations to illustrate how creative strategies in training design and delivery can lead to transformative results. The concluding section pulls together the wide range of insights proffered for strategic actions to bring about revitalised and gender-responsive AET in Africa.

## Gender and AET in Africa: A Conceptual Review

Rural women predominate as producers and providers of labour in food and agricultural systems in Africa. At the rural community level, women are also key actors in off-farm livelihood activities (FAO, 2011; Doss, 2001). Yet neither the underlying knowledge system nor the female representation in AET reflects this reality. Studies show that the low number of female scientists and students in agricultural education, research, and extension stands in stark contrast to the significant presence and predominant role women farmers play in production (Amenyah and Puplampu, 2013; FAO, 2011; Huyer and Westholm, 2007; Rathgeber, 2002). Insufficient attention to gender in agriculture has been identified in many forums as the most important factor explaining this paradox and the resulting low productivity and profitability of African agriculture.

Gender is a structural dynamic undergirding every facet of society's institutions. It refers to the socially constructed, normatively unequal position of men and women in society (World Bank, 2012). In Africa, normative gender relations disadvantage women in terms of access to and control over the means for self-development and actualization (FAO, 2011). Inequalities in male-female relations are implicated as the fundamental cause of women's under-representation in AET, and to a significant degree, in the narrow articulation of its underlying knowledge system (Jiggins, 2011; Amenyah and Puplampu, 2013).

The analysis of gender in productive activities, including agriculture, focuses on the fact that socially defined relations between women and men can configure life experiences and opportunities differently and unequally. The gender approach aims at examining these relationships in order to illuminate underlying gender-based inequities. This is a fundamental step in any purposeful action toward redressing the marginalisation of women in terms of access to productive resources in society, including access to education.

There are far-reaching implications for insufficient analysis into the sources and nature of women's under-representation in AET. Simply advocating for equal participation of women in education, training, and employment is not sufficient to address the challenges gender inequality poses. Only through clear evidence that these differences exist and an understanding of the reasons behind them are the differences likely to be taken into account in planning at all levels. Thoughtful analysis and planning ensure that steps will be taken to allow men and women to benefit equally from opportunities for self-development, such as gaining knowledge, skills, and access to technological resources and training.

Education is recognised as a powerful instrument for social transformation (Schultz, 1961). However, such transformation can only happen when there is equality of access to education for both men and women. Equality of access encompasses broad dimensions of education, including output, outcome, and value attachment (Jiggins, 2011). Studies have shown that equality cannot be understood as gender neutral. For example, while girls at primary and secondary levels of education are considered to have equal access, owing in large measure to the "education for all" policies aggressively promoted by UNESCO and the World Bank in the last decades of the twentieth century, studies have shown that unless special and deliberate measures are taken, girls are more likely to drop out of school in larger numbers, and much earlier in their schooling, than boys (Mangheni et al., 2010; Schiebinger, 2010).

A gender superstructure that ascribes lesser social value to women's roles in the public sphere perpetuates the demand for girls' labour in social reproduction activities (i.e. daily domestic labour to ensure the sustenance of household members) including food preparation and water, fodder, and fuel collection. These ascribed roles serve as disincentives to investing in girls' education and undercut two of the basic justifications that have been advanced for investing in education for development: 1) the multiplier effect of education and the social capital it creates; and 2) increased economies of scale as school cohorts move to the next level and as one generation leads to the next (Jiggins, 1994). Just as significant are the overall consequences for female representation at higher education institutions. Without adequate efforts to stimulate girls' enrolment and to stem the tide of school attrition, feeding the female recruitment pool for higher education and training will remain a challenge in developing countries.

As in many developing regions around the world, the gender dynamic in sub-Saharan Africa is deeply embedded, taken for granted, and normative. As a result, leading education institutions, especially at the higher education level, fail to take into account the underlying structural determinants of men's and women's different experiences and opportunities. By remaining oblivious to the challenges of gender inequality in the educational environment, institutions fail to facilitate actions needed to redress the problem and often end up perpetuating the status quo.

However, case studies have shown that enrolments improve dramatically where policies create enabling environments by providing services and instituting curricular innovations to include subjects of relevance and interest to female students. For example, at the College of Agricultural Studies at Sudan University of Science and Technology, policy innovations such as setting a minimum intake of 30 percent women and the construction of female housing saw female enrolment increase from 10 percent in the 1980s to 72 percent in 2007 (World Bank, 2009). Similarly, when policy makers for Tanzania's Sokoine University of Agriculture embarked on dramatic university innovation and policy changes that enhanced the climate for women, the number of women graduating increased tenfold over a decade (World Bank, 2009).

While numerical balance in male and female representation in AET is an important element of capacity building, the knowledge dimension of AET has become the subject of scrutiny in the past decade, with questions about its responsiveness to smallholder contexts. Recently there has been a call for agriculture—increasingly considered the necessary engine for growth and sustainable development—to broaden its focus beyond the conventional principle of maximising production to be more responsive to smallholders' concerns (IAASTD, 2007; Kiers et al., 2008). A more encompassing orientation that includes poverty reduction, natural resource management, and rural livelihood—domains that are largely perceived as synonymous with and more relevant to female smallholders' needs—is gaining significant interest. It is therefore necessary for this review to question whether and how the conventional knowledge system is transforming or moving beyond the traditional confines of the agricultural science paradigm to incorporate a broader understanding of knowledge. The urgency for such a shift cannot be overstated. The next section examines the intersections of gender, agricultural knowledge, and AET in Africa.

## Gender, Agricultural Knowledge, and AET in Africa

Beyond the argument for improving numerical balance and attaining gender parity in female enrolments lies a deeper concern—the putatively objective and non-gendered nature of the conventional agricultural science and knowledge system undergirding AET in Africa. The prevailing view has been that this knowledge system provides the necessary impetus to transform production and enhance people's welfare. Counter to the conventional knowledge base is an alternative, though less dominant, knowledge system characterised by a broader set of elements including food security and sustainability issues. Because these two knowledge systems hold fundamentally different implications for addressing the gender dimension of AET, we need to examine how the characteristics of each system relate to SSA's predominantly rural, female smallholder production system.

## Conventional agricultural knowledge system

The conventional agricultural knowledge system offers a technical approach focused largely on agricultural productivity. It aims to improve efficiencies along the value chain and puts considerable emphasis on scientific expertise and technological innovation (Flora, 2010; Röling, 1999), favouring commodity crops such as cereals and grains—maize, wheat, and barley—that are amenable to mono-cropping on relatively large plots (Flora, 2010).

This knowledge system has been successful in generating technologies that were widely adopted in Asia and Latin America during the Green Revolution of the 1970s. In sub-Saharan Africa, however, these technologies bypassed large numbers of female smallholder farmers who could not afford them. For many others, the technologies were simply not relevant to their farming systems, which were not based on commodity crops or large acreages (Moore et al., 2001; Flora, 2010).

Among other discernible characteristics of the conventional knowledge system are its underlying assumptions. Primary among them is the assumption that production systems are uniform and that technologies can be transferred from one context to another. Such assumptions stand in contradiction to the variability and diversity of the production niches occupied by sub-Saharan Africa's predominantly female smallholder farmers. These assumptions are now being challenged in the wider realm of agricultural research and development. A growing number of practitioners and analysts now argue that knowledge in agriculture can be "situated" in particular ecological niches and domains of production (Schiebinger, 2010; Altieri, 1998; Röling, 1999; Kloppenburg, 1991).

Because current AET is firmly grounded in the conventional knowledge system, the outputs of research (technology) and institutional arrangements, largely reflected in public extension, have typically favoured crop and livestock systems based on uniform production and external inputs. A gender analysis of this knowledge system clearly reveals the extent to which it has excluded the particular technology needs, realities, and interests of female smallholder farmers.

Women smallholders play important roles on the farm as producers and as natural resource managers. They play fundamental roles in assuring household food security through livelihood strategies that combine a myriad of non-farm entrepreneurial activities. These complex gendered relations of production, consumption, and natural resource management are not sufficiently integrated in current agricultural science curricula, partly because of the dominance of conventional agricultural knowledge. Consequently, the outputs of research and development informed by AET have had relatively little relevance to the needs of the region's predominantly female smallholder farmers.

## Gender-responsive knowledge systems

Concerns over the conventional agricultural knowledge system have led to a growing call for an agricultural knowledge base that includes agro-ecological, local, and cultural knowledge—an integrated knowledge system that more realistically mirrors the multi-functionality and diversity of Africa's production system (Jiggins, 2011; Röling, 1999; Pretty, 1999). Such a knowledge system would be one that incorporates ecological concepts in innovation and technology design and embraces a multidisciplinary, whole systems view of agricultural production. Indeed, research has illustrated the complex and adaptive nature of local agro-ecological systems within which farming has co-evolved with people over millennia.

Such an encompassing agricultural knowledge system is gender responsive in a number of significant ways:

- It recognises the plurality of knowledge and the diversity of actors, including their differently situated locations in production. This nuanced notion of knowledge is critical to understanding what types of technologies are used, how they are used, by whom, and under what conditions.
- It acknowledges the dynamically inter-connected nature of producers' social and productive activities and their real consequences for farmers. Female smallholder farmers routinely juggle household social responsibilities with agricultural production tasks, often resulting in severe labour and time constraints—insights usually not brought to bear in conventional research and development.
- It embraces a broadened view of knowledge beyond the farm gate to include knowledge niches and actors at different points along the value chain—including processing, storage and transformation. It does not assume these arenas are homogeneous and undifferentiated (Davies et al., 2008).

In addition to the ecological aspects, a gender-responsive knowledge system for AET must take into account the socially and politically differentiated nature of knowledge generation and transmission, as shown in Figure 16.1.

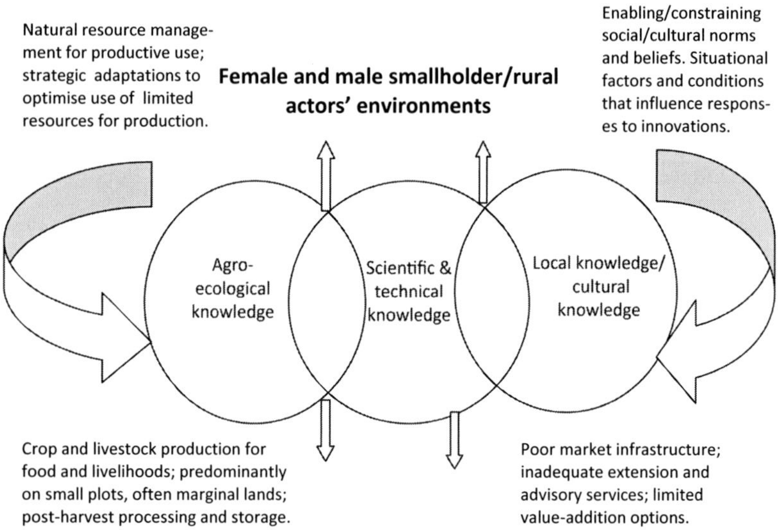


Figure 16.1. Conceptual framework of a gender-responsive knowledge system for AET in developing countries

A multi-functional orientation towards agriculture requires a pluralistic and more integrated knowledge system for AET. Beyond the classroom and laboratory, in farmers' fields, the narrow articulation of agricultural production and assumptions about who uses knowledge and technology have often led to extension training programmes targeting men for building skills in activities that, in reality, are mostly done by women (Jiggins, 2011; World Bank, 2012). A multi-functional orientation is attuned to the complex gendered relations of agricultural production and is more likely to inform technology innovations relevant to the interests and needs of smallholder farmers. Grounding AET in a gender-responsive knowledge system can help build the capacity to respond to the challenges of increasing resource scarcity and the structural challenges in smallholder agriculture.

Analysts who examine the risk implications of agricultural technologies have argued that the exclusion of the gender dimension from agricultural innovation has often led to a general failure to identify unintended social, political, and ecological consequences of different technologies in smallholder production. In Africa, just as in Asia and Latin America, substantial empirical evidence has confirmed that it is women and the poor who disproportionately bear the brunt of the unintended, often negative social consequences of modern agricultural production.

Concerns over social risks in agriculture parallel calls for a broader approach to agriculture, with a rural focus. Rural non-farming interactions, environmental and natural resource management concerns, and more recently, climate change vulnerabilities are among the elements present in this broader view. Similarly, critical health issues such as HIV and AIDS require an expanded and more integrated approach to knowledge. For continued relevance in Africa, AET needs to shift beyond a linear knowledge system that recognises only a narrow set of actors as legitimate knowledge producers to a broadened view embracing the possibility of co-producing knowledge and innovations with diverse actors situated at the rural-agricultural nexus.

The implications involved in achieving quality and effectiveness for differentiated clientele at the rural-agricultural nexus warrant a review of the current status and trends in female AET. Furthermore, such a review is necessary in order to identify and highlight where and how gender imbalances exist. Such insights are critically important to the design of strategic institutional interventions, whether at primary, secondary, or higher education levels, as well as in non-formal education. In the next section, the status and gender trends of the early twenty-first century in AET are examined.

## Gender in AET in Africa: Status and Trends

Studies have shown a high correlation between lack of schooling, food insecurity, and poverty. These relationships are clearly evident in the rural and agricultural areas that make up nearly 60 percent of sub-Saharan Africa's productive base (World Bank, 2012; Amenyah and Puplampu, 2013). Severe gender inequities in access to education and skills-building exacerbate lack of food security in many developing regions.

Between 1999 and 2007, greater gender parity has been achieved in school enrolment (Table 16.1), yet girls in sub-Saharan Africa are significantly disadvantaged relative to boys. In 2007, girls comprised 57 percent of all children outside of the schooling system (UNESCO, 2009). Analysis of trends in the region shows that girls are more likely than boys to have never attended school. Those who do go to school tend to enter late and also constitute the majority of early drop-outs.

In rural areas in particular, the multiple interlocutors of gender, poverty, and location combine to create a pernicious barrier to school entry and survival for girls. In a 2010 study involving eleven countries, of which seven were in sub-Saharan Africa, girls were found to have a 20 percent lower chance of entering school than boys (UNESCO, 2010). Thus, notwithstanding the positive trends reflected in girls' primary enrolments, the challenges remain formidable.

| | Total enrolment | | Net enrolment ratios | | Gender parity in primary | |
|---|---|---|---|---|---|---|
| | School year ending | | School year ending | | School year ending [a] | |
| | 1999 | 2007 | 1999 | 2007 | 1999 | 2007 |
| | Millions | | (%) | | F/M | |
| World | 646 | 694 | 82 | 87 | 0.92 | 0.96 |
| Developing countries | 559 | 615 | 80 | 86 | 0.91 | 0.95 |
| Countries in transition | 16 | 13 | 88 | 91 | 0.99 | 0.99 |
| Sub-Saharan Africa | 82 | 124 | 56 | 73 | 0.85 | 0.90 |
| Arab States | 35 | 41 | 78 | 84 | 0.87 | 0.90 |
| Central Asia | 7 | 6 | 88 | 92 | 0.99 | 0.98 |
| South and West Asia | 155 | 192 | 74 | 86 | 0.84 | 0.95 |

[a] Gender parity in primary education is measured by the gender parity index of gross enrolment ratios.

**Table 16.1. Primary enrolment and gender parity in selected regions Source: UNESCO 2010. Education for All Global Monitoring Report. Annex Statistical Table 5.**

While parity or numerical balance must be acknowledged as an important factor in gender equity, changes need to go beyond simple parity to ensure real transformation in opportunities for girls, especially in rural areas. By all empirical accounts, rural girls are more vulnerable than others to existing cultural norms and values. These normative elements preclude their enjoyment of equal opportunities, treatment, and outcomes in education and training. Girls are often held back for early marriage or for their labour potential. Those that beat these formidable odds and step into a classroom face subtle discrimination and unfair treatment that invariably lead to higher dropout rates (Bunyi, 2003). The consequences of this disparity for society and for sustainable agricultural development in a region whose productive activities are dominated by women include the following:

- implications for innovation and technology uptake when girls and women are denied the capacity to build the skills needed to improve rural livelihoods through lack of access to education, including numeracy and literacy: illiteracy is among the most significant barriers to women's access to extension. Lessons learned from countries in East Asia indicate that basic literacy and numeracy developed at the primary school level are key to the success of non-formal education programmes in rural and agricultural communities (UNESCO, 2010; 2009).

- a weak secondary school pool due to low enrolments and completion rates at primary school levels. This perpetuates the problem of low numbers of female entrants into secondary education, where the need to orient a proportionately higher number of girls into science and technology studies is of the greatest urgency. Analysis between 1999 and 2007 (Table 16.2) shows that instead of the continent making marked progress in redressing gender-based inequalities in enrolments at the secondary level, it has instead moved in the opposite direction, slipping from a gender parity index (GPI) of 0.82 in 1999 to 0.79 in 2007 (UNESCO, 2010).

Similarly, the gender disparity in secondary education has important implications for higher agricultural education to the extent that there needs to be a ready pool of adequately prepared females to enrol in the agricultural sciences. Secondary enrolment averages conceal significant disparities in the subject matter studied by girls as compared to boys. Studies show that fewer girls than boys study science, engineering, and mathematics at the higher secondary school level (Steady, 2004; Mangheni et al., 2010). This is due in part to cultural tendencies that discourage girls from studying science or pursuing scientific careers. Other reasons include discrimination and gender stereotyping of science and technology (Bunyi, 2003; Steady, 2004). An unfortunate consequence of this is the clustering of girls in non-science subject areas or girls being assigned to arts and humanities tracks, with very few young women following the science stream. Thus most are effectively blocked from the study of agriculture in colleges or universities.

Because the foundations of higher education in the sciences are laid at the secondary school level, it is imperative that girls receive the same level of encouragement as boys to engage in the study of science at this level. Such measures are essential to ensure that girls have an equal opportunity in higher education to engage in the study of the applied sciences, including agricultural science.

There are few gender disaggregated studies on higher education enrolments and participation in African institutions. The few that exist consistently show wide disparities in female participation in the agricultural sciences and allied disciplines when compared to male participation. A 2009 UNESCO study showed that only four out of every ten students in higher education institutions in sub-Saharan Africa were women. With a gross enrolment rate for men of 6.8 percent (which was 1.5 times higher than for women at 4.5 percent), women were clearly under-represented and appeared to face more challenges in gaining opportunities for higher education than their male counterparts.

| | **Gross enrolment ratios (GER) in secondary education [a]—%** | | | | **Gross enrolment ratios (GER) in higher education [a]—%** | | | |
|---|---|---|---|---|---|---|---|---|
| | **School year ending in** | | | | **School year ending in** | | | |
| | **1999** | | **2007** | | **1999** | | **2007** | |
| | **Total** | **GPI** | **Total** | **GPI** | **Total** | **GPI F/M** | **Total** | **GPI F/M** |
| World | 80 | 0.92 | 66 | 0.95 | 18 | 0.96 | 26 | 1.08 |
| Developing countries | 52 | 0.89 | 61 | 0.94 | 11 | 0.78 | 18 | 0.96 |
| Countries in transition | 91 | 1.01 | 90 | 0.98 | 39 | 1.21 | 58 | 1.29 |
| Sub-Saharan Africa | 24 | 0.82 | 34 | 0.79 | 4 | 0.67 | 6 | 0.66 |
| Arab States | 60 | 0.89 | 65 | 0.92 | 19 | 0.74 | 22 | 1.05 |
| Central Asia | 85 | 0.99 | 95 | 0.98 | 18 | 0.93 | 24 | 1.10 |
| South and West Asia | 45 | 0.75 | 52 | 0.85 | 7 | 0.64 | 11 | 0.77 |

[a] Note: Gender parity in secondary and higher education is measured by the gender parity index of gross enrolment ratios.

**Table 16.2. Secondary enrolment and gender parity in selected regions. Source: UNESCO 2010.**

Furthermore, evidence shows that females are less represented in the basic and agricultural sciences and are largely clustered in the so-called "feminine" fields—social sciences, humanities, and human services. The overall gender distribution in enrolments and graduation ratios mirrors this trend and shows even lower ratios for women in the agricultural sciences. Among the principal reasons for this are the consistently low pool of female students at the secondary school level who are enrolled in science subjects and the widely held cultural perception that agriculture is not a suitable area of study for girls.

A World Bank study found wide disparities in female and male enrolments across a number of AET universities and training colleges in Africa, throwing into sharp relief the severity of the gender equity challenge in the region (World Bank, 2007). In Ethiopia, women comprised only 11 percent of enrolled and 9 percent of graduate students from its twenty-five agricultural technical training centres in 2005. More recent statistics show a turnaround in female higher education enrolments in the fields of science and applied sciences, although overall, changes in the opportunities available to girls occur very slowly (Beintema and Di Marcantonio, 2009; Mangheni et al., 2010).

Higher education graduates play an important role in innovation and sustainable development at different levels of society. They constitute the stock of human capital for agricultural research, provision of extension, and other services that drive agricultural and rural development. Yet at the professional level, female representation in the stock of human capital reflects glaring gender imbalances in almost all agricultural disciplines and careers. By way of illustration (see Figure 16.2), IFPRI's Agricultural Science and Technology Indicators (ASTI) found female professionals in sub-Saharan Africa predominantly represented in the food and nutritional sciences (44 percent) compared to their representation in soil sciences (13 percent), crop sciences (18 percent), and extension (23 percent).

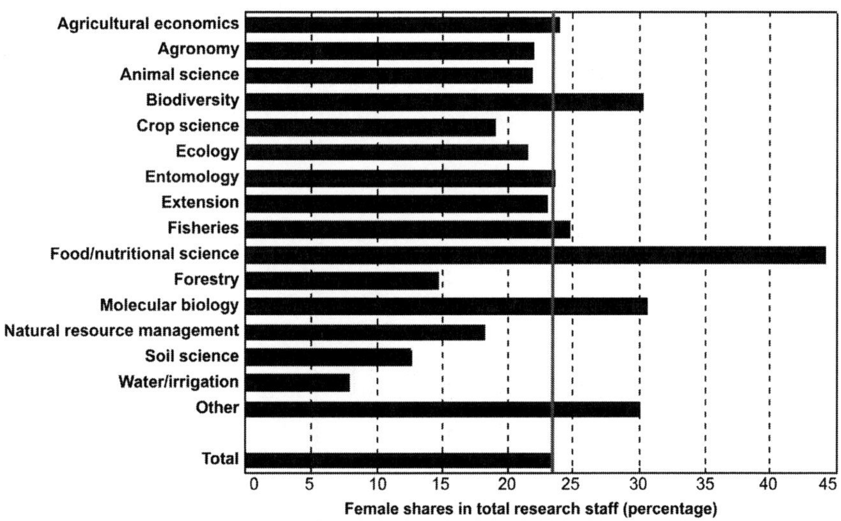


[a] Vertical bar indicates the average share of women participating across all fields of agricultural science (24 percent).

Figure 16.2. Distribution of female shares of professional staff by discipline mix, 2007–2008 in sub-Saharan Africa. Source: Beintema and Di Marcantonio, 2009.

The shortage of female professionals results in a serious dearth in diversity of ideas, nuanced analysis, and modes of interpretation that should enjoy more relevance in the region's complex, female-dominated agriculture. Female representation in national education and research systems in sampled West Africa countries in the 2007–2008 survey ranged from 9 percent in Togo and 12 percent in Burkina Faso to a significantly higher 26 percent in Nigeria. In the southern African region, these figures were similarly low, with the exceptions being Botswana, Mozambique, and South Africa, where more than 30 percent of researchers are women.

The low representation of women in agricultural research and development implies that the region is not taking advantage of the full range of human capital resources that are available and that it so critically needs. Moreover, relevance in agricultural innovation (in a region where deep-rooted cultural factors challenge conventional modes of interaction with women farmers), requires the participation of women researchers and extension agents to ensure that technologies and knowledge generated resonate with the needs and interests of both male and female farmers.

The low representation of women in the agricultural sciences is also compounded by slow career mobility. In a fifteen-country study conducted by IFPRI-ASTI, researchers found gender representation to be unbalanced, with women scientists clustered at the lower end of their professions (Beintema and Di Marcantonio, 2009). The study also found that female numbers diminished at more advanced professional levels (Figure 16.3). Significantly, the male and female enrolment and graduation ratios are shown to be widely disparate at the higher education level, with the proportion of female professionals steeply declining at later stages in terms of movement into more senior positions. Few women are found in the top level positions in agricultural institutions, whether in teaching staff, government positions, non-governmental organisations (NGOs), or the private sector. This means that in most cases there will be a lack of relevant alternative insights for agricultural policy regarding the specific needs of women producers and rural entrepreneurs.

In the experience of women professionals, it is also evident that access to higher education does not necessarily guarantee them equal opportunity to professional positions and professional agricultural careers (Rathgeber, 2002). Evidence seems to suggest that improving opportunities for professional women may be even more challenging than improving women's access to higher education. The paucity of women's representation in agricultural research and development hinders institutional capacities to work with the large numbers of female farmers and entrepreneurs who predominate in the rural sector. Transforming the gender dynamic at all levels in AET is a necessary condition for the revitalisation of Africa's economies and social fabric as encapsulated in the UN Millennium Development Goals.

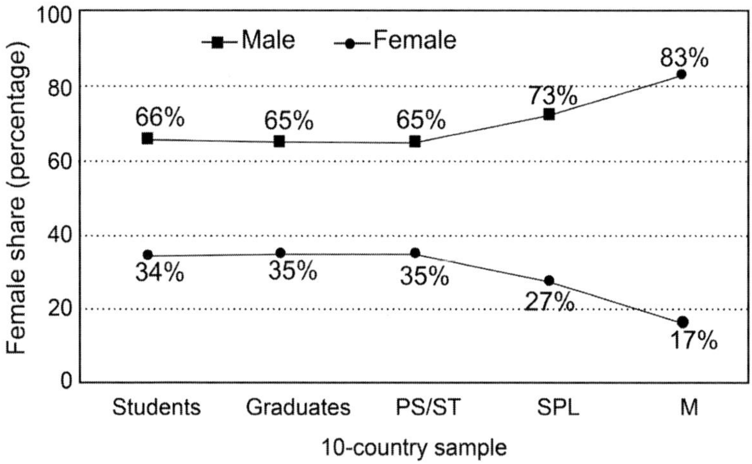


PS/ST: Professional, technical support staff; SPL: Scientists, professors, lecturers
M: Management—directors, deans, heads of department

Figure 16.3. Gender-disaggregated shares of higher education students and professional staff in agricultural science in sub-Saharan Africa. Source: Beintema & Di Marcantonio, 2009.

Investing in human capacity-building is one of the most effective means of reducing poverty and ramping up the kind of economic development likely to have long-lasting, sustainable impacts in society. The social and cultural contexts of agricultural production and natural resource management in Africa require gendered profiles of professionals in research and advisory and technical services if the quality and effectiveness of responses to the region's differentiated clientele are to be guaranteed. Historically, chronic underinvestment in the knowledge and skills of women has meant that women receive less than the proportionate share of investment in the wide range of domains in agriculture, including education, technical training, and extension.

Analysis of gender trends in the education pyramid and in the labour force is a critical entry point. By all accounts, sub-Saharan Africa lags behind other developing regions in terms of female enrolment rates in the agricultural sciences. At the higher education level, these data reflect the on-going challenges the region must address if it is to expand the pool of women who aspire to business and scientific careers in agriculture.

While recent trends reflect gradual upward movement in female enrolments in the agricultural science and technology disciplines, acknowledgement of the need to expand educational opportunities continues to lag. Several factors, ranging from historical to situational, have combined to constrain the effective integration of gender in AET in Africa.

## Factors Affecting Effective Integration of Gender in AET in Africa

Gender, as a fundamental dimension of sustainable agricultural development, appears to have received more attention at the farm level and in natural resource management, where role and responsibility analyses have helped to inform technology innovation and adoption strategies, than at the institutional level. At the level of formal institutions, the participation of women as agricultural scientists and extension agents and in broader areas of service provision has historically received less attention. Consequently, relatively few interventions for improving women's access to formal and non-formal agricultural educational systems can be identified. Several interplaying factors, examined below, limit the support of women and girls in agricultural education.

### Poverty factors and socio-cultural beliefs

At the primary school level, poverty factors and deeply entrenched social and cultural beliefs about the roles of girls combine to limit their access to basic education. In the first place, most rural parents lack the ability to meet the direct and indirect costs of primary schooling, and when forced to take their children out of school, girls are most often first among the casualties.

Moreover, because girls, like women, are primarily defined as social reproducers, they are less likely to be sent to school and often are forced into early marriages. The consequences of these effects spill over into non-formal contexts. Invariably girls who drop out of primary school lack the basic foundation for learning and, in most cases, are unable to take advantage of rural extension and other training programmes.

### Girls' participation in science at secondary levels

Studies have shown that the percentages of female secondary school learners who elect science subjects is substantially lower than those electing to take social science and humanities courses. Biases in the curriculum, syllabi, and textbooks that purvey science subjects as the domain of males have been among the contributing factors (Bunyi, 2003). Consequently young women generally are not encouraged to focus on science at these levels. This in turn is reflected in the low participation of girls in post-secondary technical and vocational training.

Unfortunately, this is where the human capacity needs for agricultural and rural development are the greatest in sub-Saharan Africa, especially in countries showing signs of sustained growth and development. Indeed, without an adequate and growing pool of young women with positive orientations toward the sciences at the post-secondary technical and vocational training levels, growing the skills base for agricultural development in the region will remain a formidable challenge. Education at the primary and secondary levels is the prerequisite for access to higher education and ultimately to professional careers in the agricultural sciences. Rather than being treated in isolation, efforts to improve women's access to and effective performance in higher agricultural education must be seen as part of the larger effort to improve girls' primary and secondary school participation.

## Higher education and professional level factors

Gender disparities in higher agricultural education can be attributed to a range of multi-faceted institutional and societal factors that culminate in a bundle of negative effects. These constraining factors include:

- poor preparation in science studies at the secondary level (the prerequisite for agricultural science courses at the college level);
- poor enabling environments including lack of infrastructure specifically designed to meet the needs of female university students;
- lack of policies to address sexual harassment and other discriminatory practices that could potentially result in hostile campus environments for female students; and
- curricula orientations that negate gender dimensions in subject matter as well as in teaching and learning.

In the professional arena, the challenges are more subtle. The consequences, however, are no less severe in terms of the extent to which they limit women's opportunities to move up the career ladder into senior research management and policy positions in agricultural research and development. A complex interplay of institutional and societal factors can be identified, including exclusionary institutional networks and conservative review and promotion committees and interview panels. The broader societal factors, such as social alienation and work/life balance challenges, combine to limit the agricultural research and development environment for women professionals. These are discussed below.

*Exclusionary institutional networks.* Women researchers face serious challenges in entering mostly male-dominated formal and informal networks. Invariably these networks are the main conduits for collaborative research, lobbying for research grants, and improving scientists' publication records. In the absence of these fundamental indicators of scientific productivity, women scientists have generally fared less well in gaining promotion to senior research leadership and management positions.

*Review and promotion committees—the gate-keeper challenge.* Recruitment and promotion committees often lack balanced gender representation owing to the convention of populating such committees with senior professionals, who are often older and male. This increases women scientists' vulnerability to deep-seated male bias and stereotyping that can work against their chances of winning competitive grants or having objective reviews (Beoku-Betts, 2005).

*Social alienation.* Workplace environments have been found to be more challenging for women scientists due to societal attitudes toward the female professional. Studies have shown that covert gender discrimination is rife in institutions where women represent a significant minority. Such discrimination can take the form of higher than normal scrutiny and inequitable reward patterns for similar levels of effort (Mangheni et al., 2010; Brush et al., 1995). This is compounded by a lack of role models and mentors, particularly in African agricultural institutions. Without a network of female peers and role models, many women find it hard to survive in a workplace characterised by discrimination and minority dynamics (Schiebinger, 2010; Rathbeger, 2002).

*Work-life balance.* Women researchers face significant challenges balancing the demands of career and family, largely because of deeply entrenched normative notions of women's roles in the public and private spheres. The prevailing perception of a woman's role as mother with primary responsibility for family maintenance imposes a double burden that severely constrains women's advancement in research institutions (Amenyah and Puplampu, 2013; Beoku-Betts, 2005; Brush et al., 1995). These challenges trigger the draining of female talent from agricultural research and development. As a consequence, many institutions suffer from a lack of female leadership and the accompanying diverse insights that could feed into developing the agricultural innovations needed to ramp up production and ensure sustainability.

Structural challenges that affect women and girls' educational opportunities are inherent in societal and institutional procedures and processes. They take time to change. Nonetheless, creative interventions found in the educational systems of other developing regions illustrate practical strategies and policy innovations that have transformed educational opportunities for girls and women. Three such creative interventions—the Bangladesh Rural Advancement Committee's (BRAC) school programme in Bangladesh; the University of Development Studies' (UDS) alternative model of higher education in Ghana; and the African Women in Agricultural Research and Development's (AWARD) career enhancement initiative in sub-Saharan Africa—are examined in the following section.

## Creative Gender Interventions in Education Systems: Case Studies

### Case study 1. Bangladesh: The Bangladesh Rural Advancement Committee (BRAC) school programme

The Bangladesh Rural Advancement Committee is an educational innovation that emerged in 1972 out of a strong need to break down deeply embedded structural barriers to the emancipation and empowerment of rural women in a country weighed down by extreme poverty and social inequity (Jiggins, 1994; Chabbott, 2006). Working through the formal education mechanisms of the country to improve women's status, the BRAC programme succeeded in enabling and motivating the system to increase participation of girls and women. BRAC's remarkable success can be attributed largely to three fundamental elements that characterised its programming: 1) its underlying philosophy and principles; 2) the rural village intervention strategy; and 3) the policy mobilisation approach.

BRAC's underlying philosophy revolves around the belief that, given the right opportunity and support, even the poorest and most vulnerable have the capacity to transform their lives through thoughtful decision making. The committee's interventions are grounded in the principle that the knowledge and experience of the poor constitute fundamental inputs in the process of identifying, shaping, and managing opportunities for self-development (Jiggins, 1994; Chabbott, 2006). Guided by these core principles and underlying philosophy, the Committee's rural education approach was designed as an incremental, reflective, and experiential learning process complementing the non-formal and formal domains of learning for Bangladeshi girls and women.

*BRACs non-formal primary education programme.* BRAC's non-formal education programme offers a second chance to dropouts at the primary school level (ages eight to ten years), of which 70 to 80 percent are girls. Designed as a strategic bridge for entry into formal schooling for children who otherwise would not have had that opportunity, the programme offers a three-year primary curriculum based on continuous assessment. With sufficient flexibility built into its structure and timing, the programme has enabled students to adapt to family demands on their time. The curriculum also largely focuses on the identified needs and interests of rural children (Jiggins, 1994; Chabbott, 2006).

By timing the three-year cycle for completion to coincide with the government school year, learners enrolled in the Bangladesh Rural Advancement Committee programme can transition immediately into formal schooling upon completion of the programme. Subsequent evaluation showed a higher percentage of girls enrolled in the programme completing their studies and moving to the next level than those who only had the formal school experience. Incrementally establishing subsequent levels both enabled the continued advancement of students from previous levels and provided a further opportunity for older children who were not able to enter previously (or who had dropped out) to continue their learning while easing their transition into formal education at a later stage.

The BRAC model uses non-formal education as the fundamental building block for formal schooling. It places emphasis on the development of numeracy and literacy skills as well as basic science concepts relevant to the rural context and experience. By placing value on non-conventional teaching and learning methods that stimulate creativity and self-confidence, the model increased girls' prospects for success in the conventional education system. While the ultimate goal for the BRAC non-formal education programme is to facilitate the entrance of the largest possible number of girls (and boys) into the formal schooling system, the programme did not mirror the approach of the formal system. Rather, it incrementally and systematically instituted carefully designed innovations at the intervention and policy levels, culminating in a set of mutually reinforcing positive educational measures, namely:

- offering an innovative three-year non-formal education curriculum that bridges to the formal primary education system;
- setting class schedules that adapt to family demands for children's labour (2.5 hours per week for the first two years; three hours a day for six days a week in the third year, for an average of 270 days per year, with flexibility to respond to local community needs and circumstances;
- inviting community participation in school planning, staff recruitment, and management;
- developing of a cadre of para-professional teachers living within the rural community;
- setting structured (not ad hoc) linkages between community-based women's training programmes and Bangladesh Rural Advancement Committee schools—joint activities to amplify the individual effects and coalesce

into larger and more systematic, socially beneficial outcomes in the community.

The impact of the programme on the formal primary and vocational education system in Bangladesh offers compelling lessons and holds significant implications for other countries that have similar poverty challenges and entrenched cultural and social barriers to harnessing the full range of girls' potential for self-development through education. Of all the world's developing regions, sub-Saharan Africa currently holds the highest percentage of out-of-school girls (UNESCO, 2004). Bold and decisive actions are needed to salvage this significant pool of potential human capacity. The majority of out-of-school girls reside in rural areas and constitute the pool of rural agricultural innovators of this century; tested and proven educational models such as BRAC can offer creative insights and strategies for adaptive innovation among this group.

The twin challenges of a highly feminised agricultural production sector and a rapid out-migration of rural male youth in predominantly agrarian SSA countries add urgency to the need for innovations that place female youth and women at the centre of rural education. The BRAC model proffers insights into strategies for establishing a foundation for skills development through non-formal learning and providing a bridge to formal education. National education systems in sub-Saharan Africa stand to gain instructive insights through prudent adaptation of such models to fit particular contexts.

## Case study 2. Ghana: The University of Development Studies

The University of Development Studies (UDS) is a multi-campus research and training institution established in northern Ghana in 1992 to serve the needs and interests of the rural communities where it is located (Kaburise, 2003). The University's compelling pro-poor vision statement—"to be the home of world class pro-poor scholarship"—explicitly drives its teaching, research, and outreach (Kaburise, 2003). Located in one of the poorest regions of Ghana, a region that suffers from extreme environmental and natural resource degradation, the University of Development Studies' vision reflects an unequivocal commitment to service relevant to the rural world in which it is located.

As in other developing African countries, women in rural Ghana constitute a majority of the poor and uneducated while contributing disproportionately to agricultural production as largely unpaid farm labourers. However, these twin elements are more intricately linked and exacerbated in rural Ghana by deep cultural and social beliefs about the place of women and girls in society. Taken together, these beliefs pose formidable barriers to women's self-development and empowerment.

The UDS's programs and interventions are modelled to respond to the particular technology and knowledge needs of northern Ghana's largely feminised rural labour force. Yet at the same time, the university is creatively modelled to address critical human capacity needs in a way that would attend to gender equity concerns at all levels—from university to community. The model is shaped around three clearly defined strategies: campus location, gender-responsive admission policies, and a practical field programme creatively embedded in the curriculum.

*Dispersed campus locations.* The University of Development Studies comprises a geographically dispersed set of college campuses across the rural areas of northern Ghana, quite the opposite of the post-colonial convention of establishing one major campus in the capital or other major metropolitan areas. Embedding campuses in the rural community serves to:

- redress the widely held perception of higher education institutions as elitist and far removed from the context of rural poverty; and
- enhance the dynamic interaction between community constituents and university actors who, over time, can elicit greater trust, empathy, and community buy-in of the UDS's innovative model from traditionally-minded leaders and opinion shapers.

*Holistic policy on gender mainstreaming.* The University's avowed commitment to addressing gender equity in student enrolments is reflected in its adoption of a gender-responsive policy that helped to dramatically increase female enrolment from 12 percent in 1997 to 29 percent in 2003 (Kaburise, 2003). However, the university's gender policy goes beyond simple expansion of female enrolment. It encompasses a broader commitment to developing the requisite capacities that can ramp up quality and relevance of technology innovations to largely poor and female producers as well as other rural actors the university was set up to serve. The policy is designed to ensure that applicants who meet the basic criteria for admission are ranked within clusters of schools from similar geographic locations rather than within a global pool that includes applicants from more endowed schools in different parts of the country. This policy was very successful in expanding admissions to northern-based applicants from the less endowed, predominantly poorer communities, and in particular, substantially increased the number of female applicants.

*Third-trimester field programme (TTFP).* The third-trimester field programme provides all students a sustained opportunity to live and work within rural communities throughout their four-year degree programs. Working in groups, students identify a research-development "problem" in a rural context and design a way to address the problem. Students work consultatively with the community and other local stakeholders, including government and non-governmental agencies. Through the third-trimester field program, student groups can experiment with an integrated research, extension, and training approach that enhances synergies across the principal components.

Every third trimester of the four-year degree programme, students stay in a rural community for several months in fulfilment of a requirement for the field program. This addresses a key gender quagmire in African technology innovation and adoption—ensuring the fit and relevance of technology to the actual users. Interactions between researchers and development workers often do not go beyond communication with the male heads of households and the older generation of producers, as sanctioned by community socio-cultural norms. The longer stay required in the third-trimester field programme encourages a more sustained pattern of interaction, which helps to lower the gender and generational barriers between community members and student researchers, thus increasing the prospects for a more nuanced gender analysis of community activities, needs, and interests.

This careful blending of a holistic gender policy with a practical field training model—integrating the tripartite goals of research, extension, and education relevant to community problems—has been hailed as a flagship model in AET. It must be noted, however, that gender policies for increasing female enrolments are not unique to the University of Development Studies. Other higher education institutions, including Makerere University in Uganda, Kenyatta University in Kenya, and The University of Dar es Salaam in Tanzania have been at the forefront in instituting gender policies aimed at increasing female enrolments.

The UDS model stands out because of its radical break with the conventional post-colonial university model in Africa. Its unequivocal embrace of a pro-poor philosophy of education and capacity building allowed the undertaking of potentially risky innovations. The cumulative effects (both in terms of output and outcomes) have vindicated the model. Stakeholder buy-in has continued to grow over the years, along with cumulative evidence of the model's effectiveness. An instructive indicator of this effectiveness is the growing demand for University of Development Studies' graduates in the Ghana labour market.

The UDS model clearly has potential to increase female enrolment in rural and agricultural education. Barring any constraints to their movements up the career ladder, these female graduates will strengthen the pool of women professionals possessing the capacity to inform gender-responsive policies in agricultural and rural development. The reality, however, is that deeply entrenched structural barriers hamper women's retention in institutions, thus endangering the gains made through innovative higher education programs. A recent intervention aimed at addressing this challenge—the African Women in Agricultural Research and Development (AWARD) project hosted by the World Agroforestry Centre in Kenya—is the next case study examined in this section.

## Case study 3: African Women in Agricultural Research and Development (AWARD)

Statistics show that in many African countries, efforts to increase female enrolment in the science disciplines, including agriculture, are beginning to bear fruit. However, the picture in the national agricultural research and education professions reflects an intractable pattern with very few women scientists effectively engaging in research and extension or moving on to senior leadership positions over a sustained period. The implicating factors cut across institutional, social, and cultural dimensions, and taken together they constitute a host of negative effects that limit women professionals' movement up the career ladder in agricultural research and development in Africa.

Launched in 2007 by the Gender and Diversity Programme of CGIAR, the AWARD project is a direct response to this challenge, addressing what has been referred to as the "leaky pipeline." The leaky pipeline refers to the high concentration of female professionals on the lower rungs of the career ladder and their significantly lower representation in senior scientific leadership and policy positions (Beintema and Di Marcantonio, 2009). AWARD works in eleven countries of the region, offering two-year fellowships aimed at boosting critical science and leadership skills while connecting women scientists more effectively with wider professional networks through structured mentoring with senior professionals in their fields.

Anchored on three critical cornerstones—mentoring, science, and leadership capacity-building—the fellowship offers a package of short-term intensive courses and participation in conferences and professional associations designed to fast-track women's careers at three critical junctures of professional development: upon completion of the bachelor's, master's, and doctoral degrees. The three cornerstones are described below.

*Mentoring and passing it forward.* In the AWARD design, each fellow is matched with a senior professional who serves as her mentor for one or two years. Through a structured and systematic process, each fellow-scientist pair sets goals and establishes a schedule to meet at least once a month. Mentoring relationships are closely monitored throughout the fellowship period to provide personalised advice, guidance, and coaching to ensure that goals are achieved. To help each fellow put into practice new skills and insights gained through her mentoring relationships, she in turn takes on a young female scientist to mentor, so that the benefits of mentoring are extended to many more women agricultural scientists in the region. Mentoring is a proven and well respected tradition in African cultures whereby female and male elders pass on historically accumulated knowledge, experience, and skills to the younger generation. In modern society, mentoring has proven to be a powerful driver for career development and for retaining women in science in particular (Brush et al., 1995).

*Science capacity building.* The AWARD programme's science cornerstone is intended to strengthen the competencies of women professionals and indirectly enhance research capacities in their institutions. Opportunities provided include membership in professional associations; support for attending science conferences; training in the use of electronic, virtual libraries; training in science, proposal writing, and presentation skills; and the option to compete for advanced science training in renowned research institutions and state-of-the-art laboratories. Fellows are thus able to hone their research skills while broadening horizons in their particular professional areas. Taken together, these opportunities provide a powerful platform for significantly improving women scientists' capacities to produce the technological innovations and advances required for positive and sustainable impacts while at the same time expanding their professional networks and increasing their visibility.

*Building leadership capacities.* This cornerstone is explicitly designed to empower African women in agricultural research and development (R&D) to serve as champions of rural women and to increase their visibility. Through a suite of leadership and management courses at different levels, fellows learn how to leverage team talents, manage conflict, and use influence appropriately. This is critical because prevailing institutional dynamics in agricultural research and development in Africa indicate that scientific excellence alone may not be sufficient to shape a successful research career or result in better research teams or institutions. Thus, through leadership training, fellows learn to manage successful research teams, navigate organisational dynamics that tend to favour men, build alliances and take risks, promote gender-sensitive policies and practices, and influence their institutions on behalf of rural women and poor farmers.

In turn, AWARD Fellows are called upon to put their leadership skills into practice by serving as role models to inspire the next generation of girls and boys to join the fight against hunger and poverty. Each Fellow organises an event in the second year of her fellowship to share her learning with the broader community. This can be in the form of a school talk, career fair, or community workshop. She talks about her work and her personal journey but also stresses the importance of agricultural research and development and the key role played by Africa's women scientists. This sharing is part of the AWARD contract and encourages Fellows to exercise their visibility while helping break down stereotypes about women and agricultural research.

A distinguishing feature of the model is the attention paid to quality in agricultural research and development; that is, African women professionals are given the opportunity for competitive initiative. The competitive thrust of the fellowship—designed to expand options for women's engagement in agricultural innovation and policy leadership—ensures benefits to those who have demonstrated the potential to capitalise on opportunities for self-empowerment and contribution to pro-poor, gender-responsive agricultural policy.

Given the recent nature of the initiative, the overall effects of the AWARD model may take some time to manifest in terms of expected behavioural changes and levels of productivity. Emerging anecdotal evidence from regional and national monitoring meetings, however, allows room for optimism with regard to the programme's potential for transformation. Still, systematic evidence is critical before there can be robust policy advocacy to scale up the model in the region.

## Implications for Gender-Responsive AET Reform and Revitalisation

The factors that negate female opportunities as compared to opportunities afforded their male counterparts for agricultural education and training as discussed in this chapter are complex and wide ranging—from socio-economic and cultural to political and institutional. While more recent evidence gives cause for optimism that these pernicious barriers are being eroded, the pace of change remains lamentably slow, and much remains to be done at all levels to sustain the momentum.

Improving gender balance in AET is not simply a matter of expanding enrolments or achieving gender parity. Fashioning a revitalisation strategy in the region's predominantly female food and agricultural production system must entail the conscious, intentional integration of mechanisms aimed at ensuring balanced opportunities, experiences, and options for female students, professionals, smallholder farmers, and rural entrepreneurs. Achieving balance also requires a fundamental transformation of the underlying, mutually reinforcing knowledge systems and normative structures that have historically informed technological development. Strategies should be directed towards both revising curriculum and expanding the number of women participants at all levels in AET. Increased participation of young women at the intermediate training level—as extension agents, laboratory technicians, and community service providers—is critical. The challenge is not limited to expanding supply but extends to ensuring balanced representation across all levels in the AET system. Key proposed strategies and policy actions are offered below.

## Primary and post-primary levels

- Establish non-formal rural education bridging programmes for out-of-school girls in rural areas to prepare them for entrance into technical and vocational training.

## Secondary and post-secondary levels

- Promote and provide special science courses that can serve the twin goals of preparing girls for higher agricultural education while at the same time helping to dispel notions of girls' relative inability to effectively learn science.
- In order to ensure girls' retention in school over a longer period of time, provide cash incentives to help poor families resist the pressure to marry off their daughters.
- Design clearly defined policies governing the treatment of girls and boys during their schooling at the primary and secondary levels to discourage stereotyping of girls as fit for particular subjects, usually outside the domain of the hard sciences.
- Establish agricultural secondary schools or agricultural courses in regular secondary schools that can feed into higher studies in agriculture.
- Review and revise admissions policies that may discriminate against, or discourage, girls from entering higher agricultural schools.
- Implement affirmative action policies to expand enrolment of girls from rural areas, and give value to on-farm experience and agricultural knowledge by considering it as an entrance requirement.
- Offer career counselling at the secondary school level that presents agriculture as a scientific enterprise with a promising professional future.
- Support non-formal extension training and rural technical training programmes at post-primary and secondary levels to provide second-chance opportunities for young females who drop out of school due to early pregnancy or poverty-related factors.

## Higher education levels

- Recognise and respond to female students' infrastructure needs, such as hostels appropriate for women who have child care needs.
- Design flexible course schedules that include evening, weekend, and part-time study opportunities to enable female students to meet family obligations.
- Enact measures to ensure a non-discriminatory work environment for women staff and students, including policies to prevent discrimination or harassment on the basis of gender and the establishment of services to deal with complaints in these areas.
- Mainstream gender in management and recruitment policies in universities, colleges, and other post-secondary training institutions.
- Target and earmark fellowships and scholarships for women, complemented by tracer studies to track supply and demand over time.
- Revise curricula to broaden subject matter areas beyond the traditional agricultural science knowledge to integrate subjects that address women's production and natural resource management niches, including interdisciplinary teaching and learning in agriculture. Introduce curriculum units and courses on the importance of women's roles and contributions to agriculture, and on working with rural women. Provide teaching materials directed specifically to the needs of women farmers.

## Professional levels

- Put in place mechanisms to ensure effectiveness and retention of female professionals.
- Establish structured mentoring programmes for newly recruited and less experienced female professionals.
- Provide professional development courses in particular areas of strategic importance for research and teaching, such as grant writing and management; course development; and facilitation of teaching and learning.
- Support competitive research, teaching, and outreach grants for women professionals.
- Provide travel grants to support women professionals' conference participation and professional networking.

Increasing access to education for women and girls can produce powerful results in nearly every dimension of development, from raising agricultural productivity and improving environmental management to lowering fertility rates through later marriages and informed decision-making. The challenge is to measure whether and to what extent policy reforms and innovations designed to increase retention of girls and encourage female enrolment in science courses actually bring about such transformations. Orienting curricula and pedagogy at the post-secondary technical, vocational, and college levels to reflect gender-responsive dimensions of knowledge is a critical aspect of the challenge. Resistance to change—a "we have always done it this way" mentality—can pose real risks to transformative change in Africa's post-colonial education culture. Moreover, the challenge to produce graduates at these levels with mind sets oriented towards serving poor and predominantly female smallholder farmer communities requires a significant shift in the philosophy inculcated into the education and training system.

While the actions proposed in the foregoing section may not be exhaustive, they present initial opportunities to bring about needed reforms at all levels of AET. In the often prevailing context of limited resources and uncertain democratic governance in African institutions, reforms will no doubt be incremental and evolving rather than radical and immediate.

## Concluding Comments

In reviewing the role of gender in agricultural education and training in sub-Saharan Africa, this chapter has also examined the consequences of its neglect in agricultural development. We have highlighted how quality, relevance of innovations, and a broadened scope for women's participation can transform agricultural and rural development in unique and powerful ways. Yet, as illuminated in the analysis, such transformations do not occur automatically. They require purposeful, committed actions undergirded by thoughtful policies.

Creating the enabling environment for change in educational institutions and rural communities is now more than ever a real possibility. National governments and donors alike are acknowledging the critical importance of agriculture as the fundamental engine for economic and social development. The challenge for sub-Saharan Africa is whether the commitment to good governance and the will to bring about fundamental change is sufficiently strong among its leaders.

## References

African Capacity Building Foundation (ACBF). 2012. Africa Capacity Indicators 2012: Capacity Development for Agricultural Transformation and Food Security. Harare, Zimbabwe.

Altieri, M. 1998. An Agro-Ecological Perspective to Guide Graduate Educational Programs in Agricultural Economics and Rural Development in Latin America of the XXI Century. Ecological Economics 27, 227–236.

Amenyah, D. and K. Puplampu. 2013. Women in Agriculture: An Assessment of the Current State of Affairs in Africa. ACBF Working Paper No. 24.

Beintema, N. and F. Di Marcantonio. 2009. Female Participation in African Agricultural Research and Higher Education: New Insights. http://www.asti.cgiar.org/gendercapacity

Beoku-Betts, J. 2005. A Lot of them Thought I Wouldn't Last there: African Women and Career Advancement in Academic Scientific Careers. Journal of Technology Transfer 30, 397–407.

Brush, E., D. Merrill-Sands, D. Gapasin, and V. Mabesa. 1995. Women Scientists and Managers in Agricultural Research in the Philippines. ISNAR Research Report No. 7. International Service for Agricultural Research, The Hague, Netherlands.

Bunyi, G. 2003. Interventions that Increase Enrolment of Women in African Tertiary Education Institutions. A Case Study Prepared for a Regional Training Conference on Improving Tertiary Education in sub-Saharan Africa: Things that Work! September 23–25, Accra.

Chabbott, C. 2006. Meeting EFA: Bangladesh Rural Advancement Committee Primary Schools (EQUIP2 Case Study). Education Quality Improvement Program 2, Academy for Educational Development (AED), Washington, DC.

Davies, K., J. Ekboir, and D. Spielman. 2008. Strengthening Agricultural Education and Training in sub-Saharan Africa from an Innovation Systems Perspective: A Case Study of Mozambique. Journal of Agricultural Education and Extension 14: 1, 35–51.

Dixon, J., A. Gulliver, and D. Gibbon. 2001. Farming Systems and Poverty: Improving Farmers' Livelihoods in a Changing World. FAO and the World Bank, Rome, Italy.

Doss, C. 2001. Designing Agricultural Technology for African Women Farmers: Lessons from 25 Years of Experience. World Development 29, 2075–2092.

Eicher, C. 2006. The Evolution of Agricultural Education and Training: Global Insights of Relevance for Africa. Staff Paper 2006–26. Department of Agricultural Economics, Michigan State University, East Lansing, MI.

FAO. 2011. The State of Food and Agriculture: Women in Agriculture. Closing the gender gap for development. FAO, UN: Rome, Italy.

—. 2012. The State of Food and Agriculture: Investing in Agriculture for a Better Future. FAO, UN: Rome, Italy.

Flora, C. 2010. Food Security in the Context of Energy and Resource Depletion: Sustainable Agriculture in Developing Countries. Renewable Agriculture and Food Systems. Cambridge University Press: UK.

Huyer, S. and G. Westholm. 2007. Gender Indicators in Science, Engineering and Technology: Information, Science and Technology for Development Series, UNESCO, Paris.

IAASTD. 2007. Report of the International Assessment of Agricultural Knowledge, Science, and Technology for Development (IAASTD). A Synthesis of the Global and sub-Global IAASTD Reports. Island Press: Washington, DC. http://www.agassessment.org/reports/IAASTD/EN/

Jiggins, J. 1994. Changing the Boundaries: Women-Centered Perspectives on Population and the Environment. Island Press: Washington, DC.

—. 2011. Gender in the Food System. Science Review SR 48: Foresight Project on Global Food and Farming Futures. http://www.bis.gov.uk/Foresight.

Kaburise, J. 2003. New Variations on the African Development University: The UDS Experience with Developing an Alternative Approach to Tertiary Education. A Case Study Prepared for a Regional Training Conference on Improving Tertiary Education in sub-Saharan Africa: Things That Work! September 23–25, Accra.

Kiers, E., R. Leakey, A. Izac, J. Heinemann, E. Rosenthal, D. Nathan, and J. Jiggins. 2008. Agriculture at the Crossroads. Science 320, 320–321.

Kloppenburg, J. 1991. Social Theory and the De/Construction of Agricultural Science: Local Knowledge for an Alternative Agriculture. Rural Sociology 56: 4, 23–32.

Mangheni, M., L. Ekirikubinza-Tibatemwa, and L. Forsythe. 2010. Gender Issues in Agricultural Education within African Universities. Gender Background Paper: Ministerial Conference on Higher Education in Agriculture in Africa. November 15–19, Kampala, Uganda.

Moore, K., S. Hamilton, P. Sarr, and S. Thiongame. 2001. Access to Technical Information and Gendered NRM Strategies: Men and Women in Rural Senegal. Agriculture and Human Value 18, 95–105.

Pretty, J. 1999. Can Sustainable Agriculture Feed Africa? New Evidence on Progress, Processes, and Impacts. Environment, Development and Sustainability 1, 253–274.

Rathgeber, E. 2002. Female and Male CGIAR Scientists in Comparative Perspective. Working Paper 37. CGIAR Gender & Diversity Program.

Röling, N. 1999. Sustainability as an Outcome of Human Interaction: Implications for Curricula in Higher Agricultural Education in Industrialized Countries. In Van de Bor, W., P. Holen, A. Wals, and W. Leals Filho, Eds. Integrating Concepts of Sustainability into Education for Agriculture and Rural Development. Peter Lang: Frankfurt.
Schultz, T. 1961. Investment in Human Capital. American Economic Review 45, 57.
Steady, F. 2004. An Investigative Framework for Gender Research in Africa in the New Millennium. In African Gender Scholarship: Concepts, Methodologies and Paradigms. CODESRIA: Dakar.
Schiebinger, L. 2010. Gender, Science and Technology. Background Paper for the Expert Group Meeting, Gender, Science and Technology. United Nations Division for the Advancement of Women. September 28–October 1, Paris, France.
UNESCO. 2004. Gender and Education for All: The Leap to Equality. Education for all Global Monitoring Report, 2003/2004. http:www.unesco.org/en/efareports/20034-gender/
—. 2009. Global Education Digest. 2009. Comparing Education Statistics across the World. www.uis.unesco.org/publication/GED 2009.
—. 2010. Education for all Global Monitoring Report. www.unesco.org/en/efareport.
World Bank. 2007. Cultivating Knowledge and Skills to Grow African Agriculture: A Synthesis of Institutional Regional and International Review. World Bank, Washington D.C.
—. 2009. Gender in Agriculture Sourcebook. World Bank: Washington D.C.
—. 2012. Gender Equality and Development. World Development Report: World Bank, Washington, D.C.

# PART FIVE

# CONCLUSION

# CHAPTER SEVENTEEN

# ON THE ROAD TO IMPACT AND RESILIENCE: TRANSFORMATIVE CHANGE IN AND THROUGH AET IN SUB-SAHARAN AFRICA

ZENDA OFIR, FRANS SWANEPOEL AND ALDO STROEBEL

Africa is changing. It has become a continent of hope. While much of the developing world is mired in financial difficulties and economic stagnation, Africa's per capita income growth has been among the highest in the world. Some therefore argue that Africa's development path should leapfrog over manufacturing and move directly from agriculture to services as its engine of growth. But this would mean that Africa would miss out on the opportunity to capitalise on the one sector that can provide it with a major competitive advantage globally—a sector that is also crucial to the well-being of Africa's people—agriculture. It has therefore been generally accepted that agriculture will remain a fundamental instrument for poverty reduction, economic growth, and environmental sustainability for Africa in the twenty-first century.

Appropriate and adequate education and training, at all levels and in many spheres, are essential for a revitalised agriculture sector. In this context, the 2010 Ministerial Conference on Higher Education in Agriculture in Africa (CHEA) emphasised the need to strengthen agricultural education and training (CHEA, 2010). In order to maximise impact, it is critical that all interventions and strategies for strengthening agricultural education and training (AET) in sub-Saharan Arica be informed by highly credible analyses and syntheses of issues and solutions, based on an understanding of practical experiences across the world and from within Africa.

This final chapter provides a synthesis of principles for systemic and transformative change in AET to advance out-of-the-box thinking, with an emphasis on the models that can help to cultivate leaders and change-makers at all levels of the agriculture sector.

## Tackling Transformative Change in the AET System

There is a striking consistency in analyses of the factors influencing agriculture and the AET system in SSA. This is true among the authors of this book—even though each has a specific area of focus—as well as elsewhere. The many positive and negative influences on and challenges to the AET system, as well as its singular position in the agricultural innovation system (AIS), appear to be well understood. As noted in chapter one, data about the status quo may in certain locations still be insufficient for credible evidence-informed decision-making. Yet reasonable and wide-ranging solutions for AET system improvements have been suggested by a range of stakeholders, including the authors contributing to this book. African leaders in business and science are seeking fresh and bold steps that can move agriculture on the continent forward towards its full potential. Political support and resources have been promised, and important coordinating mechanisms for action have been launched, with the Tertiary Education for Agriculture Mechanism in Africa (TEAM-Africa) the most pertinent among them.

Indeed, in the second decade of the millennium, the stars appear to be aligned for AET in Africa.

Importantly, the need for *transformative* change for success in the AET system is recognised also by other strategic initiatives and prominent voices from within and outside of Africa (World Bank, 2007a; World Bank, 2007b; Juma, 2011; Moock, 2011, WEF, 2013). For example, chapter 3 recalls fourteen key principles espoused by the Framework for African Agricultural Productivity (FAAP) for the "radical transformation" of AET in Africa. Six dimensions of capacity are seen as key foci for intervention. Moock (2011) notes four new "levers for transformation" for the revitalisation of higher education in the AET system. All argue in different ways for transformation that connects the mission and vision of institutions in the system with new and evolving global and African contexts.

Many mechanisms are proposed, with various forms of horizontal linkages across the AIS system; capacity-strengthening initiatives are at the centre of most. Yet it should be recognised that transformation can be achieved through grand plans as well as incremental change, as long as key levers or potential tipping points are identified. Interventions can be structured around such strategic areas.

All the proposed transformation strategies reinforce the importance of the location of AET within the agricultural innovation system. As a result of its strategic position within the larger system, transformative change in AET will in turn ripple across the AIS (Figure 17.1). The results will be felt in all parts of the system. Given the state of the agriculture sector, such changes are urgently needed.

Yet, as inferred in chapters 3, 4, and 5, institutions and vulnerable farming communities can be gripped by inertia, and some may actively resist change. Leadership, capacity, political will, and resources all need to be supportive of renewal. Across a whole system or set of systems, transformation can be elusive. Amidst a plethora of proposed interventions, it will therefore be crucial to focus on what can activate the system to undergo change that is radical and comprehensive enough to be *transformative*.

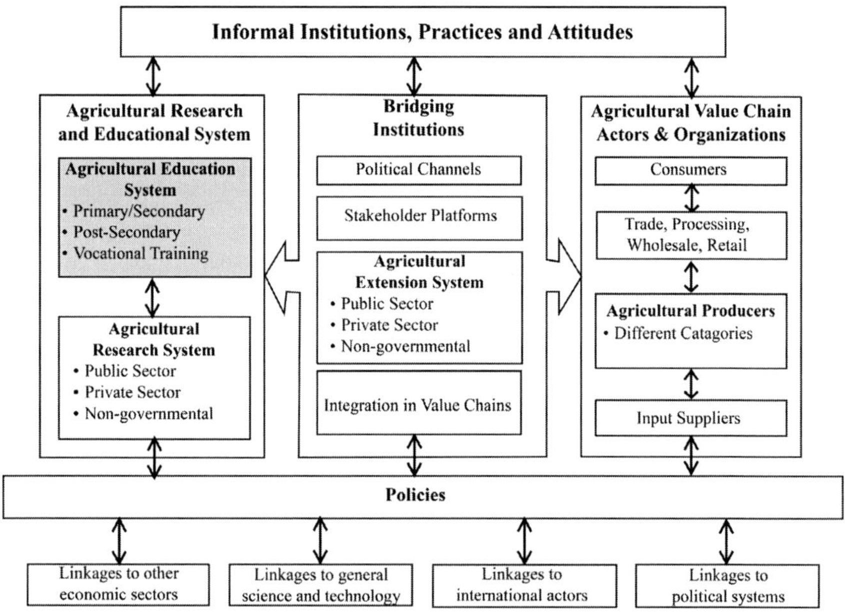


Figure 17.1. The AET system embedded in the Agricultural Innovation System
Source: Birner and Spielman (2007)

## Defining Transformative Change

Firstly, it is necessary to understand what is meant by "transformative change." Transformative change is seen as profound, fundamental, and irreversible. It is based on breakthroughs, on fundamental shifts in individual, group, institutional, or societal values and perspectives. Such shifts involve changes in viewpoint, vision, paradigm, life purpose, organisational direction, or socio-political reforms, which in turn seed fundamental shifts in behaviour or performance. These shifts bring regenerative moments and lead to radical redirections of efforts across a system (Hannum et al., 2007).

Transformative change is usually more or less unexpected, often achieved through key "levers" and sometimes through hard-to-predict tipping points. It is always more profound in consequence than developmental or episodic change. It tends toward the multidisciplinary and holistic, integrating a range of strategies that focus on people's beliefs, values, and attitudes. Strategies may also focus upon individual behaviours as well as the institutional and social systems and structures in which individuals operate.

A system can be transformed over time through a series of incremental changes; transformation may also come about as the result of a shock or strong pressure on the system. The transformation process can be accelerated by understanding what might be "transformative" and by seeking to promote interventions that have a good chance of bringing about fundamental change.

## Activating Transformative Change

It is necessary to consider on the one hand the balance between drivers and enablers (catalysts of change) and existing strengths in the AET system, and on the other hand drivers of vulnerabilities and constraints that act as impediments to change (Figure 17.2). This balance will differ by context—that is, by sub-region, country, institution, or set of institutions.

Significant or transformative change may come if the combined effect of the positive influences is more powerful and effective than the vulnerabilities and constraints in the system. If change is to happen, these two types of forces on and within the system should not be in equilibrium. Just a few strategic interventions over time may overcome the constraining forces. This is why it is important to try to recognise which interventions might be transformative for the whole system.

An understanding of what could shift the balance in critical parts of the AET system will help determine the strategies needed to bring about the desired transformative change. The challenge is to identify those factors and interventions likely to be most pivotal for this purpose and those that might be poised to result in tipping points leading to transformation. The interventions have to be combined and sequenced well for best effect and to prevent disequilibrium. If the process is not properly managed, the whole system might become ineffective or even disintegrate.

There is also a need for "best fit" solutions, i.e. solutions tailor-made for a specific set of circumstances and able to evolve as the context evolves. Thus the actual design and implementation have to be managed by leaders at all levels of the system—leaders who truly understand the context within which the changes are to take place and who are committed to working towards success over time—where necessary, in collaboration with one another.

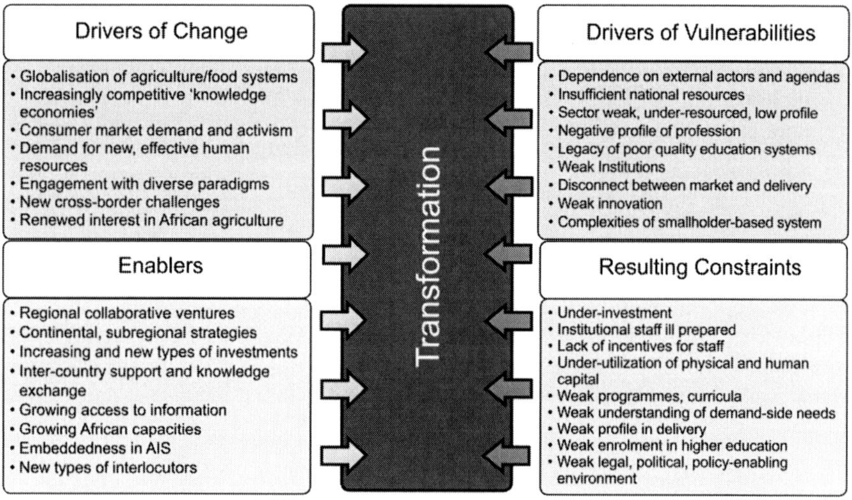


Figure 17.2. Some of the most critical forces and influences that affect the potential for transformation in the AET system

In addition, leaders should be in a position to predict, at least to some extent, the intended and unintended effects—both positive and negative—of planned interventions and ensure that capacities are in place to make fast adjustments as needed. Trajectories towards transformation are hardly predictable, and without experimentation and advanced modelling it is almost impossible to be certain that such strategies will succeed. But informed leaders in each institution or set of institutions, country, and sub-region can at the very least establish enabling conditions to improve the chances of success and develop an AET strategy that emphasises those interventions likely to make the most effective and sustainable changes.

As noted earlier, it will also be crucial to understand which actions could be catalytic for transformation, whether through "big plans" or incremental change. It is necessary to determine which interventions definitely need central or "top-down" design and activation through national or regional policies, strategies, and funding, and which may best evolve "bottom-up" as stakeholders' interests, capacities, exposure, and linkages with others grow. In this massive task, choices have to be carefully made.

Experiences in China have shown, for example, that significant progress towards transformation in agriculture can be achieved through a series of purposeful yet incremental steps, taken with consistency of purpose and investment in traditional systems (Xiaoyun et al., u.d.). The Chinese experience showed that putting smallholder farmers' needs at the centre of policies and plans is crucial, and that smallholder and large-scale agriculture can exist side by side to mutual benefit. The Chinese developed context-based strategies and facilitated transformation towards a market system by ensuring that critical inputs and services could be accessed in an economical manner. They also enabled gradual market reform to protect smallholders from a "market trap."

## Towards Transformative Change: a Focus on "Agency"

The current drivers and enablers or catalysts for change are primarily from outside the AET system, while the serious constraints tend to be from within. This warrants closer investigation.

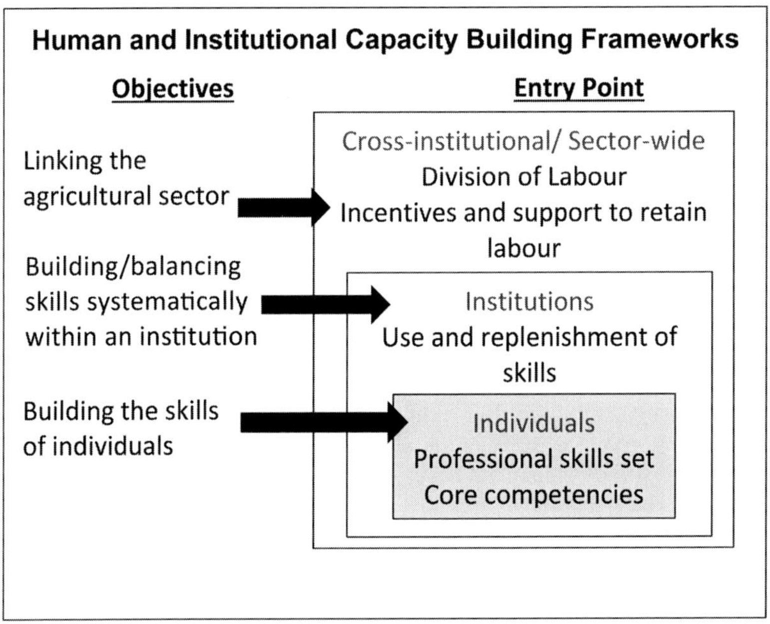


Figure 17.3. Human and institutional capacity framework. Source: Moock, 2011

The human and institutional capacity-building framework noted by Moock (2011) in Figure 17.3 is relevant both for the leaders and staff currently in the system and for new leaders the system will produce. The framework highlights the need for coherent policies and strategies that address individual, institutional, system, and sector needs. What is often implied but omitted from such frameworks is the need to focus on: 1) the *agency* of those within the system who are responsible for change; and 2) the *organisational motivation, capacities, and performance* implicit in the system.

There are many frameworks that can focus attention on key aspects for consideration when working towards transformative change. Figure 17.4 provides one simple but useful example, as it places "agency" at the centre while recognising the importance of the larger economic, environmental, and social contexts within which agency is exercised.

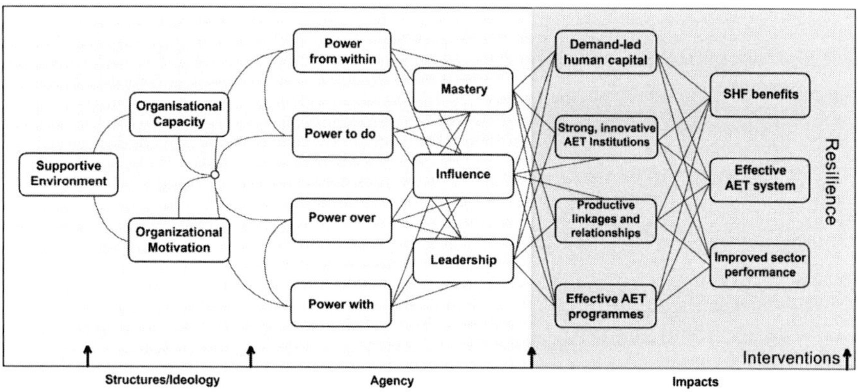


Figure 17.4. Framework highlighting key aspects for consideration in efforts to seek transformative change. Source: Adapted from Grabe, 2012

"Agency" encompasses the ability and willingness of people to make strategic choices towards desired results. If key actors in efforts to facilitate change do not have appropriate and adequate agency and incentives, the inertia in the AET system may not be sufficiently overcome for the desired transformative changes to take place.

For example, as discussed by authors in this book and summarised later in this chapter, changing higher education institutions in line with what is needed is a very challenging task. Curricula and programmes need to be restructured to unleash talent and innovative capacities. They need to reflect systems thinking, multi- and trans-disciplinary approaches, and African expertise. New attitudes, mechanisms, and relationships have to be developed for better knowledge generation and acquisition. Gender responsiveness needs to be incorporated across the system. Change agents, leaders, employees, farmers, and entrepreneurs need to be nurtured with values, knowledge, and skills for a variety of purposes and integrated into the larger AET system/agricultural higher education sector. All of this needs to take place with a focus on increasingly important but often neglected areas such as agribusiness, climate change, developing and linking into value chains, addressing the agriculture-food-nutrition nexus, and policy-making at national, regional, and global levels.

The AET system is thus presented with an extraordinary opportunity, but enormous commitment and skill will be required from the leadership, staff, and others responsible for the design and implementation of relevant policies and interventions. These actors should have not only the necessary *incentives*, but also the *agency* to lead and enable change. This means being empowered to effect change, based on four types of power that define "agency" in this context (see Table 17.1).

| | |
|---|---|
| **POWER 'FROM WITHIN'**<br>Cultivating the desire to **change**. | This type of power relates to self-knowledge, confidence, vision, and a desire to influence, i.e. *a growing motivation* to change and to effect or lead change. |
| **POWER 'TO DO'**<br>Cultivating **choice** by ensuring connections and capacities. | This type of power relates to access to financial and intellectual resources and gaining skills and knowledge, i.e. *growing capacity* to enable or lead change. |
| **POWER 'OVER'**<br>Cultivating more **control** over opportunities and resources, through influence or authority. | This type of power relates to changes in underlying resource and power constraints, and individuals' ability to overcome these, *i.e.* through *growing opportunity* to enable or lead change. |
| **POWER 'WITH'**<br>Cultivating **community** collective action for scientific, institutional or system benefit. | This type of power relates to collaboration, solidarity and joint action with others to challenge convention, *i.e. growing effort* to take collective action |

**Table 17.1. Expanding agency: the types of power needed by those responsible for transformation in the AET system. Source: Adapted from the AWARD Empowerment Framework for Leaders in Agricultural Research and Development (AWARD, 2012; Rowlands, 1997)**

The issue of agency is seldom explicitly mentioned in AET strategies, other than references to the need for capacity strengthening and mechanisms through which this can be achieved, such as networks and other forms of collaboration. There is usually a stronger focus on structure, resources, and interventions at institutional or system-wide level. While these aspects are very important, it is crucial to attend not only to individuals' motivation for change and their exposure to opportunities to do so, but also to disincentives and demotivating forces, policies, and structures in institutions. This is confirmed by Zondi (2013), who argues that individual power to expand agency is at the centre of initiating desired transformation.

## Structures and Ideologies

There are a number of contextual forces that drive institutional (organisational) performance: forces in the external environment, internal capacities—including those related to linkages with others—and internal motivation (Figure 17.5). The agency of individuals is particularly influenced by the structures and ideologies embedded in the institutions in which they work, and in the society around them. Any intervention aimed at helping to transform the AET system therefore needs to be designed and implemented with cognisance of these important factors shaping how people think and work.

"Structure" refers to structural power embedded in the way institutions work, while "ideologies" refers to the rules, norms, and values that determine how people act within the structure. Chapter 15 provides an example with a description of the subtle challenges faced by women in agricultural research and development organisations. The author speaks of a "complex interplay of institutional and societal factors" that affects women's progress and inclusion in their institutions; these may include exclusionary institutional networks, conservative review and promotion committees, social alienation, and the struggle to maintain a work-life balance. Such structural and ideological impediments might then hamper fresh efforts to remove inequalities, address disparities, and adjust curricula to reflect greater gender responsiveness, which will be essential in a revitalised AET system.

A similar situation is likely to be found in the education of extension agents. As pointed out in chapter 10, the traditional education in many institutions, as well as the norms and values that shape the curricula, may not sufficiently prepare extension agents for the required understanding of, and respect for, the knowledge and learning strategies and methods of smallholder farmers. This is exacerbated by the fact that in Africa the vast majority of smallholder farmers are women, while extension agents are mostly men.

AET strategies tend to ignore such potential barriers at ground level, where implementation has to take place. It is therefore crucial to be cognisant of and address the potential influence of *the institution itself* on transformation efforts. A useful framework developed by the International Development Research Centre and the Inter-American Development Bank (IDRC/IADB, 2002) highlights four dimensions by which an organisation is characterised (Figure 17.5).[1] *Organisational motivation*, which represents the underlying personality of the organisation, "is what drives the members of the organisation to perform." This will therefore play a major role in efforts to bring about change. Organisational motivation is determined by the organisation's evolution and history, its mission, values and vision, and in particular its culture and the incentives it offers to its leadership and staff. This dimension is usually not mentioned explicitly in strategies for change in the AET sector.

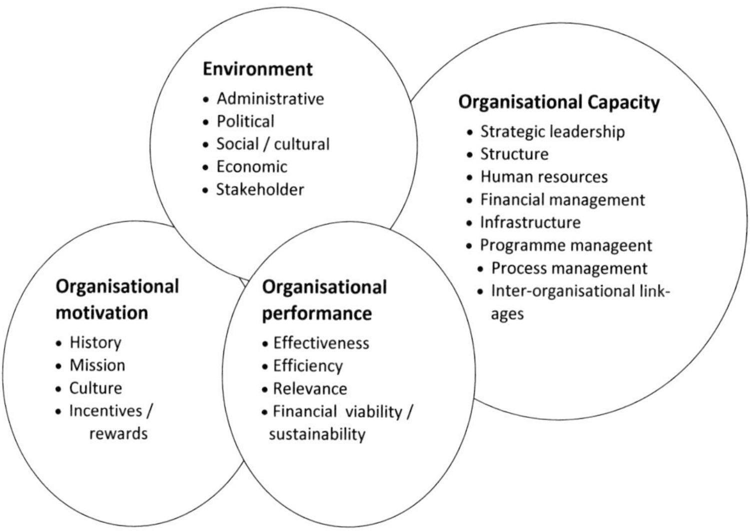


Figure 17.5. Framework for organisational assessment. Source: IDRC/IADB (2002)

Similarly, organisational capacity is critical, but more frequently mentioned in relation to improvements in the AET sector. Capacity refers to all the resources, systems, and processes the organisation develops to support its

[1] In the context of this discussion, organisational effectiveness is less important than the two dimensions mentioned here, although it is an integral part of the model. The external enabling environment is discussed later in this chapter.

work, including the systems and management practices associated with human, financial, and infrastructure resources as well as the ability of the organisation to use these resources to perform.

## Roadmap towards Transformative Change in National AET Systems

Studies cited in this book point out that there are a number of generic challenges that can be found in AET systems across sub-Saharan Africa. There are also many common drivers of vulnerabilities and drivers for change, as well as enabling and constraining factors.

Earlier in this chapter it was noted that the balance between these aspects may differ among countries and sub-regions. There may also be factors that are unique to a specific context. However, there are enough commonalities to start constructing an "ideal" roadmap that can serve as a broad indication of which routes might lead to change (Figure 17.6). The construction of this roadmap has been informed by Eicher (2006); Pal and Beyerlee (2006); Stroebel and Swanepoel (2008); and Stroebel et al. (2011).

The proposed roadmap is not intended as a blueprint. It is general by design, intended to be part of a process of rethinking, reframing, and reshaping structures and ideas to lead towards radical reform of the AET system within the framework established by the AIS. In particular, *it emphasises the interests, concerns, and needs of smallholder farmers.*

The key components of the proposed roadmap are noted below. The contributions in other chapters provide details of issues and in some cases suggestions for change related to each of the roadmap components, drawing from case studies and examples across Africa and beyond. These suggestions can be used in conjunction with other key reports to assist in efforts to develop robust national and sub-regional AET strategies.

## The Supportive Environment

In its 2007 report on AET, the World Bank described an agricultural innovation system as "a blending of institutional capacities, coordination mechanisms, communications networks, and policy incentives that foster innovation-led gains in agricultural productivity" (World Bank, 2007a). More recent work refers to the AIS as a web (chapter 3). As a subsystem of the AIS, AET has similar characteristics. It is an open system, and therefore an enabling external environment is very important. The characteristics and quality of the external environment determine the level and type of support and resources available to the individuals, institutions, collaborations, and networks in the system as well as the ease with which their work can be done.

Attention should therefore be on the economic, political, policy, socio-cultural, environmental, demographic, and technological conditions that affect institutions and their inter-relationships. These conditions will reflect the balance between the opportunities and challenges that exist in the agriculture sector and beyond. This balance is also displayed in the analysis of drivers for vulnerability and change as well as that of the enablers, catalysts, and constraints in the system (Figure 17.2). Given the many issues relating to a "supportive environment" raised in this book (and in preceding studies), it is clear that creating such an environment is of significant concern (Alemneh, 2007). Robust efforts have to be made to achieve this, and lessons have to be learned from the many challenges described in a recent progress assessment regarding implementation of the CAADP (Kimenyi et al., 2013).

Government policies, strategies, regulations, and protocols are of particular importance for transformative change given the proliferation of actors, linkages, and markets in the agricultural innovation system. The policy environment should stimulate innovation (chapters 11 and 15) and in the process support those issues perceived as critical for transformation and success. Most importantly, the real needs of smallholder farmers should be at the top of development policies and strategies—including those in AET-related plans—in tandem with rather than in opposition to or as a minor component of the sector as compared to large-scale farming.

Examples of areas for attention include gender-responsiveness, with particular emphasis on the smallholder farmer (chapter 16), entrepreneurship and intellectual property protection (chapters 6 and 15), the alignment of national incentives and policies to encourage systems-based approaches and multi- and interdisciplinary research (chapter 12), and integration between the higher education sector, industry/community, and government (chapter 15).

The paradigm shift from agriculture to agribusiness and entrepreneurial activity (described in chapters 2, 11 and 15) provides both opportunity and challenge. It requires policies, strategies, regulations, and protocols that will promote and catalyse the shift, while also providing protection against undue risk and exploitation. Participation in giant international agri-food value chain networks, co-existence and mutually beneficial collaboration between large-scale and smallholder agriculture, a bio-based green economy (chapter 2), and a focus on the agriculture-food-nutrition nexus bring new and exciting opportunities, yet are accompanied by severe power asymmetries that greatly enhance risk for the smallholder farmer.

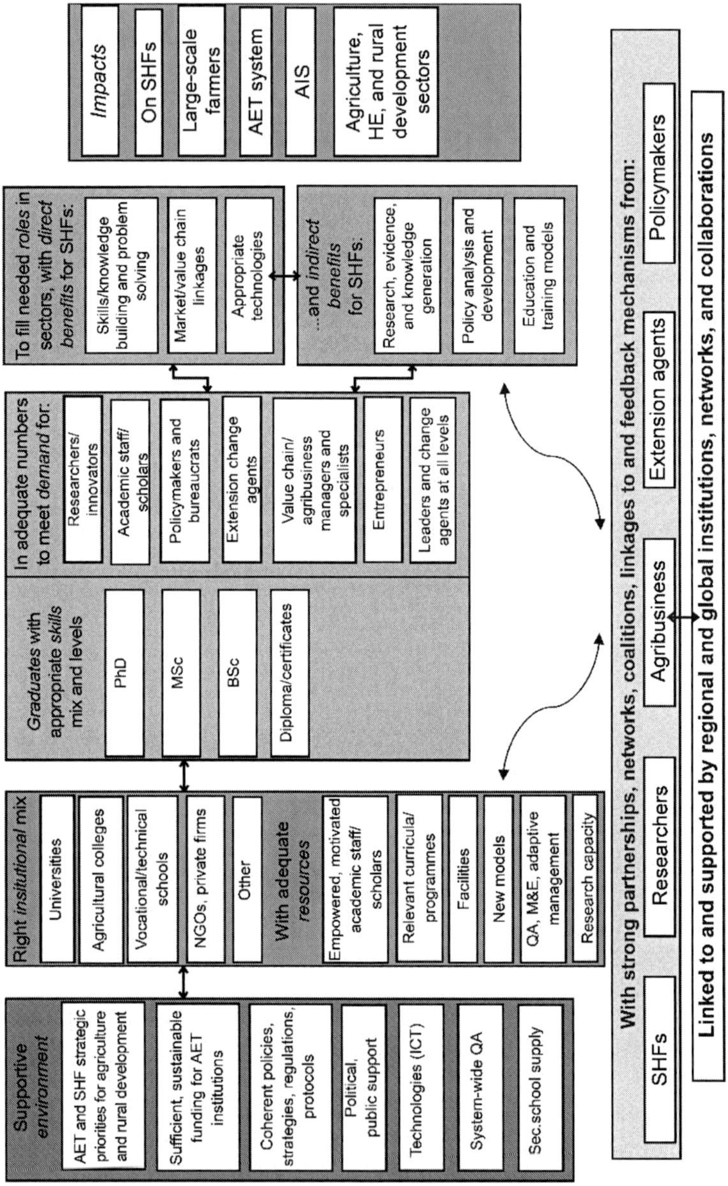


Figure 17.6. An ideal roadmap for the transformation of AET in SSA countries

Governments have to create an environment that promotes positive outcomes while guarding and protecting the agricultural sector—in particular the smallholder farmer—against any action that might hurt national, sector, or AET system interests. Institutions across the AET system are well positioned to help identify and alert governments to such opportunities and risks and should nurture relationships that will help ensure that expert opinions are sought and heard.

*Policy coherence* is a critical issue. AET-related policies have to be aligned with national investment plans and policy regimes in the higher education, agriculture, and industry sectors. *Political support* in democratic countries might be difficult to obtain if *public support* for agriculture is low. Several chapters of this book refer to the need to cultivate a positive public image of the agriculture sector and of the AET system. If stakeholders are convinced of the economic value of AET, they can help convince governments of the need to support it (chapter 14).

The government has to encourage *linkages* between education and training, research and development, and extension (chapters 1 and 13), the so-called *knowledge-triangle*, in order to allow the system to respond to demand. In chapter 14, Zuidema suggests that network analyses be used to determine to what extent the complex set of actors involved in AET delivery and financing is operating as an interlinked system. He notes the disconnect between the Ministries of Agriculture and Education in many countries as being of particular concern. This disconnect diminishes policy coherence and the efficient and effective allocation of resources, demonstrating the importance of structural linkages, i.e. land grant-style institutions (chapters 1 and 5), to encourage the appropriate linkages.

Finally, it is important to explore new investment opportunities offered by the private sector through a programmatic rather than budget-centred approach (chapter 14) and through South-South collaboration. The latter has brought with it *new models of investment,* most notably those established by BRICS—the rising global powers of Brazil, Russia, India, China and South Africa. Brazil, China, and India are all very active on the continent, including in agriculture. Although their influence and models have not been analysed in this book, they can bring fresh ideas, innovative opportunities, and new funding mechanisms to the fore (chapters 4 and 9).

## Relevant Institutions and Adequate Resources

Capacity building in Africa, especially in the higher education sector, has for a long time been a low priority for donors and governments alike. This situation has been changing in recent years; yet it will take exceptional leadership, carefully selected and innovative strategies, and new energy and commitment to ensure that the institutions in the AET system leap ahead.

Challenges facing institutions have been well documented in many chapters of this book. Fragmented resources have compounded their difficulties. The increasing diversity across the AET system in terms of pedagogies, institutions, students, expectations, and missions have to be dealt with and made to work in synergy—a very challenging task. Ministries of Agriculture do not have power over advanced agricultural education programmes. This adds to the disconnect between national goals and university programmes, with many staff members ill-prepared for the evolving challenges in the sector. Agricultural education is also expensive compared to many other areas, making it difficult to maintain research programmes and infrastructure (chapter 14). Importantly, insufficient institutional capacities continue to limit smallholder farmers' access to knowledge and technologies, hindering their efforts to thrive (chapters 4, 5, and 10).

Agricultural development needs to be a *national strategic priority*, and the important role of the AET system should be articulated in national and regional development goals. *Resource allocations* in both the agriculture and higher education sectors, and in other relevant arenas, should reflect these priorities. As several authors point out, there has consistently been severe underinvestment in AET over a number of decades. This has led to poor infrastructure and serious obstacles to cultivating scholarship, as well as insufficient and ineffective connections and learning strategies between AET actors and smallholder farmers (chapter 10). It is therefore important not only to determine whether government investment has increased, but also to ensure that the flow of funds encourage and support strategic priorities.

Funding sources are increasingly mixed, and the financing of extension training can draw from a wide variety of mechanisms outside the AET system. Catalytic and long-term investments can both be used to direct strategic priorities. There is a need for more concerted approaches and joint business plans involving donors, governments, and AET stakeholders. TEAM-Africa has been established to fulfil this function for the tertiary agricultural sector (chapter 1).

AET leaders need to enable institutions to reinvent themselves in response to changing contexts and demands by

- using monitoring and evaluation;
- understanding how to benefit from African and global expcriences;
- working with a systems orientation;
- understanding and dealing with the effects of power differentials, including those firmly embedded in newly envisaged paradigms for agricultural development;

- engaging effectively with global and national policy challenges around the environment and climate change, GMOs, and other technological and social challenges;
- cultivating resilience in the sector and among smallholder farmers; and importantly,
- delivering adaptable graduates (chapters 9 and 11).

This summary reinforces the key argument made in several chapters—that although many mechanisms exist for this purpose, *institutional capacity strengthening* programmes will have to be primarily based on the formal and informal connections between interrelated components within the AET system and the AIS. Such connections expose individuals to new information and knowledge and provide them with opportunities to test their own knowledge—whether through the support provided by the global CGIAR system (chapter 8) or other types of networks and collaborative ventures (chapter 5). There is some evidence, highlighted in chapter 5, that the *knowledge triangle* within the innovation system supported by *land-grant-style institutional structures*, where AET is closely linked to research and extension, are more successful. This is particularly well illustrated in chapter 1, referring to the land grant models in Japan and the success in building the State Agricultural University (SAU) system in India. The Vietnamese agricultural extension system also directly engages smallholder farmers in the entire process with great success.

Within these initiatives, special emphases are needed. Institutional leaders and staff members need to learn from *success factors* in past and current AET development. These can be studied to determine potential application in each specific context. Among the many examples of such efforts in this book, one that stands out is EARTH University, where the confluence of many factors brought about a successful venture (chapter 6). These include staff and students carefully selected in line with the institution's mission; sequenced course design; strategic incentives structures; student-centred experiential learning; strong linkages between the institution and its stakeholders, in particular farmers; engagement in entrepreneurial projects; targeted internships and scholarships; an active alumni network; partnerships that align programmes with national priorities; and funders willing to take risks.

Staff members need to have *agency* (Table 17.1 of this chapter) and *incentive* (chapter 9). This will help them to commit to institutions, participate in transformative change without fear or constraint, and deal with the challenge of delivering graduates who can work with success in the current and evolving AET context.

Furthermore, there should be support offered by an *adaptive management approach* that is able to respond quickly to lessons learnt and changing contexts. Adaptive management is facilitated by effective monitoring and evaluation as part of institutionalised quality assurance, i.e. "balancing quantity with quality" (chapter 11); supporting accreditation and ensuring that information for decision-making and accountability is available in a timely manner; and ensuring that curricula and programmes evolve as sector needs change (chapter 13).

*Networks and collaborative ventures* provide staff opportunities for participation in knowledge-sharing as well as regional staff development programmes in areas where there are otherwise not enough resources to support such opportunities—for example, small nations without the critical mass needed to support AET institutions, specialised technical fields, and emerging cross-disciplinary areas (chapter 14). This is further discussed below.

## Appropriately Trained Graduates

New trends and paradigms likely to influence the sector in coming decades and the resulting *demands for better and new types of graduates* are challenging the whole system (chapter 2). A new generation has to be prepared. They have to be entrepreneurs outside of and across international and local value chains; able to work effectively in systems with and as researchers, extension agents, and farmers; and adaptive enough to evolve with new demands and opportunities. Institutions therefore have a series of issues to deal with as part of the transformation of the AET system.

Ideally, if resources allow, new and innovative models should be considered, such as EARTH (*Escuela de Agricultura de la Region Tropical Humeda*) University (chapter 6). It is particularly enlightened in its unashamed and inspiring focus on *developing leaders for the AET system and for the agriculture sector*. The institution was established in a manner cognisant of the context in which it operates, with a clearly defined mission and set of values that define and inspire programmes, staff, and graduates. Chapter 16 also highlights the unique and successful approach of the African Women in Agricultural Research and Development (AWARD) programme, with a strong focus on cultivating leadership among professional women in agricultural research and development.

AET in sub-Saharan Africa must increase its understanding of African farmers' learning strategies, approaches, and methods. Farmer study groups and learning circles are examples of valuable learning approaches which allow for *farmer-centred learning*. Other levels of education and training across the spectrum—vocational, college diploma, undergraduate, and postgraduate education and training—are all critical. However, it is important to *prioritise postgraduate* education and training to invigorate research and to support national agricultural development (chapters 3, 5, 9, and 10).

Producing such graduates will entail most, if not all, of the following aspects:

- Reorienting graduates towards a multidisciplinary, systems approach (chapters 4, 6, 12);
- Educating them from within African contexts and with African experiences and solutions, but with international experience coupled to incentives to return to Africa (chapters 3, 4, 5, 7, 9, 14);
- Focusing on holism and generalism (chapter 6);
- Embedding gender, sustainability, quality assurance, and other key concepts in the underlying knowledge systems (chapters 6, 13 and 16);
- Ensuring expertise to engage with old and new user communities, agribusiness, and global value chains (chapters 2, 9, 11 and 15);
- Enabling the construction of knowledge and access to information based on multidisciplinary approaches (chapter 6) and on engendered approaches to learning, research, and work (chapter 16);
- Encouraging organisational and social entrepreneurship (chapters, 6, 11, and 15), with a mind-set that favours proactive action, risk-taking, competitiveness, autonomy, and innovation;
- Cultivating the right set of values towards people and towards the complementarity between agriculture and the environment (chapter 6);
- Understanding smallholder farmers and the rationale for technology development and use, and being able to participate in shared impact-oriented progress assessments that promote self-reliance (chapters 4 and 8);
- Developing extension agents who can act as agents of change among farmers and in their relationships with other parts of the system (chapters 6 and 10). These agents should be catalysts—initiating learning and being conduits for knowledge resources (chapters 1, 8, and 10), while respecting and activating "agency" in the farmer (chapter 4).

Institutional structures, processes and approaches need to facilitate and encourage *entrepreneurial and innovative* programmes and strategies (chapter 15). At the same time, hindrances to focusing on entrepreneurship and innovation should be removed, for example by channelling funding for systems approaches through departments as a means of stimulating such focus.

It is also necessary to attend to the *supply pipeline*—those who come through the school system as potential candidates for post-secondary studies in agricultural fields. Rapidly urbanising young people need to have the awareness and incentive to study and work in an arena that is usually seen as less attractive for a prosperous future.

The AET system needs a specific type of leader at this juncture: innovative, inspiring, willing to take calculated risks, and most importantly, committed to working on common interests with the *web of actors and institutions* in an AET system embedded in a larger national system of innovation (chapter 3).

System-wide *quality assurance and learning* through appropriate and useful external and internal monitoring and evaluation, supported by an effective accreditation system, are highlighted in chapters 7, 11, and 13; these elements are regarded as crucial for the regulation of the system. Importantly, quality assurance and learning provide information for strategic and operational decision-making at various levels within the system, and among those to which the various institutions in the system are accountable.

New mechanisms to access and share information and learning, facilitated by *information and communication technologies and social media,* are widely recognised as having significant potential to leapfrog poor infrastructure and enable better scholarship (chapter 9). Advanced ICT facilities can facilitate collaboration, for example by sharing expert scholars among institutions and drawing upon non-university experts from various spheres—government ministries, non-governmental organisations, National Agricultural Research Organisations, the CGIAR (Consultative Group on International Agricultural Research), the private sector, and think tanks—to bring their knowledge into the higher education domain.

Finally, as noted in most chapters, *curricula* need to be modernised to include:

- updated systems, in particular farming systems approaches, with a strong focus on local (social inclusion, environmental sustainability, resilience), regional, and global (trade, climate change) challenges (chapter 12);
- integrating multi- and interdisciplinary foci, thus cultivating both specialists and generalists (chapters 6, 9, 11, and 12);
- synthesising and integrating knowledge in domains such as production, environmental integrity, social benefits, consumer requirements (health and food safety), sustainability science, and others (chapter 12);
- supporting interaction with farmers—providing them with feedback for evaluation, and exposing students to practical application in the field (chapters 4 and 10);
- developing frameworks and models that Africans own, with African examples and case studies based on African agribusiness experiences (chapter 11);
- focusing on smallholder farmers, with all the characteristics expressed in this chapter and elsewhere (chapters 4, 10 and 11);

- emphasising the implications of global and regional policies and value chains as well as the interface between political and technical issues in areas such as trade, foreign direct investment, international protocols, and controversial technologies such as genetically modified organisms and food (chapters 1 and 2);
- co-creating curricula with smallholder farmers to ensure gender-responsiveness at all levels of the AET system and the agriculture sector (chapters 11 and 16);
- maintaining entrepreneurship and innovation as central tenets for working both within and outside of agricultural value chains, including work in the area of food and nutrition. Curricula for entrepreneurship and innovation require a systems orientation, multi- and interdisciplinary approaches, and experiential and work-integrated education and training. Prerequisites for success are individual and collective capabilities in innovation and entrepreneurial activities, supportive organisational cultures, external networks, and tailor-made pedagogical approaches (chapters 6 and 15). In practice it is hard for those who are not innovative and entrepreneurial to teach these characteristics and approaches to others.

## Impact on Smallholder Farmers

This book provides information and ideas about how the AET system in sub-Saharan Africa can be transformed, with a specific focus on the interests of the smallholder farmer. The underlying argument is that improvements in the AET system have to have positive, *direct benefits* for smallholders. These benefits will include 1) strengthening smallholders' knowledge and skills; 2) helping them to solve problems in their work; 3) giving them access to information; and 4) developing appropriate and active market and other value-chain linkages.

Additional, *less direct* but still important benefits are 1) an increased focus among researchers and extension agents on the real needs of farmers, including technology development—resulting in more and better technologies available to farmers; 2) policy analysis and information that serve farmers' interests; 3) greater attention to and respect for the learning approaches, knowledge, and knowledge systems of smallholder farmers; and 4) the education of researchers, extension agents, and agribusiness value-chain specialists with relevant skills and a better understanding of and respect for the context, knowledge, lives, and interests of smallholder farmers—both men and women.

Taken together and if successfully implemented, these changes should over time contribute to 1) significant and sustained improvements in the livelihoods of smallholder farmers; 2) empowerment of farmers and communities of smallholder farmers—beyond income improvement; and 3) some alleviation of the serious power asymmetries in the agriculture value chain. The recommended measures should also 5) enhance the resilience of smallholder farmers and of their groups and communities.

These are—in theory—the *long-term expected impacts* on smallholder farmers sought through transformation of the AET system (Figure 17.6). The attributes and scope of the impacts still need to be determined.

Monitoring and evaluation need to track and assess these and any other relevant outcomes and impacts. An active search for unintended negative consequences will be important. There are also many important observations and recommendations in various chapters on how these changes can best be achieved.

Bearing in mind the fact that in most African countries a large majority of farmers are women (chapters 1 and 16), the needs, demands, and interests of the *smallholder farmer* have to be at the centre of planning and implementation (chapters 4, 9, and 10) of changes. An interrelated, interdependent set of activities can be devised to stimulate research and provide appropriate, gender-responsive knowledge and technology. Respect, understanding, and the integration of farmers' knowledge and knowledge systems into current frameworks and practices are important. Support for farmers and learning approaches that resonate with them need to be encouraged in formal, non-formal, and informal settings, and farmers' effective participation in research and extension must be sought (chapters 10 and 11).

## Impact on the AET System: Towards Resilience

The transformation foreseen in this book will also have a major *impact on the AET system*. If well designed and implemented, it will lead to a more 1) effective, 2) efficient, 3) relevant, and 4) respected national AET system, with the "ideal" attributes described in the proposed roadmap. However, the scope of and balance between the desired attributes of the system will differ from country to country, depending on the vision, initial situation, and possible strategies.

The transformation should also lead in the long term to 5) a more *resilient* system. In the simplest terms, the resilience of a country, society, system, or institution depends upon its ability to be flexible and to adapt readily and effectively to slow or rapid change—or to resist such change if resistance will bring better results in the long run. Similarly, the resilience of individuals and the groups to which they belong is to a great degree determined by their ability to adapt quickly and effectively or to resist shocks or evolution in the environment.

Strategies and interventions should therefore not only transform the AET system to be more relevant, efficient, and effective in its value addition to the AIS, but also make it more respected and in the long term, more resilient.

The starting point for cultivating resilience is to identify the drivers for current vulnerabilities. Although not explicitly mentioned as such in earlier chapters, several issues have emerged that can be designated as drivers of vulnerabilities in the AET system (Figure 17.2). They show that the resilience of the system is dependent not only on forces from within the system itself, but also on external forces in the agriculture, education, and other sectors, on a national, regional, and global level.

This issue needs to be further explored. Initial analyses show that key drivers of vulnerabilities in the AET system include the following:

- Inadequate national support associated with dependence for system resources on international actors and their agendas, coupled to:
- In many countries, the relatively low public and political profile of the agriculture sector and its professionals at the national level, resulting in low prioritisation and under-resourcing;
- The legacy of poor national education systems that do not deliver an adequate number of sufficiently prepared candidates in relevant areas for the higher education system;
- In many instances, institutions weakened by years of neglect and lack of clear direction for AET efforts;
- Weak institutional and individual capacities and motivation for innovation inherent in many agricultural innovation systems;
- Disconnects and fragmentation across the AET system, including between markets and delivery, between the different institutions in the system, and between those responsible for strategic action such as the Ministries of Agriculture and Education;
- The complexities inherent in a smallholder-based system of agriculture, with its insufficient scale, great potential for asymmetric capacities and power relations, lack of bargaining power, and difficulties in achieving concerted action.

Policy- and decision-makers at regional and national sector, system, and institutional levels need to understand these drivers of vulnerabilities as well as the hurdles to resilience in the AET system, and these issues should be systematically addressed as soon as resources allow. Innovation, financial strength and access to information and knowledge are crucial. Methods are needed to predict responses to interventions and possible negative consequences during transformation, and a set of solutions should be in hand to alleviate or neutralise negative effects. There should be a will to act and to invest resources in strategies with the potential to counter the risk of negative consequences and dependence on others. Cooperation—with an open platform for communication among different parts of the system and with key external actors—is crucial. All systems should move in conjunction with one another so that they can respond systematically and in synergy to any emerging risk or failure.

A concerted effort across the continent is needed to build resilience by working to eliminate drivers of vulnerabilities and the accompanying constraints to cultivating resilience. This places responsibility on each sub-region and country working on AET strategies to attend to the issue of resilience to the best of their abilities. This will be essential if Africa is to sustain itself and flourish in today's highly competitive global context—in particular in a system that is at the core of Africa's increasingly important, yet still weak, knowledge economy.

AET is part of the AIS and interlinked with many other open systems. Its successful transformation will therefore have impacts far beyond the system itself—on agriculture value chains, on rural development, and in the higher education sector. Eventually transformations in AET will impact society itself.

Using a "theory of change" approach (Funnell and Rogers, 2011; Vogel, 2012), the contributions expected from changes in the AET system can be aligned with regional, sub-regional, and national policies and strategies. Although the contributions may be small and the impact pathways hard to trace, planning within the wider strategic frameworks helps to direct the overall approach and specific strategies towards change.

Of particular importance is the assessment and promotion of policies and strategies that can lead to mutually beneficial arrangements between smallholder and large-scale agriculture. Whether in the same value chains or existing side by side, this dual track is likely to persist. It is therefore crucial that both groups benefit from each other and that positive impacts are felt by both.

## Accelerating Transformative Change

Strategies can be designed and implemented in a top-down or bottom-up manner, or a combination of both, in order to arrive at a "best fit" solution for AET system transformation in each country. The approach in each case will depend on specific decision-making and governance systems, public and political interest, and institutional cultures and capacities, as well as on the strength of connections to regional efforts in this regard.

It will be difficult for any country to address *all* the issues proposed in the roadmap. Difficult choices will have to be made, including between strategies for incremental or radical change, and top-down or bottom-up evolution. Efforts towards transformation can be facilitated through learning from the current situation, from the past, and from others' experiences, as well as through incremental learning as new strategies and interventions unfold. Such efforts can also be bolstered through using complexity science, and if resources and expertise allow, through forecasting, modelling, and experimenting with factors that could tip the balance towards the desired transformative change.

As pointed out in chapters 11 and 13, quality assurance at various levels is needed in order to support successful higher education programming. Monitoring and evaluation can also be valuable for other purposes and can play a crucial role in accelerating transformation processes.

## Embedding Monitoring and Evaluation in the AET System

Monitoring and evaluation provide for strategic and operational learning. They are crucial to honing policies and strategies through incremental learning. They aid understanding of change logic and development trajectories, save time and resources by enabling quick adjustment after failures, and identify those aspects—incremental or radical—that are potentially most crucial for transformation. Monitoring and evaluation can also enhance the ability and motivation of key stakeholders to experiment, learn, assess, and rapidly scale up successful interventions.

Monitoring and evaluation will not only help assess progress during the transformation process itself, but will also provide evidence for the results (outcomes and impacts) that it delivers at crucial points in the AET system and beyond, both during and after transformation.

Planning, monitoring, and evaluation can be connected from the start of a strategy or programme, using state of the art techniques to combine monitoring and evaluation with change logic or "theory of change" (Funnell and Rogers, 2011; Vogel, 2012).

As transformation takes root, developmental evaluation—real-time monitoring of progress, performance and impacts—can be combined with special external or independent evaluation studies of progress, performance, impact, sustainability and resilience. This *adaptive management* approach will assist in tracking and assessing changes as they happen and will support the evolution of transformation strategies. Importantly, such real-time learning coupled to evaluation will help identify unintended consequences of interventions in a timely manner. Where such consequences are negative, monitoring and evaluation can serve as an early warning system to help prevent ineffective action and wasted resources. Monitoring and evaluation can also be a useful instrument on the complex road towards building a more resilient AET system.

Throughout all monitoring and evaluation activities, the voices of smallholder farmers—including their different groupings in society such as men and women, cooperatives, and so forth—should be heard, respected, and used. They should, after all, be at the centre of the system.

## Seeking New Models for Transformation

Models for AET development captured in this book are primarily those used in the West, in Africa, and in parts of South and Southeast Asia. While these provide many useful lessons and frameworks for action, emerging global powers from the South have excellent potential to add to current understanding and knowledge relevant to AET.

At this exciting and important stage in the evolution of agriculture and AET in Africa, it will be highly valuable to learn from as many sources and development paradigms as possible.

As pointed out (chapters 4 and 9), several of the powerful BRICS countries are already highly active in agriculture and in education and training in sub-Saharan Africa (Future Agricultures, 2013a). Brazil focused successfully on productivity and technological modernisation, yet it is debating which model(s) to follow in the future (Future Agricultures, 2013b). The Asian Green Revolution provides many lessons, including lessons in unintended negative long-term consequences of strategic interventions and investments (chapter 4). China has been successful in feeding its massive population with relatively little arable land through an incremental approach to transformation—letting large-scale and smallholder farming co-exist for mutual benefit, and combining transformative strategies such as gradual market reform with the application of innovative and often low-cost solutions (Xiaoyun et al., u.d.; Brautigam, 2009; Future Agricultures, 2013c). Although not all approaches will be relevant to Africa, understanding how they were implemented and their implications for other AET systems could yield useful insights for countries on the continent.

## Final Remarks

As more is learnt about the evolution of agriculture, framed by current debates in countries with a history of successful agricultural development that includes smallholder farmers playing a major role, new ideas and different approaches to AET will emerge. These can help inform and adjust policies and strategies as they evolve towards the transformation of AET systems in sub-Saharan Africa.

Envisioning boldly, determining values, setting priorities, and taking risks—but retaining concern for the human aspects central to AET—these are the elements for effective transformation of education and training in agriculture and its related areas of knowledge.

## References


Alemneh, T.Y. 2007. *The Ethiopian Higher Education: Creating Space for Reform*. St. Mary's UC Printing Press: Ethiopia.

AWARD. 2012. AWARD Empowerment Framework for Leaders in Agricultural Research and Development. AWARD Series: Reflections on Evidence. Nairobi, Kenya.

Birner, R. and D. Spielman. 2007. *A Benchmarking Tool for Agricultural Innovation Systems*. Presentation at the World Bank Workshop Enhancing Agricultural Innovation Systems. March 22–23. Washington, DC.

Brautigam, D. 2009. *The Dragon's Gift: The Real Story of China in Africa*. Oxford University Press. UK.

CHEA. 2010. Ministerial Communiqué issued at the Ministerial Conference on Higher Education in Agriculture in Africa (CHEA). November 15–19, Kampala, Uganda.

Eicher, C. 2006. *The Evolution of Agricultural Education and Training: Global Insights of Relevance for Africa.* Staff Paper 2006–26. East Lansing, MI, Department of Agricultural Economics, Michigan State University. http://agecon.lib.umn.edu/cgi-bin/pdf_view.pl?paperid=23691&ftype=.pdf

Funnell, S.C. and P.J. Rogers. 2011. *Purposeful Programme Theory: Effective Use of Theories of Change and Logic Models*. Jossey-Bass. San Francisco, USA.

Future Agricultures, 2013a. *South-South Cooperation in Context: Perspectives from Africa.* Working Paper 54. Future Agricultures Consortium. University of Sussex. Brighton, UK.

—. 2013b. *Narratives of Brazil-Africa Cooperation for Agricultural Development*. Working Paper 51. Future Agricultures Consortium. University of Sussex. Brighton, UK.

—. 2013c. *Narratives of China-Africa Cooperation for Agricultural Development: New Paradigms?* Working Paper 53. Future Agricultures Consortium. University of Sussex. Brighton, UK.

Grabe, A. 2012. An Empirical Examination of Women's Empowerment and Transformative Change in the Context of International Development. *American Journal of Community Psychology* 49: 233–245.

Hannum, K.M., J.W. Martineau, and C. Reinelt (eds). 2007. *The Handbook of Leadership Development Evaluation*. Center for Creative Leadership and Jossey-Bass. San Francisco, USA.

IDRC/IADB 2002. *Organisational Assessment: A Framework for Improving Performance*. International Development Research Centre/Inter-American Development Bank. Ottawa, Canada.

Juma, C. 2011. *The New Harvest: Agricultural Innovation in Africa.* Oxford University Press: New York.

Kimenyi, M.S., B. Routman, and A. Westbury. 2013. *CAADP at 10: Progress towards Agricultural Prosperity*. Africa Growth Initiative at Brookings. USA.

Moock, J. 2011. *Network Innovations: Building the Next Generation of Agricultural Scientists in Africa.* ASTI/IFPRI-FARA. Conference Working Paper 1.

Pal, S. and D. Byerlee. 2006. The Funding and Organization of Agricultural Research in India: Evolution and Emerging Policy Issues. In Pardey, P.G, J.M. Alston, and R.R. Piggott (eds.), *Agricultural R&D Policy in the Developing World*. International Food Policy Research Institute. Washington, DC.

Rowlands, J. 1997. *Questioning Empowerment*. Oxfam. Oxford, UK.

Stroebel, A. and F. Swanepoel. 2008. *Cultivating Knowledge and Skills to Grow African Agriculture—an Analysis and Synthesis.* Southern African Agricultural Education Networking and Clarification Workshop, October 3–4, Manzini, Swaziland.

Stroebel, A., F. Swanepoel, and C. Eicher. 2011. Agricultural Education and Training for Development (AET4D) in Africa. Seventh Global Confederation of Higher Education Associations for the Agricultural and Life Sciences (GCHERA) Conference, June 27–29, Beauvais, France.

Vogel, I. 2012. Theories of Change in International Development. DFID. London, UK.

World Bank. 2007a. Cultivating Knowledge and Skills to Grow African Agriculture. World Bank, Washington, DC.

—. 2007b. World Development Report 2008: Agriculture for Development. World Bank, Washington, DC.

World Economic Forum (WEF). 2013. *Achieving the New Vision for Agriculture: New Models for Action.* A report by the World Economic Forum's New Vision for Agriculture Initiative, prepared in collaboration with McKinsey & Company.

Xiaoyun, L., T. Lixia, X. Xiuli, Q. Gubo, and W Haimin. (u.d.). *What Can Africa Learn from China's Experience in Agricultural Development?* China Agricultural University. Beijing. P.R.China.

Zondi, S. 2013. *Pan Africanism and African Renaissance*. Keynote Lecture at the Africa Day Celebration. 24 May. University of the Witwatersrand. Johannesburg, South Africa.

# Biographies of Authors

**Teshome Alemneh** is Programme Officer for Africa at Higher Education for Development (HED), USA. He manages planning grants, provides leadership, and oversees the implementation of 12 higher education partnerships in Sub-Saharan African countries (Burkina Faso, Burundi, Ethiopia, Ghana, Kenya, Liberia, Malawi, Rwanda, Senegal, South Africa, South Sudan, and Uganda). He was a research fellow at the Association of Land-grant Colleges and Universities in 2008 and a New Century Fulbright Scholar from 2004 to 2005 at the State University of New York (Buffalo). His research focussed on access and quality of higher education. Dr Alemneh has researched and written extensively about the status and challenges of higher education in Africa. In Ethiopia, he has served as Vice-Dean of Ambo College of Agriculture, and Associate Professor at Hawasa University, successfully supervising a number of masters' students. As a Deputy Minister of Higher Education (2001–2004), he has led the national higher education expansion and reform strategies that resulted in the promulgation of the first higher education law in the country, the establishment of 10 new universities, the creation of the first national higher education relevance and quality assurance agency, and a higher education strategy center. He was one of the architects of the vision and framers of the strategy leading to the establishment of an additional 13 universities. Alemneh holds a BSc in Plant Sciences from Alemaya College of Agriculture in Ethiopia, an MSc in Soil Sciences and a PhD in Earth Sciences and Land Evaluation from Ghent University in Belgium.

**Ponniah Anandajayasekeram** is an independent consultant, and Adjunct Professor at the School of Environmental and Rural Sciences at the University of New England, Australia. He holds a PhD in Agricultural Economics. Ananda has over 35 years of experience with a number of international agricultural research centers and national and multi-national organizations in both developed and developing countries. His areas of expertise include agricultural marketing, farming systems and participatory research methods, research management, agricultural innovation systems, monitoring, evaluation and impact assessment, planning and priority setting, networks and partnerships and training needs assessment. He has authored and co-authored over 200 journal articles, book chapters, books and reports. He has mentored many scientists and trained hundreds of professional in Africa and Asia. He was the past President of the International Farming Systems Association and the Southern and Eastern African Farming Systems Association. Ananda has devoted his entire career to organizational change and capacity strengthening in sub-Saharan Africa.

**Irene Annor-Frempong** obtained her PhD from the University of Bristol Veterinary School, UK and her Masters from Wageningen University, The Netherlands. Her undergraduate studies in agriculture, and a Diploma in Education, was completed at the University of Cape Coast, Ghana. She has worked in research institutes in Switzerland and the Netherlands, and for more than 20 years as Lecturer and Senior Lecturer in tertiary institutions in Ghana, Nigeria and Lesotho, with extensive experience in research, education and development. She serves as a Technical Committee member of the Regional Universities Forum for Capacity Building in Agriculture (RUFORUM). She joined the Forum for Agricultural Research in Africa (FARA) in 2008, and has since coordinated continental capacity building programmes including SCARDA and UniBRAIN. She is currently the Director, Capacity Strengthening at FARA, responsible for developing and implementing FARA's capacity strengthening strategy, as well as advocacy and resource mobilization for capacity development for African agricultural research.

**Maureen Bandama** is the Agribusiness Programme Coordinator for Southern Africa at Market Matters Inc. (MM Inc.). She joined MM Inc. in 2011, responsible for developing and coordinating MM Inc.'s networking, training and capacity building programmes in Africa. In addition, Maureen is currently involved in a survey of emerging farms and agribusinesses in South Africa, carried out jointly by the National Agricultural Marketing Council and Market Matters Inc. Prior to her appointment with MM Inc. Maureen obtained an MSc in Agricultural Economics at Stellenbosch University, South Africa. She holds a BSc in Agriculture and Natural Resources with a major in Agribusiness from Africa University, Zimbabwe. Maureen spent two years as a compliance Monitor (Agricultural and Financial Services sectors) in the Exchange Control Division of the Reserve Bank of Zimbabwe. She is passionate about agribusiness entrepreneurship development within Africa.

**Ralph Christy** is the founder of Market Matters, Inc., Director of the Cornell International Institute for Food, Agriculture and Development (CIIFAD) and Professor of Emerging Markets at Cornell University in Ithaca, New York, where he teaches and conducts food marketing research and educational programmes on the economic performance of markets and distribution systems. He has advised industry leaders and public policy makers on food marketing strategies, economic development, and the organization of the global food economy. A PhD graduate of Michigan State University's Department of Agricultural Economics, Christy is past President of the American Agricultural Economics Association, and past Board Member of the Winthrop Rockefeller Foundation, Winrock International, and the Agribusiness Capital Fund. He is the author of *A Century of Service: Land Grant Colleges and Universities, 1890-1990* (1992); *Achieving Sustainable Communities in a Global Economy* (2004); and *Emerging Markets* (forthcoming).

**Luc D'Haese** is Emeritus Professor of the Department of Bio-Science Engineering of the University of Antwerp, and the Department of Agricultural Economics at the University of Ghent, Belgium. He is currently Extraordinary Professor at Stellenbosch University, and the University of Pretoria, South Africa. Previously he served as CEO of the Belgian Technical Cooperation. He managed the International Course in Food and Nutrition at the University of Ghent for 15 years. During 1977 to 1992, he as a Professor at the University of Burundi. He is an associated member of the Belgian Royal Academy of Science.

**Canagasaby Devendra** specialises in animal nutrition and animal production systems. He has worked at the University of the West Indies in Trinidad, the Malaysian Agricultural Research and Development Institute (MARDI), the Canadian International Development Research Centre (IDRC), and the International Livestock Research Institute (ILRI). He has over 30 years of international experience with research and development initiatives. Dr Devendra has written 17 books and 455 publications, most of which have been published internationally. His most recent book, *Small farms in Asia, Revitalising Agricultural Production, Food Security and Rural Prosperity* was published in 2010. He was the first Asian recipient of the International Dairy Production award from the American Dairy Science Association in 1989. In addition, he received the Medal of Honour from the Minister of Agriculture, Vietnam in 2001, the Lincoln University International Alumni Medal from New Zealand, the Asian Animal Science award from Korea in 2004, and the Excellence Award for the Development of Animal Science from Pakistan in 2009. In 2009 he also received the Rotary Gold Medal in Malaysia. He is an Academician (Senior Fellow) of the Academy of Sciences Malaysia, and serves as a member of Heifer International's Board of Directors, USA. He qualifications include BAgricSci. (N.Z.), MAgricSci (Malaysia), PhD and DSc (Nott.,UK).

**Carl Eicher's** career over the past 50 years at Michigan State University has been devoted to training students, building institutions and helping develop agricultural research capacity throughout Sub-Saharan Africa. He served as Visiting Professor at the University of Nigeria (1963-66), Stanford University (1968) and the University of Zimbabwe (1983-87). In 1963, he helped establish the Economic Development Institute at the University of Nigeria and subsequently served as the Director of the Institute from 1964 to 1966. He received the Foreign Francqui Prize, University of Leuven, Belgium in 1986 and he was awarded an Honorary Doctor of Science Degree by the University of Zimbabwe in 1998. His dedication to African development continues today through his work with the United Nations Food and Agriculture Organization, the World Bank, the Kellogg Foundation, the Bill and Melinda Gates Foundation and African organizations such as the Forum for Agricultural Research in Africa (FARA) and the New Partnership for Africa's Development (NEPAD).

**Marcus Hakutangwi** holds a Masters in Adult Education (Agricultural Education and Extension) from the former University of Rhodesia. He is currently enrolled for PhD in African leadership / community empowerment through education using study circles, at the National University of Science and Technology in Zimbabwe. He is an Adviser with the Environmental Management Agency and Senior Strategy Advisor with the Nutrition Council of Zimbabwe, and a Lecturer in Agricultural Education and Extension at Africa University and Zimbabwe Open University. He is a member of the Barefoot Education for Africa Trust (BEAT). He has a wealth of experience in the agricultural, rural and community development arenas. In addition, he has consulted for the Kellogg Foundation to design an agricultural education and training programme for grassroots farmers. He has presented papers at a number of national and international fora, and has published extensively.

**Monty Jones** is Ambassador-at-Large and Advisor to the President of Sierra Leone. He is the former Executive Director of FARA and Co-winner of the prestigious 2004 World Food Prize, awarded for his discovery of the genetic process to create the New Rice for Africa (NERICA). Prof. Jone, has spent the past 22 years of his career in Africa working on international agricultural research for development institutions. His work on NERICA has increased rice production in West, East and Central Africa and created savings for many African governments. In September 2001, he received the National Order of Merit of Côte d'Ivoire given by the Ivorian president. In 2004, he also received the Insignia of the Grand Officer of the order of the Rokel from the Sierra Leone president for his work on NERICA. Since joining FARA in 2002 as its Executive Director, Prof. Jones has mobilized key decision makers including the African Union and NEPAD, as well as major investors in

African agriculture development to work together towards the achievement of Africa's vision. He was listed as one of the 100 most influential people in the world by Time magazine for 2007.

**Johan Jordaan** is an Agricultural Economist and Senior Lecturer in Agricultural Management at the Saasveld Campus of the Nelson Mandela Metropolitan University in George, South Africa. Before joining the former Port Elizabeth Technikon as Head of Department for Agricultural Management, he practiced as agricultural economist for the South African Ministry of Agriculture. He consults on various aspects of business development, financing and entrepreneurship in the agricultural sector. He served as Project Leader for the establishment of an integrated demonstration and training business incubator to facilitate entrepreneurship education and enterprise development amongst emerging farmers. His experiences are often shared at local and international conferences and workshops. Johan holds a BSc Agric. degree from the University of the Free State, a BSc Agric (Hons) degree from the University of Pretoria and an MBA from Stellenbosch University. He is currently Programme Manager for the Agricultural Management Unit at the Saasveld Campus of the Nelson Mandela Metropolitan University.

**Margaret Kroma** is the Programme Officer, Gender and Head of Unit at the Alliance for a Green Revolution in Africa (AGRA). She was the former Project Manager of the CGIAR, African Women in Agricultural Research and Development (AWARD) project and was an Assistant Professor in the College of Agriculture and Life Sciences at Cornell University, USA (2000-2008) where she served as Faculty coordinator of various collaborative agricultural research and development projects in Ghana, Mali, Niger and Sierra Leone. Margaret has over 15 years of progressive professional experience in agricultural development, gender and rural extension and holds a PhD in Rural Sociology from Iowa State University. She has authored/co-authored over 20 publications including journal articles, book chapters, monographs, op-eds and reports. She serves on several boards and technical committees including the CTA Expert Committee on Science and Technology Capacity Building, the Agricultural Technology Adoption Initiative (ATAI), RUFORUM Technical Committee, and the International Advisory Council of Future Agricultures Consortium.

**Edward Mabaya** is an academic and a development practitioner. He is involved in several programmes that seek to improve the lives of African farmers through private enterprises. Dr Mabaya established and coordinates the Seeds of Development Programme, a business development services and networking programme for emerging seed companies in East and Southern Africa. As a Research Associate in the Department of Applied Economics and Management at Cornell University, he conducts research on food marketing and distribution, seed systems and the role of efficient agricultural markets in Africa's economic development. He is Assistant Director at the Cornell International Institute for Food, Agriculture, and Development (CIIFAD). Ed earned his MSc and PhD degrees in Agricultural Economics at Cornell University, and a BSc degree from the University of Zimbabwe. In 2007, Ed was an Archbishop Desmond Tutu Leadership Fellow at the African Leadership Institute at Oxford University. During 2010-2011 he was a Visiting Researcher at Stellenbosch University, South Africa.

**Zenda Ofir** is a South African international evaluation specialist. With a PhD in Chemistry, she initially worked as manager in a South African science council, responsible for *i.a.* national higher education/industry research programs on biotechnology and on food production and security, and later as Director Research at the University of Pretoria. Since then she has worked on assignments in nearly 40 countries on four continents, conducting evaluations, facilitating the design of monitoring and evaluation systems and helping programs and organizations clarify their change logic ('theories of change') towards impact. She has done extensive work in the agriculture sector for *i.a.* Rockefeller Foundation, the International Fund for Agricultural Development (IFAD), the Alliance for a Green Revolution in Africa (AGRA), the Consultative Group on International Agricultural Research (CGIAR) and AWARD, a Bill and Melinda Gates Foundation funded program cultivating African women leaders in agricultural research and development. Zenda is a former President of the African Evaluation Association (AfrEA) and former Vice-President of the International Organization for Cooperation in Evaluation (IOCE). She also served on the first steering committee of the Network of Networks on Impact Evaluation (NONIE) – a global network representing hundreds of donor organizations and professional evaluation networks - and the first Board member living outside North America of the (now) nearly 8,000 member American Evaluation Association (AEA).

**John Makina** is an educationist and extension specialist. He has held various positions in environmental management and conservation. For several years, he was a lecturer in the Department of Education at the University of Zimbabwe and the University of Swaziland. He has lectured on rural development, research methods, communication, philosophy of adult education, and in the Certificate and Diploma programmes in Adult Education. He moved to Tswane University of Technology, South Africa where he became the Director of Staff Development and International Links. He currently manages his own consulting firm in the education and training sector.

**Daniel Sherrard** has been with EARTH University for almost 25 years, initially as a member of the CalPoly/San Luis Obispo team that assisted in establishing the institution. He has held numerous positions in the institution including admissions coordinator, professor, coordinator of the IV year academic programme, Dean and since 2008, Provost. In addition he has coordinated various projects in the university including the "Camino al Futuro" project (Road to the Future) in collaboration with the WK Kellogg Foundation, an important step in charting EARTH's future. Dr Sherrard has been instrumental in developing relations between EARTH and numerous African institutions of higher learning in agriculture. Dr. Sherrard received his doctorate in agricultural education from Iowa State University, his MSc from CalPoly/San Luis Obispo and his BSc from Portland State University.

**Aldo Stroebel** is Executive Director International Relations and Cooperation at the National Research Foundation (NRF), South Africa, and Affiliated Associate Professor at the Centre for Sustainable Agriculture, Rural Development and Extension, at the University of the Free State (UFS), South Africa. He is the former Director International Academic Programmes in the Office of the Vice-Chancellor, UFS. He is a well-known agriculture- and rural development specialist, focusing on smallholder farming systems in Africa. He qualified with a Masters degree from the University of Ghent, Belgium before researching for his PhD at UFS and Cornell University, USA. He was a Postdoctoral Research Fellow at Wageningen University, The Netherlands. He has successfully supervised more than 15 Masters and Doctoral students, and is the author/co-author of more than 80 journal articles and publications. At the University of Pretoria, he was the Programme Coordinator of the Integrated Rural Development Programme (IRDP) in partnership with the WK Kellogg Foundation, directing the establishment of a model for human capital and economic development in nine geographical areas in six southern African countries, including Botswana, Lesotho, Mozambique, South Africa, Swaziland and Zimbabwe. He was recently elected as one of 20 Founding Members of the South African Young Academy of Science (SAYAS). His fields of specialisation include smallholder farming systems, human capital development and capacity building, and internationalisation of higher education.

**Frans Swanepoel** is Director of the African Doctoral Academy (ADA) at Stellenbosch University, South Africa. He is a Visiting Fellow at the Institute for African Development (IAD) at Cornell University, USA. He is the former

Senior Director Research Development and Professor of Sustainable Agriculture at the University of the Free State, South Africa. He holds a PhD in Animal Breeding. Previously he was Dean of Agriculture at the University of Fort Hare, South Africa, where he is now an Adjunct Professor, and Professor and Director of the Postgraduate School of Agriculture and Rural Development at the University of Pretoria, South Africa. He serves on the Board of the South African Agricultural Research Council. During 2008/2009, he was a Fulbright Visiting Fellow at the Cornell International Institute for Food, Agriculture and Development at Cornell University (CIIFAD), USA. He is a member of the Academy of Science of South Africa (ASSAf), and has recently been appointed as Panel Chair for the national consensus study to revitalise agricultural education and training. He has successfully supervised more than 30 Masters and Doctoral students, and is the author / co-author of more than 150 journal articles, monographs, reports and book chapters. His fields of specialisation include African smallholder agriculture, research and innovation management, agricultural education and training, and capacity development.

**Ranga Taruvinga** is the Director and Chief Executive Officer of the Mananga Centre for Regional Integration and Management Development, responsible for both its strategic direction and operations. The company's core business is training and management development. In addition he is involved with consultancies in organizational capacity assessment, capacity building, rural development/poverty reduction, and environmental management issues. He has a special interest in smallholder farmer development and capacity building. He has also worked as an Associate for the International Services for National Agriculture Research (ISNAR), in The Hague, The Netherlands as a training agricultural research scientist from Africa. He holds a doctorate in environmental sciences.

**Glen Taylor** is the Senior Director for Innovation and Business Development at the University of the Free State in Bloemfontein, South Africa. Prior to this position he was Head of Department for Agricultural Management at the Nelson Mandela Metropolitan University. He has also held positions as Senior Lecturer at the Central University of Technology and as lecturer at the University of the Free State in the Department of Animal Science. Previously he was engaged at the National Department of Agriculture and the Agricultural Research Council of South Africa. Glen holds a Diploma in Veterinary Technology from Pretoria Technikon, BSc Agric, BSc Agric (Hons) and MSc Agric. degrees from the University of the Free State. He obtained his PhD from the University of Pretoria. At the University of the Free State, he serves as board member of a number of companies on behalf of the University, and has developed a number of initiatives in the fields of biotechnology, bio-fuel and product development. He has supervised numerous postgraduate students and published extensively in the fields of animal science and innovation. He participated in many fora, both nationally and internationally, related to diversifying income streams in universities, laying the foundations for income from innovation, managing third stream income from technology transfer, planning, financing and resource strategies, and engaging key stakeholders and partners.

**Akke van der Zijpp** is Emeritus Professor at Wageningen University, The Netherlands and Professor Extraordinary at the University of the Free State, South Africa. She works as an international consultant and serves in various capacities, inter alia as member of the Governing Boards of the Organic Research Institute and the Animal Health Service, as President of the WPSA Netherlands branch, member of the Animal Medicines Authority for reduction of antibiotics in animal production, and other NGOs and advisory committees. Following her first degree in Animal Science at Wageningen University, she was lecturer at Reading University (UK), and Associate Professor at Wageningen University, followed by her appointment as Director of the Animal Production Research Institute, and later as Deputy-Director Research at ILRI and ICIPE, both based in Nairobi. In 1999 she was appointed as head of the Animal Production Systems Group in the Department of Animal Sciences at Wageningen University. Her PhD study focused on genetic resistance to disease in poultry, followed by a sabbatical in the USA. She supervised over 100 MSc and 20 PhD students, published close to 100 refereed papers, and (co-)edited 10 books. Her present interest is in sustainable development of animal production systems, and programmes in interdisciplinary research for development at Wageningen University.

**Johan van Rooyen** is Director of the Standard Bank Center for Agri-Leadership and Mentorship Development, and Professor in Agricultural Economics at Stellenbosch University, South Africa. He teaches agri-business strategy, agricultural and rural development, and business cases, at the Universities of Stellenbosch, Free State, Pretoria and Ghent, Belgium. He qualified with a DSc in Agricultural Economics and holds a Management Certificate (SMC) from IMD, Switzerland. He has published more than 160 peer reviewed scientific papers, and six academic books on agri-business and agri- development. He is the recipient of the Tomlinson Medal for Agricultural Economics, South Africa. Prof van Rooyen has served on a number of boards of agri-business companies and agencies, and is currently a board member of the International Food and Agri-Business Industry Management Association (IFAMA). He held various positions, incl. at the Development Bank of Southern Africa (Executive Group Manager for agriculture and rural development) the University of Pretoria (Professor and Director of the Post-graduate School for Agriculture and Rural Development, and the Absa Chair of Agribusiness Management), the Agri-business Chamber of South Africa (CEO) and at the SA Wine Industry Council (CEO).

**Larry Zuidema**, an agricultural economist, is the former Director of the International Agriculture Programme of the College of Agriculture and Life Sciences at Cornell University in Ithaca, New York, USA. At Cornell University, he coordinated several international projects in Asia, the South Pacific and Africa which focused on the strengthening of educational institutions' capacity for agricultural teaching, research and rural development. In the periods 1987-88 and 1995-98, he was a Senior Fellow at the former International Service for National Agricultural Research (ISNAR) where he focused on institutional management; linkages between agricultural research, teaching and extension; and the strengthening of university research capacity in Africa. Throughout his career, he served in Africa, Southeast Asia, South Asia and several of the newly independent republics of the former Soviet Union as a consultant for the World Bank, FAO, ADB, USAID and several consulting firms. He was a past president of the Association of US University Directors of International Agricultural Programmes.

# INDEX